Methods of Cut-Elimination

TRENDS IN LOGIC

Studia Logica Library

VOLUME 34

SCOPE OF THE SERIES

Trends in Logic is a bookseries covering essentially the same area as the journal *Studia Logica* – that is, contemporary formal logic and its applications and relations to other disciplines. These include artificial intelligence, informatics, cognitive science, philosophy of science, and the philosophy of language. However, this list is not exhaustive, moreover, the range of applications, comparisons and sources of inspiration is open and evolves over time.

Volume Editor
Heinrich Wansing

For further volumes:
http://www.springer.com/series/6645

Matthias Baaz · Alexander Leitsch

Methods of Cut-Elimination

Matthias Baaz
Vienna University of Technology
Wiedner Hauptstraße 8-10
1040 Vienna
Austria
baaz@logic.at

Alexander Leitsch
Vienna University of Technology
Favoritenstraße 9
1040 Vienna
Austria
leitsch@logic.at

ISBN 978-94-007-0319-3 e-ISBN 978-94-007-0320-9
DOI 10.1007/978-94-007-0320-9
Springer Dordrecht Heidelberg London New York

Printed on acid-free paper

Springer is part of Springer Science+Business Media (www.springer.com)

Contents

1 Preface **1**
- 1.1 The History of This Book 1
- 1.2 Potential Readers of This Book 1
- 1.3 How to Read This Book 2

2 Introduction **5**

3 Preliminaries **9**
- 3.1 Formulas and Sequents 9
- 3.2 The Calculus LK 14
- 3.3 Unification and Resolution 24

4 Complexity of Cut-Elimination **39**
- 4.1 Preliminaries 39
- 4.2 Proof Complexity and Herbrand Complexity 44
- 4.3 The Proof Sequence of R. Statman 51

5 Reduction and Elimination **63**
- 5.1 Proof Reduction 63
- 5.2 The Hauptsatz 73
- 5.3 The Method of Tait and Schütte 85
- 5.4 Complexity of Cut-Elimination Methods 93

6 Cut-Elimination by Resolution **105**
- 6.1 General Remarks 105
- 6.2 Skolemization of Proofs 106
- 6.3 Clause Terms 111
- 6.4 The Method CERES 114
- 6.5 The Complexity of CERES 127
- 6.6 Subsumption and p-Resolution 133

6.7 Canonic Resolution Refutations 139
6.8 Characteristic Terms and Cut-Reduction 146
6.9 Beyond $\mathcal{R}$: Stronger Pruning Methods 157
6.10 Speed-Up Results 159

7 Extensions of CERES **163**
7.1 General Extensions of Calculi 163
7.2 Equality Inference 169
7.3 Extension by Definition 172

8 Applications of CERES **175**
8.1 Fast Cut-Elimination Classes 175
8.2 CERES and the Interpolation Theorem 189
8.3 Generalization of Proofs 209
8.4 CERES and Herbrand Sequent Extraction 212
8.5 Analysis of Mathematical Proofs 214
8.5.1 Proof Analysis by Cut-Elimination 214
8.5.2 The System ceres 215
8.5.3 The Tape Proof 216
8.5.4 The Lattice Proof 221

9 CERES in Nonclassical Logics **229**
9.1 CERES in Finitely Valued Logics 230
9.1.1 Definitions 230
9.1.2 Skolemization 238
9.1.3 Skolemization of Proofs 238
9.1.4 CERES-m 241
9.2 CERES in Gödel Logic 250
9.2.1 First Order Gödel Logic and Hypersequents 251
9.2.2 The Method hyperCERES 255
9.2.3 Skolemization and de-Skolemization 256
9.2.4 Characteristic Hyperclauses and Reduced Proofs . . . 259
9.2.5 Hyperclause Resolution 265
9.2.6 Projection of Hyperclauses into **HG**-Proofs 267

10 Related Research **271**
10.1 Logical Analysis of Mathematical Proofs 271
10.2 New Developments in CERES 272

References **275**

Index **283**

Chapter 1

Preface

1.1 The History of This Book

This book comprises 10 years of research by Matthias Baaz and Alexander Leitsch on the topic of cut-elimination. The aim of this research was to consider computational aspects of cut-elimination, the most important method for analyzing formal first-order proofs. During this period a new method of cut-elimination, cut-elimination by resolution (CERES), has been developed which is based on the refutations of formulas characterizing the cut-structure of the proofs. This new method connects automated theorem proving with classical proof theory, allowing the development of new methods and more efficient implementations; moreover, CERES opens a new view on cut-elimination in general. This field of research is evolving quite fast and we expect further results in the near future (in particular concerning cut-elimination in higher-order logic and in nonclassical logics).

1.2 Potential Readers of This Book

This book is directed to graduate students and researchers in the field of automated deduction and proof theory. The uniform approach, developed by Alexander Leitsch, serves the purpose of importing mathematical techniques from automated deduction to proof theory, to facilitate the implementation and derivation of complexity bounds for basically indeterministic methods. Matthias Baaz has been responsible for proof theoretic considerations and for the extension of CERES to nonclassical logics.

M. Baaz, A. Leitsch, *Methods of Cut-Elimination*, Trends in Logic 34,

DOI 10.1007/978-94-007-0320-9_1,

1.3 How to Read This Book

The book can be read from a computer-science or from a proof-theoretic perspective as the diagram below indicates.

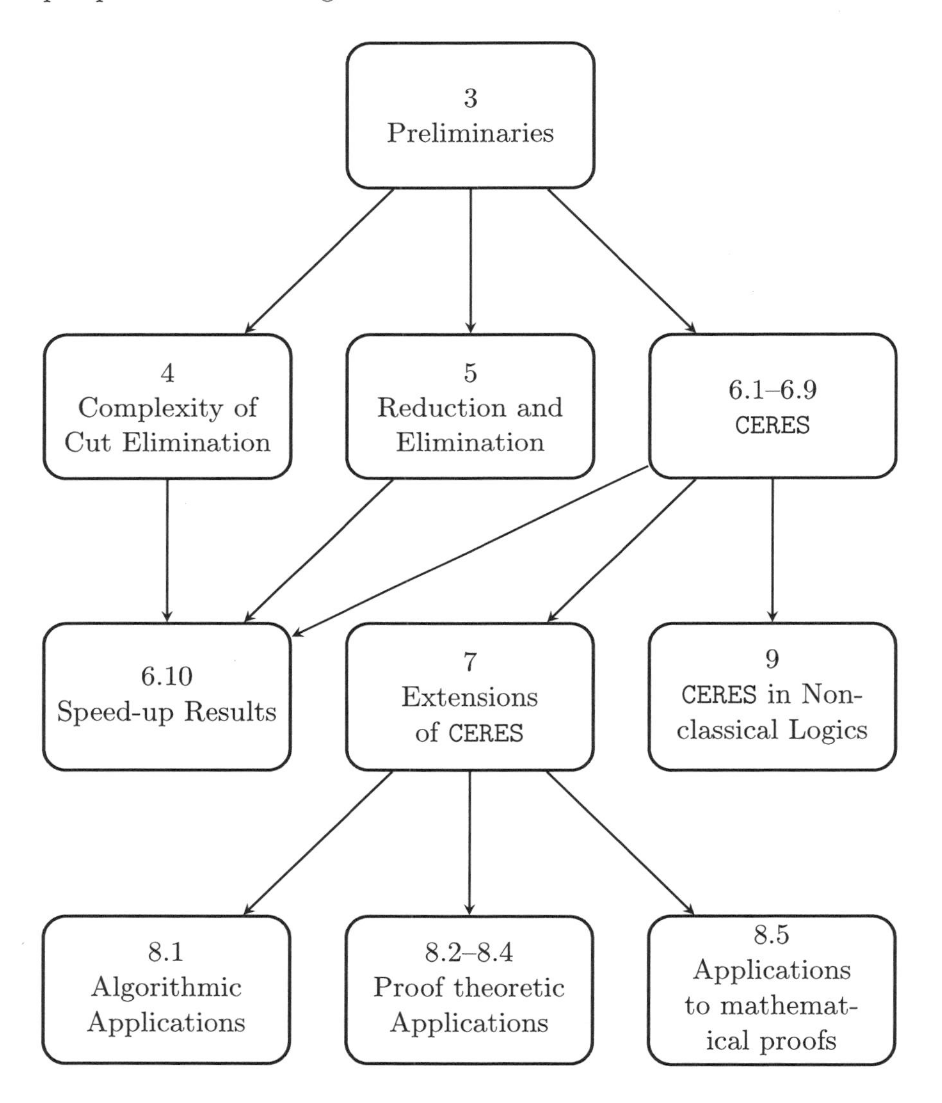

Acknowledgments We thank Daniele Mundici for his encouragement to write a book on this topic and for his steady interest in our research during the last 15 years. We also are grateful to the Austrian Science Fund for supporting the research on cut-elimination by funding the projects P16264, P17995, and P19875.

The research on this topic began with the authors and Alessandra Carbone during the time of her Lise Meitner fellowship. In the course of the following FWF research projects P16264, P17995, and P19875, the Ph.D. students Stefan Hetzl, Clemens Richter, Hendrik Spohr, Daniel Weller, and Bruno Woltzenlogel-Paleo contributed substantially to the theoretic and, especially, to the practical development of the CERES method. The extension of the method to Gödel logic has been carried out together with Agata Ciabattoni and Chris Fermüller.

Our special thanks go to Tomer Libal, Daniel Weller and Bruno Woltzenlogel for their careful and critical reading of the text. Their comments and suggestions have been integrated in the text and have resulted in a substantial improvement of the book.

We are very grateful to the reviewer for his numerous critical comments and suggestions for improvements which had a substantial impact on the final version of text.

Chapter 2

Introduction

Gottfried Wilhelm Leibniz called a proof analytic iff[1] the proof is based on concepts contained in the proven statement (praedicatum inest subjecto [59]). His own example [60] shows that this notion is significant, as it is connected to the distinction between inessential derivation steps (mostly formulated as definitions) and derivation steps which may or may not be based on concepts contained in the result:

(a) $4 = 2 + 2$ (the result)

(b) $3 + 1 = 2 + 2$ (by the definition $4 = 3 + 1$)

(c) $(2 + 1) + 1 = 2 + 2$ (by the definition $3 = 2 + 1$)

(d) $2 + (1 + 1) = 2 + 2$ (by associativity)

(e) $2 + 2 = 2 + 2$ (by the definition $2 = 1 + 1$)

The interest in the notion of analytic proof and analytic provability is twofold:

- First, the reduction of the concepts constituting a proof to the concepts contained in the (desired) result is essential to construct a proof by an analysis of the result (This was the main aim of Leibniz). Therefore analytic proofs in a suitable definition are the core of any approach to automated theorem proving.
- Second, analytic proofs allow control not only of the result but also the means of the proof and admit the derivation of additional information

[1] If and only if.

M. Baaz, A. Leitsch, *Methods of Cut-Elimination*, Trends in Logic 34,
DOI 10.1007/978-94-007-0320-9_2,

related to the result from the proof. In other words: a theorem with an analytic proof can be strengthened by looking at the proof.

In mathematics, the obvious counterpart to the notion of analytic proof is the notion of elementary proof. What elementary means, however, changes in time (from avoiding arguments on complex numbers in the Prime Number Theorem [37] to omitting arguments on p-adic numbers and more recently ergodic theory). In a more modern expression analyticity relates to the distiction between soft and hard analysis by Terence Tao [76].
David Hilbert introduced the concept of purity of methods (Reinheit der Methode) as an emphasis on analytic provability (and not so much on analytic proofs). He discussed for the first time whether, for a given mathematical theory, all provable statements are in fact analytically provable (this line of thought is already present in *Grundlagen der Geometrie* [50]).

The social value of mathematics (and of science in general) is connected to the establishment of verified statement i.e. theorems which can be applied without (using the) knowledge of its proofs. It is not necessary to understand the proof of the central limit theorem for working with normal distributions.

This principle also applies within mathematics w.r.t. the intermediary statements i.e. lemmata. In terms of propositional reasoning this is expressed by the rule of modus ponens

$$\frac{A \quad A \to B}{B}$$

which is the historically primal example of a cut rule. The presence of such rules in a proof, however, might hide valuable information such as an implicit constructive content.

The introduction of cut-free derivations in the sequent calculus **LK** (**LJ**) in Gerhard Gentzen's seminal papers *Über das logische Schliessen I+II* [38] provided a stable notion of analytic proof for classical (intuitionistic) first-order logic based on the subformula property. The structural rules represent the obvious derivation steps not necessarily related to the result. Gentzen was the first to actually prove, that everything derivable can be derived analytically (the Hauptsatz).

In this book we focus on cut-elimination for classical logic from a procedural point of view. In the tradition of proof theory, the emphasis is on cut-free provability with restricted means, not on the actual elimination of cuts from proofs. We develop a more radical form of cut-elimination using the fact, that the cuts after cancellation of other parts of the proof can be considered as contradictions. The method (called CERES[2] – cut-elimination by resolution) works as follows:

- extract from the parts of the axioms, leading to cuts, a set of clauses (in the sense of the resolution calculus) which is refutable. The set of clauses can be represented by clause terms, which are algebraic objects.

- For every clause, there exists a cut-free part of the original proof (the projection), which derives the original end sequent extended by the clause.

- Refute the set of clauses using resolution, construct a ground resolution proof and augment the clauses with the associated (substitution instances) of projections.

By the method CERES an essentially cut-free proof is obtained. The remaining atomic cuts are easily removable in the presence of logical axioms. This is even not necessary as they do not interfere with the extraction of desired information implicitly contained in proofs as Herbrand disjunctions, interpolants etc. To apply CERES, it is necessary to reduce compound logical axioms to atomic ones and to replace strong quantifiers in the end-sequent by adequate Skolem functions without increasing the complexity of the proof. The elimination of the Skolem functions from a cut-free proof is of at most exponential expense.
CERES simulates the usual cut elimination methods of Gentzen and Schütte/-Tait, here formulated nondeterministically. On the other hand there are sequences of proofs, whose cut-free normal forms according to Gentzen and Schütte/Tait grow nonelementarily w.r.t.[3] the cut-free normal forms according to CERES. The reason is, that usual cut-elimination methods are local in the sense that only a small part of the proof is analyzed, namely the derivation corresponding to the introduction of the uppermost logical connective. As a consequence many types of redundancies in proofs are left undetected leading to a bad computational behaviour.
The strong regularity properties of cut-free normal forms obtained by CERES (the proofs are composed from the projections) together with the simulation results (reductive methods can be simulated by CERES) allow the formulation of negative results also for the traditional methods. For example no cut-free proof, whose Herbrand disjunction is not composed from substitution

[2]http://www.logic.at/ceres
[3]With respect to.

instances of the Herbrand disjunctions of the projections can be obtained by Gentzen or Schütte/Tait cut-elimination.

As intended, CERES is used to extract structural information implicit in proofs with cuts such as interpolants etc. It serves as a tool for the generalization of proofs (justifying the Babylonian reasoning by examples). Furthermore we demonstrate how to apply CERES to the analysis of mathematical proofs using two straightforward examples. CERES relates these proofs with cuts to the spectrum of all cut-free proofs obtainable in a reasonable way. By analyzing CERES itself, we establish easy-to-describe classes of proofs, which admit fast (i.e. elementary) cut elimination. Possibilities and limits of the extension of CERES-like methods to the realm of nonclassical, especially intermediate logics are discussed using the example of first-order Gödel-Dummett logic (i.e. the logic of linearly ordered Kripke structures with constant domains).

We finally stress that the proximity of CERES to the resolution calculus facilitates its implementation (and thereby the implementation of the traditional cut-elimination methods) using state-of-the-art automated theorem proving frameworks. Furthermore, resolution strategies might be employed to express knowledge about cut formulas obvious to mathematicians but usually algorithmically difficult to represent. This includes the difference between the proved lemma (positive occurrence of the cut formula) and its application (negative occurrence of the cut-formula).

Chapter 3

Preliminaries

3.1 Formulas and Sequents

In this chapter we present some basic concepts which will be needed throughout the whole book. We assume that the the reader is familiar with the most basic notions of predicate logic, like terms, formulas, substitutions and interpretations.

We denote predicate symbols by P, Q, R, function symbols by f, g, h, constant symbols by a, b, c. We distinguish a set of free variables V_f and a set of bound variables V_b (both sets are assumed to be countably infinite).

Remark: The distinction between free and bound variables is vital to proof transformations like cut-elimination, where whole proofs have to be instantiated. $\diamond$

We use α, β for free variables and x, y, z for bound ones. Terms are defined as usual with the restriction that they may not contain bound variables.

Definition 3.1.1 (semi-term, term) We define the set of semi-terms inductively:

- bound and free variables are semi-terms,
- constants are semi-terms,
- if $t_1, \ldots, t_n$ are semi-terms and f is an n-place function symbol then $f(t_1, \ldots, t_n)$ is a semi-term.

$\diamond$

Semi-terms which do not contain bound variables are called terms.

M. Baaz, A. Leitsch, *Methods of Cut-Elimination*, Trends in Logic 34,

DOI 10.1007/978-94-007-0320-9_3,

Example 3.1.1 $f(\alpha, \beta)$ is a term. $f(x, \beta)$ is a semi-term. $P(f(\alpha, \beta))$ is a formula. ◇

Replacement on positions play a central role in proof transformations. We first introduce the concept of position for terms.

Definition 3.1.2 (position) We define the positions within semi-terms inductively:

- If t is a variable or a constant symbol then ϵ is a position in t and $t.\epsilon = t$
- Let $t = f(t_1, \ldots, t_n)$ then ϵ is a position in t and $t.\epsilon = t$. Let μ be a position in a t_j (for $1 \leq j \leq n$), $\mu = (k_1, \ldots, k_l)$ and $t_j.\mu = s$; then ν, for $\nu = (j, k_1, \ldots, k_l)$, is a position in t and $t.\nu = s$.

◇

Positions serve the purpose to locate sub-semi-terms in a semi-term and to perform replacements on sub-semi-terms. A sub-semi-term s of t is just a semi-term with $t.\nu = s$ for some position ν in t. Let $t.\nu = s$; then $t[r]_\nu$ is the term t after replacement of s on position ν by r, in particular $t[r]_\nu.\nu = r$. Let P be a set of positions in t; then $t[r]_P$ is defined from t by replacing all $t.\nu$ with $\nu \in P$ by r.

Example 3.1.2 Let $t = f(f(\alpha, \beta), a)$ be a term. Then

$$\begin{aligned} t.\epsilon &= t, \\ t.(1) &= f(\alpha, \beta), \\ t.(2) &= a, \\ t.(1,1) &= \alpha, \\ t.(1,2) &= \beta, \\ t[g(a)].(1,1) &= f(g(a), \beta). \end{aligned}$$

◇

Positions in formulas can be defined in the same way (the simplest way is to consider all formulas as terms).

Definition 3.1.3 (substitution) A *substitution* is a mapping from $V_f \cup V_b$ to the set of *semi-terms* s.t. $\sigma(v) \neq v$ for only finitely many $v \in V_f \cup V_b$.

If σ is a substitution with $\sigma(x_i) = t_i$ for $x_i \neq t_i$ $(i = 1, \ldots, n)$ and $\sigma(v) = v$ for $v \notin \{x_1, \ldots, x_n\}$ then we denote σ by $\{x_1 \leftarrow t_1, \ldots, x_n \leftarrow t_n\}$. We call the set $\{x_1 \leftarrow t_1, \ldots, x_n \leftarrow t_n\}$ the *domain* of σ and denote it by $dom(\sigma)$. Substitutions are written in postfix, i.e. we write $F\sigma$ instead of $\sigma(F)$. $\diamond$

Substitutions can be extended to terms, atoms and formulas in a homomorphic way.

Definition 3.1.4 A substitution σ is called *more general* than a substitution ϑ $(\sigma \leq_s \vartheta)$ if there exists a substitution μ s.t. $\vartheta = \sigma\mu$. $\diamond$

Example 3.1.3 Let $\vartheta = \{x \leftarrow a, y \leftarrow a\}$ and $\sigma = \{x \leftarrow y\}$. Then $\sigma\mu = \vartheta$ for $\mu = \{y \leftarrow a\}$ and thus $\sigma \leq_s \vartheta$. Note that for the identical substitution we get $\emptyset \leq_s \lambda$ for all substitutions λ. $\diamond$

Definition 3.1.5 (semi-formula, formula) $\top$ and $\bot$ are formulas.
If $t_1, \ldots, t_n$ are terms and P is an n-place predicate symbol then $P(t_1, \ldots, t_n)$ is an (atomic) formula.

- If A is a formula then $\neg A$ is a formula.
- If A, B are formulas then $(A \rightarrow B)$, $(A \wedge B)$ and $(A \vee B)$ are formulas.
- If $A\{x \leftarrow \alpha\}$ is a formula then $(\forall x)A, (\exists x)A$ are formulas.

Semi-formulas differ from formulas in containing free variables in V_b. $\diamond$

Example 3.1.4 $P(f(\alpha, \beta))$ is a formula, and so is $(\forall x)P(f(x, \beta))$. $P(f(x, \beta))$ is a semi-formula. $\diamond$

Definition 3.1.6 (logical complexity of formulas) If F is a formula in PL then the complexity $comp(F)$ is the number of logical symbols occurring in F. Formally we define

$comp(F) = 0$ if F is an atomic formula,

$comp(F) = 1 + comp(A) + comp(B)$ if $F \equiv A \circ B$ for $\circ \in \{\wedge, \vee, \rightarrow\}$,

$comp(F) = 1 + comp(A)$ if $F \equiv \neg A$ or $F \equiv (Qx)A$ for $Q \in \{\forall, \exists\}$ and $x \in V_b$.

$\diamond$

Gentzen's famous calculus **LK** is based on so called sequents; sequents are structures with sequences of formulas on the left and on the right hand side of a symbol which does not belong to the syntax of formulas. We call this symbol *the sequent sign* and denote it by $\vdash$.

Definition 3.1.7 (sequent) Let Γ and Δ be finite (possibly empty) sequences of formulas. Then the expression $S\colon \Gamma \vdash \Delta$ is called a *sequent*. Γ is called the *antecedent* of S and Δ the *consequent* of S. ◇

Let

$$\bigwedge_{i=1}^{1} A_i = A_1, \quad \bigwedge_{i=1}^{n+1} A_i = A_{n+1} \wedge \bigwedge_{i=1}^{n} A_i \text{ for } n \geq 1,$$

and analogous for $\bigvee$.

Definition 3.1.8 (semantics of sequents) Semantically a sequent

$$S\colon A_1, \ldots, A_n \vdash B_1, \ldots, B_m$$

stands for

$$F(S)\colon \bigwedge_{i=1}^{n} A_i \to \bigvee_{j=1}^{m} B_j.$$

In particular we define $\mathcal{M}$ to be an interpretation of S if $\mathcal{M}$ is an interpretation of $F(S)$. If $n = 0$ (i.e. there are no formulas in the antecedent of S) we assign $\top$ to $\bigwedge_{i=1}^{n} A_i$, if $m = 0$ we assign $\bot$ to $\bigvee_{j=1}^{m} B_j$. Note that the empty sequent is represented by $\top \to \bot$ which is equivalent to $\bot$ and represents falsum. We say that S is true in $\mathcal{M}$ if $F(S)$ is true in $\mathcal{M}$. S is called *valid* if $F(S)$ is valid. ◇

Example 3.1.5

$$S\colon\ P(a), (\forall x)(P(x) \to P(f(x))) \vdash P(f(a))$$

is a sequent. The corresponding formula

$$F(S)\colon\ (P(a) \wedge (\forall x)(P(x) \to P(f(x)))) \to P(f(a))$$

is valid; so S is a valid sequent. ◇

Definition 3.1.9 A sequent $A_1, \ldots, A_n \vdash B_1, \ldots, B_m$ is called *atomic* if the A_i, B_j are atomic formulas. ◇

Definition 3.1.10 (composition of sequents) If $S = \Gamma \vdash \Delta$ and $S' = \Pi \vdash \Lambda$ we define the composition of S and S' by $S \circ S'$, where $S \circ S' = \Gamma, \Pi \vdash \Delta, \Lambda$. ◇

Definition 3.1.11 Let Γ be a sequence of formulas. Then we write $\Gamma - A$ for Γ after deletion of all occurrences of A. Formally we define

$$\begin{array}{rcl} (A_1, \ldots A_n) - A & = & (A_2, \ldots A_n) - A \text{ for } A = A_1, \\ & = & A_1, ((A_2, \ldots A_n) - A) \text{ for } A \neq A_1, \\ \epsilon - A & = & \epsilon. \end{array}$$

◇

Definition 3.1.12 (permutation of sequents) Let S be the sequent $A_1, \ldots, A_n \vdash B_1, \ldots, B_m$, π be a permutation of $\{1, \ldots, n\}$, and π' be a permutation of $\{1, \ldots, m\}$. Then the sequent

$$S' \colon A_{\pi(1)}, \ldots, A_{\pi(n)} \vdash B_1, \ldots, B_m$$

is called a *left permutation* of S (based on π), and

$$S'' \colon A_1, \ldots, A_n \vdash B_{\pi'(1)}, \ldots, B_{\pi'(m)}$$

is called a *right permutation* of S (based on π'). A *permutation* of S is a left permutation of a right permutation of S. ◇

Definition 3.1.13 (subsequent) Let S, S' be sequents. We define $S' \sqsubseteq S$ if there exists a sequent S'' s.t. $S' \circ S''$ is a permutation of S and call S' a *subsequent* of S. ◇

Example 3.1.6 S': $P(b) \vdash Q(a)$ is a subsequent of

$$S\colon\ P(a), P(b), P(c) \vdash Q(a), Q(b).$$

S'' has to be defined as $P(a), P(c) \vdash Q(b)$. Then clearly

$$S' \circ S'' = P(b), P(a), P(c) \vdash Q(a), Q(b).$$

The left permutation (12) then gives S. ◇

By definition of the semantics of sequents, every sequent is implied by all of its subsequents. The empty sequent (which stands for falsum) implies every sequent.

Definition 3.1.14 Substitutions can be extended to sequents in an obvious way. If $S = A_1, \ldots, A_n \vdash B_1, \ldots, B_m$ and σ is a substitution then

$$S\sigma = A_1\sigma, \ldots, A_n\sigma \vdash B_1\sigma, \ldots, B_m\sigma.$$

$\diamond$

Definition 3.1.15 (polarity) Let λ be an occurrence of a formula A in B. If $A \equiv B$ then λ is a positive occurrence in B. If $B \equiv (C \wedge D), B \equiv (C \vee D), B \equiv (\forall x)C$ or $B \equiv (\exists x)C$ and λ is a positive (negative) occurrence of A in C (or in D respectively) then the corresponding occurrence λ' of A in B is positive (negative). If $B \equiv (C \to D)$ and λ is a positive (negative) occurrence of A in D then the corresponding occurrence λ' in B is positive (negative); if, on the other hand, λ is a positive (negative) occurrence of A in C then the corresponding occurrence λ' of A in B is negative (positive). If $B \equiv \neg C$ and λ is a positive (negative) occurrence of A in C then the corresponding occurrence λ' of A in B is negative (positive). If there exists a positive (negative) occurrence of a formula A in B we say that A is of positive (negative) polarity in B. $\diamond$

Definition 3.1.16 (strong and weak quantifiers)
If $(\forall x)$ occurs positively (negatively) in B then $(\forall x)$ is called a strong (weak) quantifier. If $(\exists x)$ occurs positively (negatively) in B then $(\exists x)$ is called a weak (strong) quantifier. $\diamond$

Note that (Qx) may occur several times in a formula B; thus it may be strong and weak at the same time. If confusion might arise we refer to the specific position of (Qx) in B. In particular we may replace every formula A by a logically equivalent "variant" A' s.t. every (Qx) (for $Q \in \{\forall, \exists\}$ and $x \in V$) occurs at most once in A'. In this case the term "(Qx) is a strong (weak) quantifier" has a unique meaning.

Definition 3.1.17 A sequent S is called *weakly quantified* if all quantifier occurrences in S are weak. $\diamond$

3.2 The Calculus LK

Like most other calculi Gentzen's **LK** is based on axioms and rules.

Definition 3.2.1 (axiom set) A (possibly infinite) set $\mathcal{A}$ of sequents is called an *axiom set* if it is closed under substitution, i.e., for all $S \in \mathcal{A}$ and for all substitutions θ we have $S\theta \in \mathcal{A}$. If $\mathcal{A}$ consists only of atomic sequents we speak about an *atomic axiom set.* $\diamond$

Remark: The closure under substitution is required for proof transformations, in particular for cut-elimination. ◇

Definition 3.2.2 (standard axiom set) Let $\mathcal{A}_T$ be the smallest axiom set containing all sequents of the form $A \vdash A$ for arbitrary atomic formulas A. $\mathcal{A}_T$ is called the *standard axiom set.* ◇

Definition 3.2.3 (LK) Basically we use Gentzen's version of **LK** (see [38]) with the exception of the permutation rule. There are two groups of rules, the logical and the structural ones. All rules with the exception of cut have left and right versions; left versions are denoted by $\xi\colon l$, right versions by $\xi\colon r$. Every logical rule introduces a logical operator on the left or on the right side of a sequent. Structural rules serve the purpose of making logical inferences possible (e.g. permutation) or to put proofs together (cut). A and B denote formulas, $\Gamma, \Delta, \Pi, \Lambda$ sequences of formulas. In the rules there are introducing or *auxiliary formulas* (in the premises) and introduced or *principal formulas* in the conclusion. We indicate these formulas for all rules. In particular we mark the auxiliary formula occurrences by $+$ and the principal ones by $\star$. We frequently say auxiliary (main) *formula* instead of auxiliary (main) formula occurrence.

The logical rules:

- $\wedge$-introduction:

$$\frac{A^+, \Gamma \vdash \Delta}{(A \wedge B)^\star, \Gamma \vdash \Delta}\ \wedge\colon l_1 \qquad \frac{B^+, \Gamma \vdash \Delta}{(A \wedge B)^\star, \Gamma \vdash \Delta}\ \wedge\colon l_2 \qquad \frac{\Gamma \vdash \Delta, A^+ \quad \Gamma \vdash \Delta, B^+}{\Gamma \vdash \Delta, (A \wedge B)^\star}\ \wedge\colon r$$

- $\vee$-introduction:

$$\frac{A^+, \Gamma \vdash \Delta \quad B^+, \Gamma \vdash \Delta}{(A \vee B)^\star, \Gamma \vdash \Delta}\ \vee\colon l \qquad \frac{\Gamma \vdash \Delta, A^+}{\Gamma \vdash \Delta, (A \vee B)^\star}\ \vee\colon r_1 \qquad \frac{\Gamma \vdash \Delta, B^+}{\Gamma \vdash \Delta, (A \vee B)^\star}\ \vee\colon r_2$$

- $\rightarrow$-introduction:

$$\frac{\Gamma \vdash \Delta, A^+ \quad B^+, \Pi \vdash \Lambda}{(A \rightarrow B)^\star, \Gamma, \Pi \vdash \Delta, \Lambda}\ \rightarrow\colon l \qquad \frac{A^+, \Gamma \vdash \Delta, B^+}{\Gamma \vdash \Delta, (A \rightarrow B)^\star}\ \rightarrow\colon r$$

- $\neg$-introduction:

$$\frac{\Gamma \vdash \Delta, A^+}{\neg A^\star, \Gamma \vdash \Delta}\ \neg\colon l \qquad \frac{A^+, \Gamma \vdash \Delta}{\Gamma \vdash \Delta, \neg A^\star}\ \neg\colon r$$

- $\forall$-introduction:

$$\frac{A\{x \leftarrow t\}^+, \Gamma \vdash \Delta}{(\forall x)A^\star, \Gamma \vdash \Delta}\ \forall\colon l$$

where t is an arbitrary *term.*

$$\frac{\Gamma \vdash \Delta, A\{x \leftarrow \alpha\}^+}{\Gamma \vdash \Delta, (\forall x)A^\star}\ \forall\colon r$$

where α is a free variable which may not occur in Γ, Δ, A. α is called an *eigenvariable.*

- The logical rules for $\exists$-introduction (the variable conditions for $\exists : l$ are the same as those for $\forall\colon r$, and similarly for $\exists : r$ and $\forall\colon l$):

$$\frac{A\{x \leftarrow \alpha\}^+, \Gamma \vdash \Delta}{(\exists x)A^\star, \Gamma \vdash \Delta}\ \exists\colon l \qquad \frac{\Gamma \vdash \Delta, A\{x \leftarrow t\}^+}{\Gamma \vdash \Delta, (\exists x)A^\star}\ \exists\colon r$$

The structural rules:

- *permutation*

$$\frac{S}{S'}\ \pi\colon l \qquad \frac{S}{S''}\ \pi'\colon r$$

where S' is a left permutation of S based on π, and S'' is a right permutation of S based on π' . In $(: l\pi) : l$ all formulas on the left side of S' are principal formulas and all formulas on the left side of S are auxiliary formulas; similarly for $p(\pi) : r$. Mostly we write the rules in the form

$$\frac{S}{S'}\ p\colon l \qquad \frac{S}{S''}\ p\colon r$$

when we not interested in specifying the particular permutation.

- *weakening*:

$$\frac{\Gamma \vdash \Delta}{\Gamma \vdash \Delta, A^\star}\ w\colon r \qquad \frac{\Gamma \vdash \Delta}{A^\star, \Gamma \vdash \Delta}\ w\colon l$$

- *contraction:*

$$\frac{A^+, A^+, \Gamma \vdash \Delta}{A^\star, \Gamma \vdash \Delta}\ c\colon l \qquad \frac{\Gamma \vdash \Delta, A^+, A^+}{\Gamma \vdash \Delta, A^\star}\ c\colon r$$

- The *cut rule:* Let us assume that A occurs in Δ and in Π. Then we define

$$\frac{\Gamma \vdash \Delta \quad \Pi \vdash \Lambda}{\Gamma, \Pi^* \vdash \Delta^*, \Lambda}\ cut(A)$$

where Π^* is Π after deletion of at least one occurrence of A, and Δ^* is Δ after deletion of at least one occurrence of A. The formula A is the auxiliary formula of $cut(A)$ and there is no principal one. If $\Pi^* = \Pi - A$ and $\Delta^* = \Delta - A$, i.e. we delete all occurrences of A in Π and Δ we speak about a *mix*. If A is not an atomic formula we call the cut *essential*, and *inessential* if A is an atom.

The cut rule can be simulated by mix and other structural rules. Indeed let ψ be the proof

$$\frac{\begin{matrix}(\psi_1) & (\psi_2)\\ \Gamma \vdash \Delta & \Pi \vdash \Lambda\end{matrix}}{\Gamma, \Pi^* \vdash \Delta^*, \Lambda}\ cut(A)$$

Then the proof ψ':

$$\cfrac{\cfrac{\begin{matrix}(\psi_1) & (\psi_2)\\ \Gamma \vdash \Delta & \Pi \vdash \Lambda\end{matrix}}{\Gamma, \Pi - A \vdash \Delta - A, \Lambda}\ mix(A)}{\Gamma, \Pi^* \vdash \Delta^*, \Lambda}\ w^* + p^*$$

is a derivation of the same end sequent. The number of additional weakenings is bounded by the number of occurrences of A in Π and Δ. At most two permutations are necessary to obtain the desired end sequent.

Note that the version of cut we are defining here is more general than the cut and mix rules in Gentzen's original paper. If we delete only one occurrence of A in Π and Δ we obtain the cut rule (according to Gentzen's terminology); if we delete all occurrences in Π and Δ we get a mix (which corresponds to Gentzen's terminology). As we are dealing with classical logic only this version of cut does not lead to problems and makes the analysis of cut-elimination more comfortable.

◇

Definition 3.2.4 Let

$$\frac{S_1 \quad S_2}{S}\ \xi$$

be a binary rule of **LK** and let S', S_1', S_2' be instantiations of the schema variables in S, S_1, S_2. Then (S_1', S_2', S') is called an *instance* of ξ. The instance of a unary rule is defined analogously. ◇

Example 3.2.1 Consider the rule

$$\frac{\Gamma \vdash \Delta, A^+ \quad \Gamma \vdash \Delta, B^+}{\Gamma \vdash \Delta, (A \land B)^\star} \land : r$$

Then

$$\frac{(\forall x)P(x), (\forall x)Q(x) \vdash P(a)^+ \quad (\forall x)P(x), (\forall x)Q(x) \vdash Q(b)^+}{(\forall x)P(x), (\forall x)Q(x) \vdash (P(a) \land Q(b))^\star} \land : r$$

is an instance of $\land : r$. ◇

Definition 3.2.5 (LK-derivation) An **LK**-*derivation* is defined as a finite directed labeled tree where the nodes are labelled by sequents (via the function *Seq*) and the edges by the corresponding rule applications. The label of the root is called the *end-sequent*. Sequents occurring at the leaves are called *initial sequents* or *axioms*. We give a formal definition:

- Let ν be a node and $Seq(\nu) = S$ for an arbitrary sequent S. Then ν is an **LK**-derivation and ν is the root node (and also a leaf).

- Let φ be a derivation tree and ν be a leaf in φ. Let (S_1, S_2, S) be an instance of the binary **LK**-rule ξ. We extend φ to φ' by appending the edges $e_1 : (\nu, \mu_1)$, $e_2 : (\nu, \mu_2)$ to ν s.t. $Seq(\mu_1) = S_1$, $Seq(\mu_2) = S_2$, and the label of e_1, e_2 is ξ. Then φ' is an **LK**-derivation with the same root as φ. μ_1, μ_2 are leaves in φ', but ν is not. ν is called a ξ-node in φ'.

- Let φ be a derivation tree and ν be a leaf in φ. Let (S', S) be an instance of a unary **LK**-rule ξ. We extend φ to φ' by appending the edge $e : (\nu, \mu)$ to ν s.t. $Seq(\mu) = S'$, and the label of e is ξ. Then φ' is an **LK**-derivation with the same root as φ. μ is a leaf in φ', but ν is not. Again ν is called a ξ-node in φ'.

We write

$$\begin{matrix} (\psi) \\ S \end{matrix}$$

to express that ψ is an **LK**- derivation with end sequent S. ◇

Definition 3.2.6 Let φ be an **LK**-derivation with initial sequent S and end sequent S' s.t. all edges are labelled by unary structural rules (these are all structural rules with the exception of cut). Then we may represent φ by

$$\frac{S}{S'}\ s^*$$

Moreover, if the structural rules are only weakenings we may write w^* instead of s^*, for weakenings and permutations $(w+p)^*$, for arbitrary weakenings and one permutation w^*+p. This notation applies to any combination of unary structural rules, where w stands for weakening, p for permutation and c for contraction. ◇

Example 3.2.2 Let φ be the **LK**-derivation

$$\cfrac{\cfrac{\nu_1\colon P(a) \vdash P(a)}{\nu_2\colon (\forall x)P(x) \vdash P(a)}\ \forall\colon l \qquad \cfrac{\nu_3\colon P(a) \vdash Q(a)}{\nu_4\colon P(a) \vdash (\exists x)Q(x)}\ \exists\colon r}{\cfrac{\nu_5\colon (\forall x)P(x) \vdash (\exists x)Q(x)}{\nu_6\colon\ \vdash (\forall x)P(x) \to (\exists x)Q(x)}\ \to\colon r}\ cut$$

The ν_i denote the nodes in φ. The leaf nodes are ν_1 and ν_3, the end node is ν_6. $Seq(\nu_2) = (\forall x)P(x) \vdash P(a)$. In practice the representation of nodes is omitted in writing down **LK**-proofs. ◇

Definition 3.2.7 (cut-complexity) Let φ be an **LK**-derivation with cuts and $\mathcal{C}$ be the set of all cut-formulas occurring in φ . Then $\max\{comp(A) \mid A \in \mathcal{C}\}$ is called the *cut-complexity* of φ and is denoted by $cutcomp(\varphi)$. If φ is cut-free (i.e. $\mathcal{C} = \emptyset$) we define $cutcomp(\varphi) = -1$ ◇

Example 3.2.3 Let φ be the **LK**-derivation in Example 3.2.2. Then

$$cutcomp(\varphi) = 0.$$

In fact the only cut formula in φ is $P(a)$ which is atomic. ◇

Definition 3.2.8 Let $\mathcal{A}$ be an axiom set. An **LK**-*proof* φ of S from $\mathcal{A}$ is an **LK**-derivation of S with initial sequents in $\mathcal{A}$. If $\mathcal{A}$ is the standard axiom set we simply call φ a proof of S. The set of all **LK**-proofs from $\mathcal{A}$ is denoted by $\Phi^{\mathcal{A}}$. If the axiom set $\mathcal{A}$ is clear from the context we frequently write Φ. For all $i \geq 0$ we define:

$$\Phi^{\mathcal{A}}_i \quad = \quad \{\varphi \mid \varphi \in \Phi^{\mathcal{A}},\ cutcomp(\varphi) \leq i\}.$$

The set of cut-free proofs is denoted by $\Phi^{\mathcal{A}}_{\emptyset}$. ◇

Example 3.2.4 Let $\mathcal{A} = \{P(a) \vdash P(a),\ P(a) \vdash Q(a)\}$. Then $\mathcal{A}$ is an axiom set (indeed there are no variables in the sequents of $\mathcal{A}$). The **LK**-derivation φ, defined in Example 3.2.2, is an **LK**-proof of $Seq(\nu_6)$ from $\mathcal{A}$, i.e. $\varphi \in \Phi^{\mathcal{A}}$. Moreover $\varphi \in \Phi_0^{\mathcal{A}}$. Note that $\mathcal{A}$ is not a subset of the standard axiom set. ◇

Definition 3.2.9 (path) Let π: $\mu_1, \ldots, \mu_n$ be a sequence of nodes in an **LK**-derivation φ s.t. for all $i \in \{1, \ldots, n-1\}$ (μ_i, μ_{i+1}) is an edge in φ. Then π is called a *path* from μ_1 to μ_n in φ of *length* $n-1$ (denoted by $lp(\pi) = n-1$). If $n = 1$ and $\pi = \mu_1$ then ψ is called a trivial path. π is called a *branch* if μ_1 is the root of φ and μ_n is a leaf in φ. We use the terms *predecessor* and *successor* contrary to the direction of edges in the tree: if there exists a path from μ_1 to μ_2 then μ_2 is called a *predecessor* of μ_1. The successor relation is defined in a analogous way. E.g. every initial sequent is a predecessor of the end sequent. ◇

Example 3.2.5 Let $\varphi =$

$$\dfrac{\dfrac{\nu_1\colon P(a) \vdash P(a)}{\nu_2\colon (\forall x)P(x) \vdash P(a)}\ \forall\colon l \qquad \dfrac{\nu_3\colon P(a) \vdash Q(a)}{\nu_4\colon P(a) \vdash (\exists x)Q(x)}\ \exists\colon r}{\dfrac{\nu_5\colon (\forall x)P(x) \vdash (\exists x)Q(x)}{\nu_6\colon\ \vdash (\forall x)P(x) \to (\exists x)Q(x)}\ \to\colon r}\ cut$$

as in Example 3.2.2. $\nu_6, \nu_5, \nu_2, \nu_1$ is a path in φ which is also a branch. ν_2 is a predecessor of ν_6. ν_1 is not a predecessor of ν_4. ◇

Definition 3.2.10 (subderivation) Let φ' be the subtree of an **LK**-derivation φ with root node ν (where ν is a node in φ). Then φ' is called a *subderivation* of φ and we write $\varphi' = \varphi.\nu$.
Let ρ be an (arbitrary) **LK**-derivation of $Seq(\nu)$. Then we write $\varphi[\rho]_\nu$ for the deduction φ after the replacement of the subderivation $\varphi.\nu$ by ρ on the node ν in φ (under the restriction that $\varphi.\nu$ and ρ have the same end-sequent). ◇

Example 3.2.6 Let $\varphi =$

$$\dfrac{\dfrac{\nu_1\colon P(a) \vdash P(a)}{\nu_2\colon (\forall x)P(x) \vdash P(a)}\ \forall\colon l \qquad \dfrac{\nu_3\colon P(a) \vdash Q(a)}{\nu_4\colon P(a) \vdash (\exists x)Q(x)}\ \exists\colon r}{\dfrac{\nu_5\colon (\forall x)P(x) \vdash (\exists x)Q(x)}{\nu_6\colon\ \vdash (\forall x)P(x) \to (\exists x)Q(x)}\ \to\colon r}\ cut$$

$\varphi.\nu_4 =$

$$\dfrac{\nu_3\colon P(a) \vdash Q(a)}{\nu_4\colon P(a) \vdash (\exists x)Q(x)}\ \exists\colon r$$

Let $\rho =$

$$\cfrac{\cfrac{\nu_8\colon P(a), P(a) \vdash Q(a)}{\nu_9\colon\ P(a), P(a) \vdash (\exists x)Q(x)}\ \exists\colon r}{\nu_{10}\colon\ P(a) \vdash (\exists x)Q(x)}\ c\colon l$$

Then $\varphi[\rho]_{\nu_4} =$

$$\cfrac{\cfrac{\cfrac{\nu_1\colon P(a) \vdash P(a)}{\nu_2\colon (\forall x)P(x) \vdash P(a)}\ \forall\colon l \qquad \cfrac{\cfrac{\nu_8\colon P(a), P(a) \vdash Q(a)}{\nu_9\colon\ P(a), P(a) \vdash (\exists x)Q(x)}\ \exists\colon r}{\nu_{10}\colon\ P(a) \vdash (\exists x)Q(x)}\ c\colon l}{\nu_5\colon (\forall x)P(x) \vdash (\exists x)Q(x)}\ cut}{\nu_6\colon\ \vdash (\forall x)P(x) \rightarrow (\exists x)Q(x)}\ \rightarrow\colon r$$

Note that $\varphi[\rho]_{\nu_4}$ is an **LK**-proof from the axiom set

$$\{P(a) \vdash P(a);\ P(a), P(a) \vdash Q(a)\}.$$

◇

Definition 3.2.11 (depth) Let φ be an **LK**-derivation and ν be a node in φ. Then the *depth* of ν (denoted by depth(ν)) is defined by the maximal length of a path from ν to a leaf of $\varphi.\nu$. The depth of any leaf in φ is zero. ◇

Definition 3.2.12 (regularity) An **LK**-derivation φ is called *regular* if

- all eigenvariables of quantifier introductions $\forall\colon r$ and $\exists\colon l$ in φ are mutually different.
- If an eigenvariable α occurs as an eigenvariable in a proof node ν then α occurs only above ν in the proof tree.

◇

There exists a straightforward transformation from **LK**-derivations into regular ones: just rename the eigenvariables in different subderivations. The necessity of renaming variables was the main motivation for changing Hilbert's linear format to the tree format of **LK**. From now on we assume, without mentioning the fact explicitly, that all **LK**-derivations we consider are regular.

The formulas in sequents on the branch of a deduction tree are connected by a so-called *ancestor relation*. Indeed if A occurs in a sequent S and A is marked as principal formula of a, let us say binary, inference on the sequents S_1, S_2, then the auxiliary formulas in S_1, S_2 are *immediate ancestors* of A (in S). If A occurs in S_1 and is not an auxiliary formula of an inference then A occurs also in S; in this case A in S_1 is also an immediate ancestor of A in S. The case of unary rules is analogous. General ancestors are defined via reflexive and transitive closure of the relation.

Example 3.2.7 Instead of using special symbols for formula occurrences we mark the occurrences of a formula in different sequents by numbers. Let $\varphi =$

$$\dfrac{\dfrac{\nu_1\colon P(a)^4 \vdash P(a)}{\nu_2\colon (\forall x)P(x)^5 \vdash P(a)}\;\forall\colon l \qquad \dfrac{\nu_3\colon P(a) \vdash Q(a)^1}{\nu_4\colon P(a) \vdash (\exists x)Q(x)^2}\;\exists\colon r}{\dfrac{\nu_5\colon (\forall x)P(x)^6 \vdash (\exists x)Q(x)^3}{\vdash (\forall x)P(x) \to (\exists x)Q(x)^7}\;\to\colon r}\;cut$$

1 is ancestor of 2, 2 is ancestor of 3, 3 is ancestor of 7. 1 is ancestor of 3 and of 7. 4 is ancestor of 5, 5 of 6 and 6 of 7. 4 is ancestor of 7, but not of 2. ◇

Definition 3.2.13 (ancestor path) A sequence $\bar{\alpha}\colon (\alpha_1, \ldots, \alpha_n)$ for formula occurrences α_i in an **LK**-derivation φ is called an *ancestor path* in φ if for all $i \in \{1, \ldots, n-1\}$ α_i is an immediate ancestor of α_{i+1}. If $n = 1$ then α_1 is called a (trivial) ancestor path. ◇

Example 3.2.8 In Example 3.2.7 the sequence $4, 5, 6, 7$ is an ancestor path. ◇

Definition 3.2.14 Let Ω be a set of formula occurrences in an **LK**-derivation φ and ν be a node in φ. Then $S(\nu, \Omega)$ is the subsequent of $Seq(\nu)$ obtained by deleting all formula occurrences which are not ancestors of occurrences in Ω. ◇

Example 3.2.9 Let $\varphi =$

$$\dfrac{\dfrac{\nu_1\colon P(a) \vdash P(a)}{\nu_2\colon (\forall x)P(x) \vdash P(a)}\;\forall\colon l \qquad \dfrac{\nu_3\colon P(a) \vdash Q(a)}{\nu_4\colon P(a) \vdash (\exists x)Q(x)}\;\exists\colon r}{\dfrac{\nu_5\colon (\forall x)P(x) \vdash (\exists x)Q(x)}{\nu_6\colon \vdash (\forall x)P(x) \to (\exists x)Q(x)}\;\to\colon r}\;cut$$

and α the left occurrence of the cut formula in φ, and β the right occurrence. Let $\Omega = \{\alpha, \beta\}$. Then

$$\begin{array}{rcl} S(\nu_1, \Omega) & = & \vdash P(a), \\ S(\nu_3, \Omega) & = & P(a) \vdash . \end{array}$$

◇

Remark: If Ω consists just of the occurrences of all cut formulas which occur "below" ν then $S(\nu, \Omega)$ is the subsequent of $Seq(\nu)$ consisting of all formulas which are ancestors of a cut. These subsequents are crucial for the definition of the characteristic set of clauses and of the method CERES in Chapter 6. ◇

Definition 3.2.15 The *length* of a proof φ is defined by the number of nodes in φ and is denoted by $l(\varphi)$. ◇

Definition 3.2.16 (cut-derivation) Let ψ be an **LK**-derivation of the form

$$\frac{\begin{array}{cc}(\psi_1) & (\psi_2) \\ \Gamma_1 \vdash \Delta_1 & \Gamma_2 \vdash \Delta_2\end{array}}{\Gamma_1, \Gamma_2^* \vdash \Delta_1^*, \Delta_2} \; cut(A)$$

Then ψ is called a *cut-derivation*; note that ψ_1 and ψ_2 may contain cuts. If the cut is a mix we speak about a *mix-derivation*. ψ is called *essential* if $comp(A) > 0$ (i.e. if the cut is essential). ◇

Definition 3.2.17 (rank, grade) Let ψ be a cut-derivation of the form

$$\frac{\begin{array}{cc}(\psi_1) & (\psi_2) \\ \Gamma_1 \vdash \Delta_1 & \Gamma_2 \vdash \Delta_2\end{array}}{\Gamma_1, \Gamma_2^* \vdash \Delta_1^*, \Delta_2} \; cut(A)$$

Then we define the *grade* of ψ as $comp(A)$.
Let μ be the root node of ψ_1 and ν be the root node of ψ_2. An A-right path in ψ_1 is a path in ψ_1 of the form $\mu, \mu_1, \ldots, \mu_n$ s.t. A occurs in the consequents of all $Seq(\mu_i)$ (note that A clearly occurs in Δ_1). Similarly an A-left path in ψ_2 is a path in ψ_2 of the form $\nu, \nu_1, \ldots, \nu_m$ s.t. A occurs in the antecedents of all $Seq(\nu_j)$. Let P_1 be the set of all A-right paths in ψ_1 and P_2 be the set of all A-left paths in ψ_2. Then we define the *left-rank* of ψ ($\mathrm{rank}_l(\psi)$) and the right-rank of ψ ($\mathrm{rank}_r(\psi)$) as

$$\begin{array}{rcl}\mathrm{rank}_l(\psi) & = & \max\{lp(\pi) \mid \pi \in P_1\} + 1, \\ \mathrm{rank}_r(\psi) & = & \max\{lp(\pi) \mid \pi \in P_2\} + 1.\end{array}$$

The *rank* of ψ is the sum of right-rank and left-rank, i.e. $\mathrm{rank}(\psi) = \mathrm{rank}_l(\psi) + \mathrm{rank}_r(\psi)$. ◇

Example 3.2.10 Let $\varphi =$

$$
\dfrac{\dfrac{\dfrac{\nu_1\colon P(a) \vdash P(a)}{\nu_2\colon (\forall x)P(x) \vdash P(a)}\ \forall\colon l \qquad \dfrac{\nu_3\colon P(a) \vdash Q(a)}{\nu_4\colon P(a) \vdash (\exists x)Q(x)}\ \exists\colon r}{\nu_5\colon (\forall x)P(x) \vdash (\exists x)Q(x)}\ cut}{\nu_6\colon\ \vdash (\forall x)P(x) \to (\exists x)Q(x)}\ \to\colon r
$$

Then the only cut-derivation in φ is ψ:

$$
\dfrac{\dfrac{\nu_1\colon P(a) \vdash P(a)}{\nu_2\colon (\forall x)P(x) \vdash P(a)}\ \forall\colon l \qquad \dfrac{\nu_3\colon P(a) \vdash Q(a)}{\nu_4\colon P(a) \vdash (\exists x)Q(x)}\ \exists\colon r}{\nu_5\colon (\forall x)P(x) \vdash (\exists x)Q(x)}\ cut
$$

The grade of ψ is 0 as the cut is atomic, $\text{rank}_l(\psi) = 2$, $\text{rank}_r(\psi) = 2$ and $\text{rank}(\psi) = 4$. ◇

3.3 Unification and Resolution

Definition 3.3.1 (unifier) Let $\mathcal{A}$ be a nonempty set of atoms and σ be a substitution. σ is called a *unifier* of $\mathcal{A}$ if the set $\mathcal{A}\sigma$ contains only one element. σ is called a *most general unifier* (or m.g.u.) of $\mathcal{A}$ if σ is a unifier of $\mathcal{A}$ and for all unifiers λ of $\mathcal{A}$ $\sigma \leq_s \lambda$. ◇

Sometimes we have to unify not only a single set of atoms but several sets of atoms simultaneously. We call such a problem a simultaneous unification problem.

Definition 3.3.2 Let $W = (\mathcal{A}_1, \ldots, \mathcal{A}_n)$ where the $\mathcal{A}_i$ are nonempty sets of atoms for $i = 1, \ldots, n$. A substitution ϑ is called a simultaneous unifier of W if ϑ unifies all $\mathcal{A}_i$. σ is called a most general simultaneous unifier of W if σ is a simultaneous unifier of W and $\sigma \leq_s \vartheta$ for all simultaneous unifiers ϑ of W. ◇

Most general unification was the key novel feature of the resolution principle by J.A. Robinson in 1965 (see [69]). He proved that for all unifiable sets there exists also a most general unifier (making the computation of other unifiers superfluous). Because most general unification plays an important role in the complexity analysis of CERES (Section 6.5) and in the generalization of proofs (Section 8.3) we present unification in more detail; in particular we define a unification algorithm UAL , prove the unification theorem and give a complexity analysis of UAL. We largely follow the presentation in [61].

Example 3.3.1 Let $\mathcal{A} = \{P(x, f(y)), P(x, f(x)), P(x', y')\}$, and

$$\begin{aligned} \sigma &= \{y \leftarrow x, x' \leftarrow x, y' \leftarrow f(x)\} \text{ and} \\ \sigma_t &= \{x \leftarrow t, y \leftarrow t, x' \leftarrow t, y' \leftarrow f(t)\} \end{aligned}$$

All substitutions σ, σ_t are unifiers of $\mathcal{A}$. Moreover, we see that the unifier σ plays an exceptional role. Indeed,

$$\sigma\{x \leftarrow t\} = \sigma_t, \quad \text{i.e.,} \quad \sigma \leq_s \sigma_t.$$

It is easy to verify that for all unifying substitutions ϑ (including those with $dom(\vartheta) - V(\mathcal{A}) \neq \emptyset$) we obtain $\sigma \leq_s \vartheta$. σ is more general than all other unifiers of W, it is indeed "most" general. However, σ is not the only most general unifier; for the unifier $\lambda : \{y \leftarrow x', x \leftarrow x', y' \leftarrow f(x')\}$ we get

$\lambda \leq_s \sigma$, $\sigma \leq_s \lambda$, and $\lambda \leq_s \vartheta$ for all unifiers ϑ of $\mathcal{A}$.

◇

Example 3.3.2 Let

$$W = (\{P(x), P(a)\}, \{P(y), P(f(x))\}, \{Q(x), Q(z)\}).$$

Then $\sigma = \{x \leftarrow a, y \leftarrow f(a), z \leftarrow a\}$ is a simultaneous unifier of W; it is also a most general one. Indeed, $\{P(x), P(a)\}\sigma = \{P(a)\}$, $\{P(y), P(f(x))\}\sigma = \{P(f(a))\}$, $\{Q(x), Q(z)\}\sigma = \{Q(a)\}$. Note that the union of all three sets in W is not unifiable. ◇

The question remains, whether there is always an m.g.u. of $\mathcal{A}$ in case $\mathcal{A}$ is unifiable. We will give a positive answer and design an algorithm which always computes an m.g.u. if $\mathcal{A}$ is unifiable and stops otherwise. Before we define such an algorithm we show that the problem of unifying an arbitrary finite set of two or more atoms (or of solving a simultaneous unification problem) can be reduced to unifying a set $\{t_1, t_2\}$ consisting of two terms only.

Let $\mathcal{A} = \{A_1, \ldots, A_n\}$ be a set of atoms and $n \geq 2$. Let $\{P_1, \ldots, P_m\}$ be the set of all predicate symbols appearing in W. For every P_i we introduce a new function symbol f_i (which is not contained in $\mathcal{A}$) of the same arity. Then we translate

$$P_i(S_1^i, \ldots, S_{k_i}^i) \quad \text{into} \quad f_i(S_1^i, \ldots, S_{k_i}^i).$$

Let $\mathcal{T}$ be the set of all translated expressions.

Clearly ϑ is a unifier of $\mathcal{A}$ iff ϑ is a unifier of $\mathcal{T}$. Thus let $\mathcal{T} = \{t_1, \ldots, t_n\}$ be the set of terms corresponding to $\mathcal{A}$ and let f be an arbitrary n-ary function symbol.
Then $t_1\sigma = \ldots = t_n\sigma$ iff $f(t_1, \ldots, t_n)\sigma = f(t_i, \ldots, t_i)\sigma$ for some $i \in \{1, \ldots, n\}$, and $\mathcal{T}$ is unifiable by unifier ϑ iff

$$\{f(t_1, \ldots, t_n), f(t_i, \ldots, t_i)\}$$

is unifiable by unifier ϑ.
Now let $W = (\mathcal{A}_1, \ldots, \mathcal{A}_k)$ be a simultaneous unification problem. First we translate this problem into a simultaneous unification problem for sets of terms

$$W' = (\mathcal{T}_1, \ldots \mathcal{T}_k).$$

Let $\mathcal{T}_j = \{t_1^j, \ldots, t_{m_j}^j\}$ for $j = 1, \ldots, k$. Then σ is an m.g.u. of W' iff σ unifies the two terms s_1, s_2 for

$$\begin{aligned} s_1 &= g(f(t_1^1, \ldots, t_{m_1}^1), \ldots, f(t_1^k, \ldots, t_{m_k}^k)), \\ s_2 &= g(f(t_{i_1}^1, \ldots, t_{i_1}^1), \ldots, f(t_{i_k}^k, \ldots, t_{i_k}^k)), \end{aligned}$$

for $i_j \in \{1, \ldots, m_i\}$.
This completes the reduction of the unification problem for W to a unification problem of $\{t_1, t_2\}$, t_1 and t_2 being terms. The translation of the unification problems to the unification of two terms is in fact linear. We first define the measure of symbolic complexity in a general way:

Definition 3.3.3 Let X be a syntactic expression. Then $\|X\|$ = number of symbol occurrences in X. In particular we define for terms

$$\begin{aligned} \|t\| = 1 \quad &\text{for} \quad t \in \mathrm{CS} \cup V, \\ \|f(t_1, \ldots, t_n)\| \quad &= \quad 1 + \|t_1\| + \cdots + \|t_n\|. \end{aligned}$$

The definition of $\|A\|$ for atoms is analogous. If $W = \{w_1, \ldots, w_k\}$ we define

$$\|W\| = \sum_{i=1}^{k} \|w_i\|.$$

$\|W\|$ for $W = (w_1, \ldots, w_n)$ is defined in the same way. ◇

For the complexity of the translations of the unification problems we obtain

Proposition 3.3.1 *Let $\mathcal{A}$ be a unification problem of a set of atoms. Then there exists an equivalent unification problem for two terms t_1, t_2 s.t. $\|\{t_1, t_2\}\| \leq 4 * \|\mathcal{A}\|$. If W is a simultaneous unification problem then there exists an equivalent one for two terms s_1, s_2 s.t. $\|\{s_1, s_2\}\| \leq 5 * \|W\|$.*

Proof: If $\mathcal{A} = \{A_1, \ldots, A_n\}$ the translation to a problem $\mathcal{T} = \{t_1, \ldots, t_n\}$ does not increase the size of the problem. For

$$\mathcal{T}' = \{f(t_1, \ldots, t_n), f(t_i, \ldots, t_i)\}$$

we select a t_i s.t. $\|t_i\|$ is minimal. Then

$$\|\mathcal{T}'\| \leq 2 + 2 * \|\mathcal{T}\| \leq 4 * \|\mathcal{T}\|.$$

For the simultaneous unification problem we get a unification problem $\{s_1, s_2\}$ of the form

$$\begin{aligned} s_1 &= g(f(t_1^1, \ldots, t_{m_1}^1), \ldots, f(t_1^k, \ldots, t_{m_k}^k)), \\ s_2 &= g(f(t_{i_1}^1, \ldots, t_{i_1}^1), \ldots, f(t_{i_k}^k, \ldots, t_{i_k}^k)). \end{aligned}$$

For an appropriate choice of the $t_{i_j}^j$ we obtain

$$\|\{s_1, s_2\}\| \leq 2 * (2 * \|W\| + 1) \leq 5 * \|W\|.$$

□

Example 3.3.3 $\mathcal{A} = \{P(x, f(y)), P(x, f(x)), P(u, v)\}$.
We transfer the unification problem to a unification problem for two terms.

$$\mathcal{T} = \{h(x, f(y)), h(x, f(x)), h(u, v)\} \text{ for } h \in FS_2.$$

Now let $i \in FS_3$. We define

$$\mathcal{T}' = \{i(h(x, f(y)), h(x, f(x)), h(u, v)), i(h(x, f(y)), h(x, f(y), h(x, f(y))\}.$$

Then
$\sigma = \{y \leftarrow x, u \leftarrow x, v \leftarrow f(x)\}$ (an m.g.u. of $\mathcal{T}$) is also an m.g.u. of $\mathcal{T}'$.
Indeed
$\mathcal{T}'\sigma = \{i(h(x, f(x)), h(x, f(x)), h(x, f(x)))\}$.

◇

By the transformations described above it is justified to reduce unification to "binary" unification. We now investigate the syntactical structures within $\{t_1, t_2\}$ which characterize unifiability. Let

$$\mathcal{T} = \{f(t_1, \ldots, t_n), f(s_1, \ldots, s_n)\}.$$

By elementary properties of substitutions we obtain

$$f(t_1, \ldots, t_n)\sigma = f(t_1\sigma, \ldots, t_n\sigma),$$
$$f(s_1, \ldots, s_n)\sigma = f(s_1\sigma, \ldots, s_n\sigma).$$

Then $\mathcal{T}$ is unifiable iff there exists a substitution σ such that

$$s_1\sigma = t_1\sigma, \ldots, s_n\sigma = t_n\sigma.$$

We observe that the unifiability of $\mathcal{T}$ is equivalent to the simultaneous unification problem $(\{s_1, t_1\}, \ldots, \{s_n, t_n\})$. The s_i or the t_i may be again of the form $g(u_1, \ldots, u_m)$ and the property of decomposition holds recursively. This leads to the following definition:

Definition 3.3.4 (corresponding pairs) Let t_1, t_2 be two terms. The set of *corresponding pairs* $\mathrm{CORR}(t_1, t_2)$ is defined as follows:

1. $(t_1, t_2) \in \mathrm{CORR}(t_1, t_2)$
2. If $(s_1, s_2) \in \mathrm{CORR}(t_1, t_2)$ such that $s_1 = f(r_1, \ldots, r_n)$ and $s_2 = f(w_1, \ldots, w_n)$, where $f \in FS$, then $(r_i, w_i) \in \mathrm{CORR}(t_1, t_2)$ for all $i = 1, \ldots, n$.
3. Nothing else is in $\mathrm{CORR}(E_1, E_2)$.

A pair $(s_1, s_2) \in \mathrm{CORR}(t_1, t_2)$ is called *irreducible* if the leading symbols of s_1, s_2 are different. ◇

There are two different types of irreducible pairs. Take for example the pairs $(x, f(y)), (x, f(x))$ and $(f(x), g(a))$. All these pairs are irreducible. But for $\sigma = \{x \leftarrow f(y)\}$ the pair $(x, f(y))\sigma = (f(y), f(y))$ is reducible and even identical; thus there exists a substitution which removes the irreducibility of $(x, f(y))$. We show now that no such substitutions exist for the pairs $(f(x), g(a))$: For arbitrary substitutions λ we have the property

$$(f(x), g(a))\lambda = (f(x\lambda), g(a))$$

and thus $(f(x), g(a))\lambda$ is irreducible for all λ.

Let us consider the pair $(x, f(x))$ and the substitution $\lambda = \{x \leftarrow f(y)\}$ then $(x, f(x))\lambda = (f(y), f(f(y)))$; $(f(y), f(f(y)))$ is reducible but reduction yields $(y, f(y))$. Because, for all substitutions $\lambda, x\lambda$ is properly contained in $f(x)\lambda$ the set $\{x, f(x)\}$ is not unifiable.

Definition 3.3.5 We call a pair of terms (t_1, t_2) *unifiable* if the set $\{t_1, t_2\}$ is unifiable. $\diamond$

In this terminology, $(x, f(y))$ is irreducible but unifiable, but $(x, f(x))$ and $(f(x), g(a))$ are irreducible and nonunifiable. It is easy to realize that (t_1, t_2) is only unifiable if all irreducible elements in $\mathrm{CORR}(t_1, t_2)$ are (separately) unifiable; note that this property is necessary but not sufficient.

Example 3.3.4

$$\begin{aligned} t_1 &= f(x, f(x, y)), \\ t_2 &= f(f(u, u), v), \\ \mathrm{CORR}(t_1, t_2) &= \{(t_1, t_2), (x, f(u, u)), (f(x, y), v)\}. \end{aligned}$$

We eliminate the irreducible pair $(x, f(u, u))$ by applying the substitution

$$\sigma_1 = \{x \leftarrow f(u, u)\}.$$

Applying σ_1 we obtain $(t_1\sigma_1, t_2\sigma_1)$ and

$$\mathrm{CORR}(t_1\sigma_1, t_2\sigma_1) = \{(f(f(u, u), f(f(u, u), y)), \quad f(f(u, u), v)), \\ (f(u, u), f(u, u)), (f(f(u, u), y), v), (u, u)\}.$$

In $\mathrm{CORR}(t_1\sigma_1, t_2\sigma_1)$ the only irreducible pair is $(f(f(u, u), y), v)$ which can be eliminated by the substitution $\sigma_2 = \{v \leftarrow f(f(u, u), y)\}$.
Now it is easy to see that $t_1\sigma_1\sigma_2 = t_2\sigma_1\sigma_2$ and that $\mathrm{CORR}(t_1\sigma_1, \sigma_2, t_2\sigma_1\sigma_2)$ consists of identical pairs only. $\sigma_1\sigma_2$ is clearly a unifier of $\{t_1, t_2\}$ (it is even the m.g.u.). For the unification above it was essential that all irreducible pairs were unifiable. $\diamond$

For an algorithmic treatment of unification the reducible and identical corresponding pairs are irrelevant; it suffices to focus on irreducible pairs. We are led to the following definition:

Definition 3.3.6 (difference set) The set of all irreducible pairs in $\mathrm{CORR}(t_1, t_2)$ is called the *difference set* of (t_1, t_2) and is denoted by $\mathrm{DIFF}(t_1, t_2)$. $\diamond$

Example 3.3.5 Let $(t_1, t_2) = (f(x, f(x, y)), f(f(u, u), v))$ as in Example 3.3.4. Then

$$\mathrm{DIFF}(t_1, t_2) = \{(x, f(u, u)), (f(x, y), v)\}.$$

By application of $\sigma_1 = \{x \leftarrow f(u,u)\}$ we first obtain the pair $(f(u,u), f(u,u))$ which is in $\text{CORR}(t_1\sigma_1, t_2\sigma_1)$, but not in $\text{DIFF}(t_1\sigma_1, t_2\sigma_1)$. Thus we obtain $\text{DIFF}(t_1\sigma_1, t_2\sigma_1) = \{(f(f(u,u), y), v)\}$. ◇

We have already mentioned that $\{t_1, t_2\}$ is unifiable only if all corresponding pairs are unifiable. By definition of $\text{DIFF}(t_1, t_2)$ we get the following necessary condition for unifiability: For all pairs $(s,t) \in \text{DIFF}(t_1, t_2)$, (s,t) is unifiable. The following proposition shows that the unification problem for (single) pairs in $\text{DIFF}(t_1, t_2)$ is very simple.

Proposition 3.3.2 *Let t_1, t_2 be terms and (s,t) be a pair in $\text{DIFF}(t_1, t_2)$. Then (s,t) is unifiable iff the following two conditions hold:*

(a) *$s \in V$ or $t \in V$,*

(b) *If $s \in V$ ($t \in V$) then s does not occur in t (t does not occur in s).*

Proof: (s,t) is unifiable $\Rightarrow$:
By definition of $\text{DIFF}(t_1, t_2)$ the pair (s,t) is irreducible; because it is unifiable, s or t must be a variable (otherwise s and t are terms with different head symbols and thus are not unifiable). Suppose now without loss of generality that $s \in V$. If s occurs in t then $s\lambda$ (properly) occurs in $t\lambda$ for all $\lambda \in \text{SUBST}$; in this case (s,t) is not unifiable. We have shown that (a) and (b) both hold.
(a), (b) $\Rightarrow$
Suppose without loss of generality that $s \in V$. Define $\lambda = \{s \leftarrow t\}$; then $s\lambda = t$ and, because s does not occur in t, $t\lambda = t$. It follows that λ is a unifier of (s,t). □

The idea of the unification algorithm shown in Figure 3.1 is the following: construct the difference set D. If there are nonunifiable pairs in D then stop with failure; otherwise eliminate a pair (x,t) in D by the substitution $\{x \leftarrow t\}$ and construct the next difference set.
Note that UAL is a nondeterministic algorithm, because the selection of a unifiable pair (s,t) is nondeterministic. UAL can be transformed into different deterministic (implementable) versions by choosing appropriate search strategies. But even if the pairs (s,t) are selected from left to right (according to their positions in (t_1, t_2)), both s and t may be variables and thus $\{s \leftarrow t\}$ and $\{t \leftarrow s\}$ can both be used in extending the substitution ϑ.
UAL is more than a decision algorithm in the usual sense, because in case of a positive answer (termination without failure) it also provides an m.g.u. for $\{t_1, t_2\}$.

```
 algorithm UAL {input is a pair of terms (t1, t2)};
begin
  ϑ ← ε;
  while DIFF(t1ϑ, t2ϑ) ≠ ∅ do
    if DIFF(t1ϑ, t2ϑ) contains a nonunifiable pair
    then failure
    else
      select a unifiable pair (s, t) ∈ DIFF(t1ϑ, t2ϑ)
      if s ∈ V
      then α := s; β := t
      else α := t; β := s
      end if
      ϑ := ϑ{α ← β}
    end if
  end while
  {ϑ is m.g.u.}
end.
```

Figure 3.1: Unification algorithm.

Theorem 3.3.1 (unification theorem) UAL *is a decision algorithm for the unifiability of two terms. In particular the following two properties hold:*

(a) *If $\{t_1, t_2\}$ is not unifiable then UAL stops with failure.*

(b) *If $\{t_1, t_2\}$ is unifiable then UAL stops and ϑ (the final substitution constructed by UAL) is a most general unifier of $\{t_1, t_2\}$.*

Proof:

(a) If (t_1, t_2) is not unifiable then for all substitutions λ $t_1\lambda \neq t_2\lambda$. Thus for every ϑ defined in UAL we get $\mathrm{DIFF}(t_1\vartheta, t_2\vartheta) \neq \emptyset$. In order to prove termination we have to show that the while-loop is not an endless loop: In every execution of the while loop a new substitution ϑ is defined as
$\vartheta = \vartheta'\{x \leftarrow t\}$, where ϑ' is the substitution defined during the execution before and (x, t) (or (t, x)) is a pair in $\mathrm{DIFF}(t_1\vartheta', t_2\vartheta')$ with $x \in V$. Because $x \notin V(t)$ (otherwise UAL terminates with failure before), the pair $(t_1\vartheta, t_2\vartheta)$ does not contain x anymore; we conclude

$$|V(\{(t_1\vartheta', t_2\vartheta')\})| > |V(\{(t_1\vartheta, t_2\vartheta)\})|.$$

It follows that the number of executions of the while-loop must be $\leq k$ for $k = |V(\{t_1, t_2\})|$. We see that, whatever result is obtained, UAL must terminate. Because UAL terminates and (by nonunifiability) $\text{DIFF}(t_1\vartheta, t_2\vartheta) \neq \emptyset$ for all ϑ, it must stop with failure. *

(b) In the k-th execution of the while loop (provided termination with failure does not take place) the k-th definition of ϑ via $\vartheta := \vartheta\{\alpha \leftarrow \beta\}$ is performed. We write ϑ_k for the value of ϑ defined in the k-th execution.

Suppose now that η is an arbitrary unifier of $\{t_1, t_2\}$. We will show by induction on k, that for all ϑ_k there exist substitutions λ_k such that $\vartheta_k\lambda_k = \eta$. We are now in a position to conclude our proof as follows:

Because UAL terminates (see part (a) of the proof), there exists a number m such that the m-th execution of the while-loop is the last one.

From $\vartheta_m\lambda_m = \eta$ we get $\vartheta_m \leq_s \eta$. Moreover ϑ_m must be a unifier: Because the m-th execution is the last one, either $\text{DIFF}(t_1\vartheta_m, t_2\vartheta_m) = \emptyset$ or there is a nonunifiable pair $(s, t) \in \text{DIFF}(t_1\vartheta_m, t_2\vartheta_m)$;

but the second alternative is impossible, as $\eta = \vartheta_m\lambda_m$ and λ_m is a unifier of $(t_1\vartheta_m, t_2\vartheta_m)$. Because ϑ_m is a unifier, η is an arbitrary unifier and $\vartheta_m \leq_s \eta$, ϑ_m is an m.g.u. of $\{t_1, t_2\}$ (note that m and ϑ_m depend on (t_1, t_2) only, but λ_m depends on η).

Therefore it remains to show that the following statement A(k) holds for all $k \in N$:

A(k): Let ϑ_k be the substitution defined in the k-th execution of the while-loop. Then there exists a substitution λ_k such that $\vartheta_k\lambda_k = \eta$.

We proceed by induction on k:

A(0): $\vartheta_0 = \epsilon$.
We choose $\vartheta_0 = \eta$ and obtain $\vartheta_0\lambda_0 = \eta$.

(IH) Suppose that A(k) holds.

If ϑ_{k+1} is not defined by UAL (because it stops before) the antecedent of A($k+1$) is false and thus A($k+1$) is true. So we may assume that ϑ_{k+1} is defined by UAL.

Then $\vartheta_{k+1} = \vartheta_k\{x \leftarrow t\}$ where $x \in V$, $t \in T$ and $(x,t) \in \mathrm{DIFF}(t_1\vartheta_k, t_2\vartheta_k)$ or $(t,x) \in \mathrm{DIFF}(t_1\vartheta_k, t_2\vartheta_k)$.

By the induction hypothesis (IH) we know that there exists a λ_k such that $\vartheta_k\lambda_k = \eta$. Our aim is to find an appropriate substitution λ_{k+1} such that $\vartheta_{k+1}\lambda_{k+1} = \eta$.

Because λ_k is a unifier of $(t_1\vartheta_k, t_2\vartheta_k)$ it must unify the pair (x,t), i.e., $x\lambda_k = t\lambda_k$. Therefore λ_k must contain the element $x \leftarrow t\lambda_k$. We define

$$\lambda_{k+1} = \lambda_k - \{x \leftarrow t\lambda_k\}.$$

The substitution λ_{k+1} fulfils the property

$$(*) \quad \{x \leftarrow t\}\lambda_{k+1} = \lambda_{k+1} \cup \{x \leftarrow t\lambda_k\}.$$

To prove $(*)$ it is sufficient to show that

$$v\{x \leftarrow t\}\lambda_{k+1} = v(\lambda_{k+1} \cup \{x \leftarrow t\lambda_k\})$$

holds for all $v \in dom(\lambda_k)$ (note that $dom(\lambda_{k+1}) \subseteq dom(\lambda_k)$). If $v \neq x$ then $v\{x \leftarrow t\} = v$ and $v\{x \leftarrow t\}\lambda_{k+1} = v\lambda_{k+1}$. If $v = x$ then

$$\begin{array}{lcl} x\{x \leftarrow t\}\lambda_{k+1} & = & t\lambda_{k+1} \text{ and} \\ x(\lambda_{k+1} \cup \{x \leftarrow t\lambda_k\}) & = & t\lambda_k. \end{array}$$

By definition of UAL, ϑ_{k+1} is only defined if $x \notin V(t)$. But $x \notin V(t)$ implies $t\lambda_{k+1} = t\lambda_k$ and

$$x\{x \leftarrow t\}\lambda_{k+1} = x(\lambda_{k+1} \cup \{x \leftarrow t\lambda_k\}).$$

We see that $(*)$ holds.

We obtain

$$\vartheta_k\lambda_k = \vartheta_k(\lambda_{k+1} \cup \{x \leftarrow t\lambda_k\}) = \vartheta_k\{x \leftarrow t\}\lambda_{k+1} = \vartheta_{k+1}\lambda_{k+1}.$$

This concludes the proof of $A(k+1)$. □

Example 3.3.6 Let $t_1 = f(x, h(x,y), z)$ and $t_2 = f(u, v, g(v))$. We compute an m.g.u. of $\{t_1, t_2\}$ by using UAL.

$\vartheta_0 = \epsilon$.

$$\mathrm{DIFF}(t_1, t_2) = \{(x,u), (h(x,y), v), (z, g(v))\}.$$

All pairs in $\mathrm{DIFF}(t_1, t_2)$ are unifiable.

$$\vartheta_1 = \vartheta_0\{z \leftarrow g(v)\} = \{z \leftarrow g(v)\}.$$

$$\mathrm{DIFF}(t_1\vartheta_1, t_2\vartheta_1) = \{(x,u), (h(x,y), v)\}.$$

Again all pairs in the difference set are unifiable.

$$\vartheta_2 = \vartheta_1\{x \leftarrow u\} = \{z \leftarrow g(v),\ x \leftarrow u\}$$
$$\mathrm{DIFF}(t_1\vartheta_2, t_2\vartheta_2) = \{(h(u,y), v)\}.$$

As v does not occur in $h(u,y)$ we continue and obtain

$$\vartheta_3 = \vartheta_2\{v \leftarrow h(u,y)\} = \{z \leftarrow g(h(u,y)),\ x \leftarrow u,\ v \leftarrow h(u,y)\}$$

Now

$$\mathrm{DIFF}(t_1\vartheta_3, t_2\vartheta_3) = \emptyset,$$

and (due to Theorem 3.3.1) ϑ_3 is an m.g.u. of $\{t_1, t_2\}$. The expression $t_1\vartheta_3 (= t_2\vartheta_3)$ obtained by unification is $f(u, h(u,y), g(h(u,y)))$. If we replace t_2 by $t'_2 = f(v, v, g(v))$ then we obtain

$$\mathrm{DIFF}(t_1, t'_2) = \{(x, v), (h(x,y), v), (z, g(v))\}.$$

By defining $\vartheta'_1 = \{x \leftarrow v\}$ we get

$$\mathrm{DIFF}(t_1\vartheta'_1, t_2\vartheta'_1) = \{(h(v,y), v), (z, g(v))\}.$$

Because the pair $(h(v,y), v)$ is not unifiable we stop with failure. ◇

UAL is exponential if the unified expression is constructed explicitly. Below we show that the complexity of UAL is also at most exponential.

Theorem 3.3.2 *Let $\mathcal{T} = \{t_1, t_2\}$ be a unifiable set of terms and σ be an m.g.u. computed by* UAL. *Then*

$$\|\mathcal{T}\sigma\| \leq \|\mathcal{T}\| * 2^{\|\mathcal{T}\|}.$$

Proof: UAL unifies pairs of the difference set $\mathrm{DIFF}(t_1, t_2)$. In the beginning the difference set is of the form

$$\{(x_1, t^1_1), \ldots, (x_1, t^1_{k_1}), \ldots, (x_n, t^n_1), \ldots, (x_n, t^n_{k_n})\}.$$

As $\mathcal{T}$ is unifiable the terms t^n_i do not contain x_i. Now let us apply $\vartheta_1\colon \{x_1 \leftarrow t^1_1\}$. Then $\mathrm{DIFF}(t_1\vartheta_1, t_2\vartheta_2)$ contains pairs of the following form:

- $(x_i, t^i_j\{x_1 \leftarrow t^1_1\})$ (for $x_1 \in V(t^i_j)$) or
- (x_i, s^i_j), or
- (y_i, w^i_j),

for $i = 2\ldots, n$; the s^i_j are subterms of some t^1_j and the y_i are variables in $\mathcal{T}$ which appear in the terms t^1_j. Now assume that after k steps of the algorithm you have irreducible pairs of the form

- $(x_i, t^i_j\{x_m \leftarrow t^m_{r_m}\} \cdots \{x_k \leftarrow t^k_{r_k}\})$

for $i = k+1, \ldots, n$ and $m \in \{1, \ldots, m\}$. Then we apply the substitution

$$\{x_{k+1} \leftarrow t^{k+1}_1\}\{x_m \leftarrow t^m_{r_m}\} \cdots \{x_k \leftarrow t^k_{r_k}\}$$

and get new irreducible pairs of the form

$$(\star)\ (x_i, t^i_j\{x_m \leftarrow t^m_{r_m}\} \cdots \{x_k \leftarrow t^k_{r_k}\}\{x_{k+1} \leftarrow t^{k+1}_1\{x_m \leftarrow t^m_{r_m}\}) \cdots \{x_k \leftarrow t^k_{r_k}\}).$$

Repetition of substitutions of the form $\{x_i \leftarrow t_i\}$ can be avoided as, by unifiability, there is no cycle in the problem. By reordering the substitutions appropriately we can thus drop multiple occurrences. Therefore $(\star)$ can be rewritten to a form

$$(x_i, t^i_j\{y_1 \leftarrow s_1\} \cdots \{y_k \leftarrow s_k\}\{y_{k+1} \leftarrow s_{k+1}\})$$

where the s_i are terms in $\mathcal{T}$. So if m is the number of steps in UAL the m.g.u. $\sigma = \theta_m$ is of the form

$$\begin{array}{l}\{x_1 \leftarrow s_1, \ldots, x_n \leftarrow s_n\},\\ s_i = \{x_{i_1} \leftarrow t_{i_1}\} \cdots \{x_{i_k} \leftarrow t_{i_k},\end{array}$$

where the terms t_{i_j} occur in the original problem $\mathcal{T}$ and, in particular,

$$\|t_{i_1}\| + \cdots \|t_{i_k}\| \leq \|\mathcal{T}\|.$$

Therefore we have the problem of maximizing

$$\|s_i\| \leq \|t_1\| * \cdots * \|t_n\|$$

where $\|t_1\| + \cdots \|t_n\| \leq \|\mathcal{T}\|$. It is easy to see that for all s_i

$$(I)\ \|s_i\| \leq 2^{\|\mathcal{T}\|}.$$

Let $r = \max\{\|s_i\| \mid i = 1, \ldots, n\}$. Then clearly $\|\mathcal{T}\sigma\| \leq \|\mathcal{T}\| * r$. Using (I) we finally obtain

$$\|\mathcal{T}\sigma\| \leq \|\mathcal{T}\| * 2^{\|\mathcal{T}\|}. \ \Box$$

Definition 3.3.7 (clause) A *clause* is an atomic sequent. ◇

Definition 3.3.8 (contraction normalization) Let C be a clause. A *contraction normalization* of C is a clause D obtained from C by omitting multiple occurrences of atoms in C_+ and in C_-. ◇

Definition 3.3.9 (factor) Let C be a clause and D be a nonempty subclause of C_+ or of C_- and let σ be an m.g.u. of the atoms of D. Then a contraction normalization of $C\sigma$ is called a *factor* of C. ◇

Definition 3.3.10 (resolvent) Let C and D be clauses of the form

$$\begin{array}{rcl} C & = & \Gamma \vdash \Delta_1, A_1, \ldots, \Delta_n, A_n, \Delta_{n+1}, \\ D & = & \Pi_1, B_1, \ldots, \Pi_m, B_m, \Pi_{m+1} \vdash \Lambda \end{array}$$

s.t. C and D do not share variables and the set $\{A_1, \ldots, A_n, B_1, \ldots, B_m\}$ is unifiable by a most general unifier σ. Then the clause

$$R\colon\ \Gamma\sigma, \Pi_1\sigma, \ldots \Pi_{m+1}\sigma \vdash \Delta_1\sigma, \ldots, \Delta_{n+1}\sigma, \Lambda\sigma$$

is called a *resolvent* of C and D.
If $m = 1$ and Γ is empty we speak about a PRF-resolvent (Positive Restricted Factoring). The (single) atom in $\{A_1, \ldots, A_n, B_1, \ldots, B_m\}\sigma$ is called the *resolved atom.* ◇

Remark: There are several ways to define the concept of resolvent (see e.g. [31, 61, 62]). We chose the original concept defined in [69] which combines unification, contraction and cut in a single rule. ◇

Definition 3.3.11 (p-resolvent) Let $C = \Gamma \vdash \Delta, A^m$ and $D = A^n, \Pi \vdash \Lambda$ with $n, m \geq 1$. Then the clause $\Gamma, \Pi \vdash \Delta, \Lambda$ is called a *p-resolvent* of C and D. ◇

Remark: The p-resolvents of C and D are just sequents obtained by applying the cut rule to C and D. Thus resolution of clauses is a cut combined with most general unification. ◇

In order to resolve two clauses C_1, C_2 we must ensure that C_1 and C_2 are variable disjoint. This can always be achieved by renaming variables by permutation of variables.

Definition 3.3.12 Let C be a clause and π be a permutation substitution (i.e. π is a binary function $V \to V$). Then $C\sigma$ is called a *variant* of C. ◇

Definition 3.3.13 (resolution deduction) A *resolution deduction* γ is a labelled tree like an **LK**-derivation with the exception that it is (purely) binary and all edges are labelled by the resolution rule. If we replace the resolutions by p-resolutions we speak about a *p-resolution derivation.* If γ is a p-resolution deduction and all clauses are variable-free we call γ a *ground resolution deduction.* Let $\mathcal{C}$ be a set of clauses. If all initial sequents (initial clauses) in γ are variants of clauses in $\mathcal{C}$ and D is the clause labelling the root, then γ is called a *resolution derivation of D from $\mathcal{C}$*. If $D = \vdash$ then γ is called a *resolution refutation* of $\mathcal{C}$. $\diamond$

Definition 3.3.14 (ground projection) Let γ' be a ground resolution deduction which is an instance of a resolution deduction γ. Then γ' is called a *ground projection* of γ. $\diamond$

Example 3.3.7 Let

$$\mathcal{C} = \{\vdash P(x), P(a);\ P(y) \vdash P(f(y));\ P(f(f(a))) \vdash\}.$$

Then the following derivation γ is a resolution refutation of $\mathcal{C}$:

$$\cfrac{\cfrac{\cfrac{\vdash P(x), P(a) \quad P(y) \vdash P(f(y))}{\vdash P(f(a))} \quad P(z) \vdash P(f(z))}{\vdash P(f(f(a)))} \quad P(f(f(a))) \vdash}{\vdash}$$

The following instantiation γ' of γ

$$\cfrac{\cfrac{\cfrac{\vdash P(a), P(a) \quad P(a) \vdash P(f(a))}{\vdash P(f(a))} \quad P(f(a)) \vdash P(f(f(a)))}{\vdash P(f(f(a)))} \quad P(f(f(a))) \vdash}{\vdash}$$

is a ground resolution refutation of $\mathcal{C}$ and a ground projection of γ .

$\diamond$

Remark: A p-resolution derivation γ is an **LK**-derivation with atomic sequents, where the only rule in γ is cut. Paths, the ancestor relation and ancestor paths for resolution derivations can be defined exactly like for **LK**-derivations. $\diamond$

Chapter 4

Complexity of Cut-Elimination

4.1 Preliminaries

Our aim is to compare different methods of cut-elimination. For this aim we need logic-free axioms. The original formulation of **LK** by Gentzen also served the purpose of simulating Hilbert-type calculi and deriving axiom schemata within fixed proof length. Below we show that there exists a polynomial transformation from an **LK**-proof with arbitrary axioms of type $A \vdash A$ to atomic ones.

Lemma 4.1.1 *Let S be the sequent $A \vdash A$ for an arbitrary formula A. Then there exists a cut-free* **LK***-proof $\pi(A)$ of S from the standard axiom set $\mathcal{A}_T$ with $l(\pi(A)) \leq 4 * comp(A) + 1$.*

Proof: We proceed by induction on $comp(A)$.

(IB) $comp(A) = 0$: then $A \vdash A$ is an axiom in $\mathcal{A}_T$ and we simply define

$$\pi(A) = A \vdash A.$$

Obviously $l(\pi(A)) = 4 * comp(A) + 1 = 1$.

(IH) assume that, for all formulas A with $comp(A) \leq k$, there are cut-free proofs $\pi(A)$ of $A \vdash A$ from $\mathcal{A}_T$ s.t. $l(\pi(A)) \leq 4 * comp(A) + 1$.

Now let A be a formula with $comp(A) = k + 1$. We have to distinguish several cases:

M. Baaz, A. Leitsch, *Methods of Cut-Elimination*, Trends in Logic 34,

DOI 10.1007/978-94-007-0320-9_4,

(a) $A \equiv \neg B$. Then $comp(B) = k$ and, by (IH), there exists a proof $\pi(B)$ of $B \vdash B$ from $\mathcal{A}_T$ s.t. $l(\pi(B)) \leq 4 * k + 1$. We define $\pi(A)$ as

$$\cfrac{\cfrac{\cfrac{\begin{array}{c}(\pi(B))\\ B \vdash B\end{array}}{\vdash B, \neg B}\ \neg\colon r}{\vdash \neg B, B}\ p\colon r}{\neg B \vdash \neg B}\ \neg\colon l$$

and thus

$$l(\pi(A)) = l(\pi(B)) + 3 \leq 4 * k + 1 + 3 < 4 * (k + 1) + 1 = 4 * comp(A) + 1.$$

(b) $A \equiv B \wedge C$. Then $comp(B) + comp(C) = k$ and, by (IH), there exists proofs $\pi(B)$ of $B \vdash B$ from $\mathcal{A}_T$ and $\pi(C)$ of $C \vdash C$ from $\mathcal{A}_T$ s.t.

(I) $l(\pi(B)) \leq 4 * comp(B) + 1,\ l(\pi(C)) \leq 4 * comp(C) + 1.$

We define $\pi(A) =$

$$\cfrac{\cfrac{\begin{array}{c}\pi(B)\\ B \vdash B\end{array}}{B \wedge C \vdash B}\ \wedge\colon l_1 \qquad \cfrac{\begin{array}{c}\pi(C)\\ C \vdash C\end{array}}{B \wedge C \vdash C}\ \wedge\colon l_2}{B \wedge C \vdash B \wedge C}\ \wedge\colon r$$

Then

$$l(\pi(A)) = l(\pi(B)) + l(\pi(C)) + 3 \leq_{(\mathrm{I})} 4 * comp(B) + 4 * comp(C) + 5 = 4 * (comp(B) + comp(C) + 1) + 1 = 4 * comp(A) + 1.$$

(c) $A \equiv B \vee C$. Symmetric to (b).

(d) $A \equiv B \rightarrow C$. Then $comp(B) + comp(C) = k$ and, by (IH), there exists proofs $\pi(B)$ of $B \vdash B$ from $\mathcal{A}_T$ and $\pi(C)$ of $C \vdash C$ from $\mathcal{A}_T$ s.t.

(II) $l(\pi(B)) \leq 4 * comp(B) + 1,\ l(\pi(C)) \leq 4 * comp(C) + 1.$

We define $\pi(A) =$

$$\cfrac{\cfrac{\cfrac{\begin{array}{cc}(\pi(B)) & (\pi(C))\\ B \vdash B & C \vdash C\end{array}}{B \rightarrow C, B \vdash C}\ \rightarrow\colon l}{B, B \rightarrow C \vdash C}\ p\colon l}{B \rightarrow C \vdash B \rightarrow C}\ \rightarrow\colon r$$

Then

$$l(\pi(A)) = l(\pi(B)) + l(\pi(C)) + 3 \leq_{(\mathrm{II})} 4 * comp(B) + 4 * comp(C) + 5 \leq 4 * (comp(B) + comp(C) + 1) + 1 = 4 * comp(A) + 1.$$

(e) $A \equiv (\forall x)B(x)$. Let α be a free variable not occurring in A; then $comp(B(\alpha)) = k$ and, by (IH), there exists a proof $\pi(B(\alpha))$ of $B(\alpha) \vdash B(\alpha)$ from $\mathcal{A}_T$ s.t.

$$\text{(III) } l(\pi(B(\alpha)) \leq 4 * k + 1.$$

We define $\pi(A) =$

$$\cfrac{\cfrac{\begin{array}{c}(\pi(B(\alpha)))\\ B(\alpha) \vdash B(\alpha)\end{array}}{(\forall x)B(x) \vdash B(\alpha)}\ \forall\colon l}{(\forall x)B(x) \vdash (\forall x)B(x)}\ \forall\colon r$$

Then

$$l(\pi(A)) = l(\pi(B(\alpha))) + 2 \leq_{\text{(III)}} 4 * k + 3 < 4 * (k + 1) + 1 = 4 * comp(A) + 1.$$

(f) $A \equiv (\exists x)B(x)$. Symmetric to (e). □

Statman's proof sequence to be defined in Section 4.3 expresses proofs in combinatory logic with equality being the only predicate symbol. In formalizing a proof in predicate logic with equality we have several choices: (1) we add appropriate equality axioms to the antecedent of the end-sequents, (2) we add atomic equality axioms to the leaves of the proofs, and (3) we extend **LK** by equational rules. We choose (2) because it is most appropriate for complexity analysis.[1] Alternative (3) would be closer to mathematical reasoning, but we would have to extend the calculus **LK**; in Chapter 7 we will explore alternative (3) as a tool to analyze real mathematical proofs.

Definition 4.1.1 We define an axiom system for equality $\mathcal{A}_e$ which contains the standard axiom set $\mathcal{A}_T$ and the following axioms:

- (ref) $\vdash s = s$ for all terms s,
- (symm) $s = t \vdash t = s$ for all terms s, t,
- (trans) $s = t, t = r \vdash s = r$ for all terms s, t, r.
- (subst) For all atoms A and sets of positions Λ in A, and for all terms s, t we add the axioms

$$s = t, A[s]_\Lambda \vdash A[t]_\Lambda.$$

[1] Note that, in [16] we chose alternative (1).

Note that the substitution axioms admit the replacement of the term s by the term t either everywhere in the atoms, or only on specified places. (subst) corresponds to the presence of inessential cut in Takeuti's calculus LKe [74]. A specific subset of (subst) are the axioms

$$s = t, A[x]\{x \leftarrow s\} \vdash A[x]\{x \leftarrow t\}.$$

which we also denote by $s = t, A(s) \vdash A(t)$. ◇

For non-atomic formulas A the principle $s = t, A(s) \vdash A(t)$ is derivable from $\mathcal{A}_e$ by proofs linear in $comp(A)$.

Lemma 4.1.2 *Let $A(x)$ be an arbitrary formula (possibly) containing the free variable x and s, t arbitrary terms. Then there exists a proof $\lambda(A, s, t)$ from $\mathcal{A}_e$ of $s = t, A(s) \vdash A(t)$ with only atomic cuts s.t.*

$l(\lambda(A, s, t)) \leq 11 * comp(A) + 1.$

Proof: We construct the proofs $\lambda(A, s, t)$ by induction on $comp(A)$.
(IB) Let A be an atom and s, t arbitrary terms. Then we define

$$\lambda(A, s, t) \equiv s = t, A(s) \vdash A(t).$$

Indeed, $\lambda(A, s, t)$ is an axiom in $\mathcal{A}_e$ and thus a (cut-free) proof from $\mathcal{A}_e$. Moreover

$$l(\lambda(A, s, t)) = 1 = 11 * comp(A) + 1.$$

(IH) Assume that for all A with $comp(A) \leq k$ and for all terms s, t we have proofs $\lambda(A, s, t)$ of $s = t, A(s) \vdash A(t)$ from $\mathcal{A}_e$ with only atomic cuts and $l(\lambda(A, s, t)) \leq 11 * comp(A) + 1$.
Now let A be a formula with $comp(A) = k + 1$ and s, t be arbitrary terms. We have to distinguish the following cases:

(a) $A \equiv \neg B$. Then, by (IH) there exists a proof $\lambda(B, t, s)$ of $t = s, B(t) \vdash B(s)$ with only atomic cuts and $l(\lambda(B, t, s)) \leq 11 * comp(B) + 1$.

We define $\lambda(A, s, t) =$

$$\dfrac{\dfrac{\text{symm}}{s = t \vdash t = s} \quad \dfrac{\dfrac{\dfrac{\dfrac{\dfrac{(\lambda(B,t,s))}{t = s, B(t) \vdash B(s)}}{\neg B(s), t = s, B(t) \vdash}\ \neg{:}\,l}{B(t), \neg B(s), t = s \vdash}\ p_1{:}\,l}{\neg B(s), t = s \vdash \neg B(t)}\ \neg{:}\,r}{t = s, \neg B(s) \vdash \neg B(t)}\ p_2{:}\,l}{s = t, \neg B(s) \vdash \neg B(t)}\ cut$$

Then $\lambda(A,s,t)$ is a proof from $\mathcal{A}_e$ with only atomic cuts and we have $l(\lambda(A,s,t)) = l(\lambda(B,t,s)) + 6 \leq 11 * comp(B) + 7 < 11 * comp(A) + 1$.

(b) $A \equiv B \wedge C$. Then $comp(B) + comp(C) = k$ and, by (IH), we have proofs $\lambda(B,s,t)$, $\lambda(C,s,t)$ from $\mathcal{A}_e$ with at most atomic cuts and

$$l(B,s,t) \leq 11 * comp(B) + 1,\ l(C,s,t) \leq 11 * comp(C) + 1.$$

We define $\lambda(A,s,t) =$

$$\dfrac{\dfrac{\dfrac{\dfrac{\dfrac{\dfrac{\dfrac{\dfrac{(\lambda(B,s,t))}{s=t, B(s) \vdash B(t)}}{B(s), C(s), s=t \vdash B(t)}\ w{:}\,l + p{:}\,l \quad \dfrac{\dfrac{(\lambda(C,s,t))}{s=t, C(s) \vdash C(t)}}{B(s), C(s), s=t \vdash C(t)}\ w{:}\,l + p{:}\,l}{B(s), C(s), s=t \vdash B(t) \wedge C(t)}\ \wedge{:}\,r}{B(s) \wedge C(s), C(s), s=t \vdash B(t) \wedge C(t)}\ \wedge{:}\,l}{B(s) \wedge C(s), B(s) \wedge C(s), s=t \vdash B(t) \wedge C(t)}\ p{:}\,l + \wedge{:}\,l}{B(s) \wedge C(s), s=t \vdash B(t) \wedge C(t)}\ c{:}\,l}{s=t, B(s) \wedge C(s) \vdash B(t) \wedge C(t)}\ p{:}\,l}{s=t, (B \wedge C)(s) \vdash (B \wedge C)(t)}\ \text{(def)}$$

Clearly $\lambda(A,s,t)$ is a proof from $\mathcal{A}_e$ with at most atomic cuts and

$$l(\lambda(A,s,t)) = l(\lambda(B,s,t)) + l(\lambda(C,s,t)) + 10 \leq$$
$$11 * comp(B) + 11 * comp(C) + 12 =$$
$$11 * (comp(B) + comp(C) + 1) + 1 = 11 * comp(A) + 1.$$

(c) $A \equiv B \vee C$: symmetric to (b).

(d) $A \equiv B \to C$. Then $comp(B) + comp(C) = k$ and, by (IH), we have proofs $\lambda(B,t,s)$, $\lambda(C,s,t)$ from $\mathcal{A}_e$ with at most atomic cuts and

$$l(\lambda(B,t,s)) \leq 11 * comp(B) + 1,\ l(\lambda(C,s,t)) \leq 11 * comp(C) + 1.$$

We define $\lambda(A,s,t) =$

$$\dfrac{\dfrac{\dfrac{\dfrac{\text{symm}}{s=t \vdash t=s} \quad \dfrac{\dfrac{\dfrac{(\lambda(B,t,s))}{t=s, B(t) \vdash B(s)} \quad \dfrac{\dfrac{(\lambda(C,s,t))}{s=t, C(s) \vdash C(t)}}{C(s), s=t \vdash C(t)}\ p{:}\,l}{B(s) \to C(s), t=s, B(t), s=t \vdash C(t)}\ \to{:}\,l}{t=s, s=t, B(s) \to C(s) \vdash B(t) \to C(t)}\ p{:}\,l + \to{:}\,r}{s=t, s=t, B(s) \to C(s) \vdash B(t) \to C(t)}\ cut}{s=t, B(s) \to C(s) \vdash B(t) \to C(t)}\ c{:}\,l}{s=t, (B \to C)(s) \vdash (B \to C)(t)}\ \text{(def)}$$

Then $\lambda(A,s,t)$ is a proof from $\mathcal{A}_e$ with at most atomic cuts and

$$\begin{aligned} & l(\lambda(A,s,t)) = l(\lambda(B,t,s)) + l(\lambda(C,s,t)) + 7 \leq \\ & \quad 11 * comp(B) + 11 * comp(C) + 9 < \\ & \quad 11 * (comp(B) + comp(C) + 1) + 1 = 11 * comp(A) + 1. \end{aligned}$$

(e) $A \equiv (\forall x)B(x)$. Let $A \equiv A(\alpha)$. Then $(\forall x)B(x) \equiv (\forall x)B(\alpha,x)$. Let β be a free variable not occurring in $A(s), A(t)$. Then, by (IH), there exists a proof $\lambda(B(\alpha,\beta),s,t)$ of $s = t, B(s,\beta) \vdash B(t,\beta)$ from $\mathcal{A}_e$ with at most atomic cuts and $l(\lambda(B(\alpha,\beta),s,t)) \leq 11 * comp(B(\alpha,\beta)) + 1$. Note that $comp(B(\alpha,\beta)) = comp(B(\alpha,x)) = k$.

We define $\lambda(A,s,t) =$

$$\cfrac{\cfrac{\cfrac{\begin{array}{c}(\lambda(B(\alpha,\beta),s,t))\\ s = t, B(s,\beta) \vdash B(t,\beta)\end{array}}{s = t, (\forall x)B(s,x) \vdash B(t,\beta)}\; p{:}\,l + \forall{:}\,l + p{:}\,l}{s = t, (\forall x)B(s,x) \vdash (\forall x)B(t,x)}\; \forall{:}\,r}{s = t, ((\forall x)B(\alpha,x))(s) \vdash ((\forall x)B(\alpha,x))(t)}\; (\mathrm{def})$$

Clearly $\lambda(A,s,t)$ is a proof from $\mathcal{A}_e$ with at most atomic cuts and

$$\begin{aligned} l(\lambda(A,s,t)) &= l(\lambda(B(\alpha,\beta),s,t)) + 4 \leq \\ & \quad 11 * comp(B(\alpha,\beta)) + 5 < 11 * comp(A) + 1. \end{aligned}$$

(f) $A \equiv (\exists x)B(x)$: symmetric to (e). □

4.2 Proof Complexity and Herbrand Complexity

The content of the area of proof complexity in propositional logic is the analysis of recursive relations between formulas and their proofs. As first-order logic is undecidable there exists no recursive bound on the lengths of proofs. Therefore, in predicate logic another measure which is independent of the calculus is needed, which can be used as reference measure to compare proof theoretic transformations. The most natural measure for the complexity of formulas of sequents comes from Herbrand's theorem, namely the minimal size of a Herbrand disjunction (or more general of a Herbrand sequent). This minimal size (the so-called Herbrand complexity) can be considered as a kind of *logical length* of a first-order formula to which proof complexities of various calculi can be related (see [15, 16]).

Definition 4.2.1 Let $S: A_1, \ldots, A_n \vdash B_1, \ldots, B_m$ be a weakly quantified sequent. Let A_i^-, B_j^- be the formulas A_i, B_j after omission of the quantifier occurrences. For every i, j let $\vec{A_i}, \vec{B_j}$ be sequences of instances of A_i^- and B_j^-, respectively. Then any permutation of the sequent

$$S': \vec{A_1}, \ldots, \vec{A_n} \vdash \vec{B_1}, \ldots, \vec{B_m}$$

is called an *instantiation sequent* of S. ◇

Example 4.2.1 Let $S = P(a), (\forall x)(P(x) \to P(f(x))) \vdash (\exists y)P(f(f(y)))$. Then

$$S': P(a), P(a) \to P(f(a)) \vdash P(f(f(x))), P(f(f(a)))$$

is an instantiation sequent of S. ◇

Definition 4.2.2 Let $\mathcal{A}$ be an axiom set (see Definition 3.2.1). A sequent S is called $\mathcal{A}$-valid if $\mathcal{A} \models S$. ◇

Definition 4.2.3 (Herbrand sequent) Let $\mathcal{A}$ be an axiom set and let S be a weakly quantified $\mathcal{A}$-valid sequent. An instantiation sequent S' of S is called an *$\mathcal{A}$-Herbrand sequent* of S if S' is $\mathcal{A}$-valid. If $\mathcal{A}$ is the standard axiom set then S' is called a *Herbrand sequent* of S. ◇

Example 4.2.2 Let $S = P(a), (\forall x)(P(x) \to P(f(x))) \vdash (\exists y)P(f(f(y)))$ be as in Example 4.2.1. Then

$$S': P(a), P(a) \to P(f(a)) \vdash P(f(f(x))), P(f(f(a)))$$

is an instantiation sequent of S, but not a Herbrand sequent of S.

$$S': P(a), P(a) \to P(f(a)), P(f(a)) \to P(f(f(a))) \vdash P(f(f(a)))$$

is a Herbrand sequent of S. ◇

Example 4.2.3 Let $\mathcal{T}$ be the set of all terms, $\mathcal{A} = \mathcal{A}_e \cup \{e \circ t = t \mid t \in \mathcal{T}\}$ (for $\mathcal{A}_e$ see Definition 4.1.1), and

$$S = P(a), (\forall x)(P(x) \to P(f(x))) \vdash P(f(e \circ f(a))).$$

S is $\mathcal{A}$-valid, but not valid (so there is no Herbrand-sequent of S). But

$$S': P(a), P(a) \to P(f(a)), P(f(a)) \to P(f(f(a))) \vdash P(f(e \circ f(a)))$$

is an $\mathcal{A}$-Herbrand sequent of S. ◇

As a consequence of Herbrand's theorem [44] every valid weakly quantified sequent has a Herbrand sequent. A semantic proof of this theorem requires König's lemma and thus a weak form of the axiom of choice. The completeness proofs in automated deduction are based on this semantic proof. However, there exists a constructive method to obtain Herbrand sequents S' from **LK**-proofs φ of S, provided the cut formulas in φ are quantifier-free. These Herbrand sequents can be directly obtained from proofs of arbitrary weakly quantified sequents; see [16] and for more efficient algorithms [79]. To simplify the construction of Herbrand sequents we restrict it to prenex sequents only.

Definition 4.2.4 A sequent $S\colon A_1, \ldots, A_n \vdash B_1, \ldots, B_m$ is called *prenex* if all A_i, B_j are prenex formulas. ◇

Remark: If S is prenex and weakly quantified then the A_i are universal prenex forms and the B_j are existential prenex forms. ◇

In his famous paper [38] G. Gentzen proved the so-called *midsequent theorem* which yields a construction method for Herbrand sequents S' of S from cut-free proofs of S. Here we are not interested in specific normal forms of proofs but only in the Herbrand sequent itself; for this reason we define another more direct method for its construction (see also [79]).

Definition 4.2.5 Let φ be a proof of a prenex weakly quantified sequent S and let A be a formula occurring at position μ in S. If A contains quantifiers we define $q(\varphi, \mu)$ as a sequence of all ancestors B of A in φ s.t. B is quantifier-free and is the auxiliary formula of a quantifier inference (i.e. we locate the "maximal" non-quantified ancestors of A in φ). If such an ancestor B does not exist (some quantified ancestors might have been introduced by weakening) we define $q(\varphi, \mu)$ as the empty sequence. If A is quantifier-free we define $q(\varphi, \mu) = A$. ◇

Definition 4.2.6 Let φ be a proof of a prenex weakly quantified sequent S for $S = A_1, \ldots, A_n \vdash B_1, \ldots, B_m$, where the μ_i are the occurrences of A_i and ν_j the occurrences of B_j in S. We define

$$S^\star(\varphi) = q(\varphi, \mu_1), \ldots, q(\varphi, \mu_n) \vdash q(\varphi, \nu_1), \ldots, q(\varphi, \nu_n).$$

From $S^\star(\varphi)$ we construct an instantiation sequent $H^\star(\varphi)$ by deleting double occurrences of formulas in $S^\star(\varphi)$ and then by ordering the remaining formulas on both sides of the sequent lexicographically. Note that, by Definition 4.2.1, $H^\star(\varphi)$ is indeed an instantiation sequent of S. ◇

Example 4.2.4 We give a proof $\varphi \in \Phi_0^{\mathcal{A}}$ of S defined in Example 4.2.3 and extract $H^\star(\varphi)$ from φ.
Let $\varphi =$

$$
\dfrac{\dfrac{P(a) \vdash P(a) \quad \dfrac{\dfrac{\dfrac{P(f(a)) \vdash P(f(a)) \quad \overset{(\psi)}{P(f(f(a))) \vdash P(f(e \circ f(a)))}}{P(f(a)) \to P(f(f(a))), P(f(a)) \vdash P(f(e \circ f(a)))}\ \to: l}{(\forall x)(P(x) \to P(f(x))), P(f(a)) \vdash P(f(e \circ f(a)))}\ \forall: l}{P(f(a)), (\forall x)(P(x) \to P(f(x))) \vdash P(f(e \circ f(a)))}\ p: l}{\dfrac{\dfrac{P(a) \to P(f(a)), (\forall x)(P(x) \to P(f(x))), P(a) \vdash P(f(e \circ f(a)))}{(\forall x)(P(x) \to P(f(x))), (\forall x)(P(x) \to P(f(x))), P(a) \vdash P(f(e \circ f(a)))}\ \forall: l}{(\forall x)(P(x) \to P(f(x))), P(a) \vdash P(f(e \circ f(a)))}\ c: l}\ \to: l
$$

$$
\dfrac{(\forall x)(P(x) \to P(f(x))), P(a) \vdash P(f(e \circ f(a)))}{P(a), (\forall x)(P(x) \to P(f(x))) \vdash P(f(e \circ f(a)))}\ p: l
$$

and $\psi =$

$$
\dfrac{\overset{(\psi_1)}{\vdash f(f(a)) = f(e \circ f(a))} \quad f(f(a)) = f(e \circ f(a)), P(f(f(a))) \vdash P(f(e \circ f(a)))}{P(f(f(a))) \vdash P(f(e \circ f(a)))}\ cut
$$

$\psi_1 =$

$$
\dfrac{\vdash f(a) = e \circ f(a) \quad \overset{(\psi_{1,1})}{f(a) = e \circ f(a) \vdash f(f(a)) = f(e \circ f(a))}}{\vdash f(f(a)) = f(e \circ f(a))}\ cut
$$

$\psi_{1,1} =$

$$
\dfrac{\vdash f(f(a)) = f(f(a)) \quad \dfrac{f(a) = e \circ f(a), f(f(a)) = f(f(a)) \vdash f(f(a)) = f(e \circ f(a))}{f(f(a)) = f(f(a)), f(a) = e \circ f(a) \vdash f(f(a)) = f(e \circ f(a))}\ p: l}{f(a) = e \circ f(a) \vdash f(f(a)) = f(e \circ f(a))}\ cut
$$

We get

$$
\begin{aligned}
q(\varphi, \mu_1) &= P(a), \\
q(\varphi, \mu_2) &= P(a) \to P(f(a)), P(f(a)) \to P(f(f(a))), \\
q(\varphi, \nu_1) &= P(f(e \circ f(a))) \\
S^\star(\varphi) &= H^\star(\varphi) = \\
& \quad P(a), P(a) \to P(f(a)), P(f(a)) \to P(f(f(a))) \vdash P(f(e \circ f(a))).
\end{aligned}
$$

◇

The following theorem shows that the size of Herbrand sequents define a lower bound for the size of cut-free proofs.

Theorem 4.2.1 *Let $\mathcal{A}$ be an axiom set and $\varphi \in \Phi_0^{\mathcal{A}}$ be a proof of a prenex weakly quantified sequent S. Then (1) $H^\star(\varphi)$ is an $\mathcal{A}$-Herbrand sequent of S and*
(2) $\|H^\star(\varphi)\| \leq \|\varphi\|$.

Proof: We prove (1) by induction on $l(\varphi)$.

$l(\varphi) = 1$. Then the proof consists only of the root labeled by an axiom S in $\mathcal{A}$. By Definition 4.2.6 $H^\star(\varphi)$ is constructed from S by omitting multiple occurrences and then by ordering. So we have

$$\frac{S}{H^\star(\varphi)}\ s^*$$

As S is (trivially) $\mathcal{A}$-valid and the structural rules are sound $H^\star(\varphi)$ is $\mathcal{A}$-valid, too.

(IH) Assume that the theorem holds for all proofs φ with $l(\varphi) \leq n$.

Now let $\varphi \in \Phi_0^{\mathcal{A}}$ and $l(\varphi) = n + 1$. We distinguish several cases:

- The last inference in φ is a unary structural rule ξ. Then φ is of the form

  $$\frac{\begin{matrix}(\varphi')\\ \Gamma \vdash \Delta\end{matrix}}{\Gamma' \vdash \Delta'}\ \xi$$

  Then, by (IH), $H^\star(\varphi')$ is an $\mathcal{A}$-Herbrand sequent of $\Gamma \vdash \Delta$ and thus is $\mathcal{A}$-valid. If ξ is a contraction- or a permutation rule then, by definition of $H^\star(\varphi)$, we have $H^\star(\varphi) = H^\star(\varphi')$ and so $H^\star(\varphi)$ is $\mathcal{A}$-valid. If ξ is a weakening rule there are two cases: (1) the main formula A (occurring on the position μ) is quantified; in this case $q(\varphi, \mu) = \emptyset$ and $H^\star(\varphi) = H^\star(\varphi')$. If A is not quantified then $H^\star(\varphi)$ can be obtained from $H^\star(\varphi')$ by weakening, contractions and permutations (which are all sound rules); so $H^\star(\varphi)$ is $\mathcal{A}$-valid too.

- The last inference in φ is a unary propositional rule. We consider only the rule $\vee\colon r_1$; the proof for the other rules $\vee\colon r_2$, $\wedge\colon l_1$, $\wedge\colon l_2$, $\neg\colon l$ and $\neg\colon r$ is completely analogous.

  So φ is of the form

  $$\frac{\begin{matrix}(\varphi')\\ \Gamma \vdash \Delta, (A)_{\nu'}\end{matrix}}{\Gamma \vdash \Delta, (A \vee B)_\nu}\ \vee\colon r_1$$

  As $\Gamma \vdash \Delta, A \vee B$ is a prenex sequent A and $A \vee B$ must be quantifier-free(!); therefore $q(\varphi', \nu') = A$ and $q(\varphi, \nu) = A \vee B$. Now let $\Gamma^* \vdash$

Δ^*, A be a permutation variant of $H^\star(\varphi')$; then $\Gamma^* \vdash \Delta^*, A \vee B$ is a permutation variant of $H^\star(\varphi)$. So we can obtain $H^\star(\varphi)$ from $H^\star(\varphi')$ by the following derivation:

$$\cfrac{\cfrac{\cfrac{H^\star(\varphi')}{\Gamma^* \vdash \Delta^*, A}\ s^*}{\Gamma^* \vdash \Delta^*, A \vee B}\ \vee{:}\, r_1}{H^\star(\varphi)}\ s^*$$

All rules in the derivation above are sound. By (IH) $H^\star(\varphi')$ is $\mathcal{A}$-valid and so $H^\star(\varphi)$ is $\mathcal{A}$-valid, too.

- The last rule in φ is a quantifier rule (it must be either $\exists{:}\, r$ or $\forall{:}\, l$ as S is weakly quantified). We only consider the case $\exists{:}\, r$, the proof for $\forall{:}\, l$ is completely analogous. Hence φ is of the form

$$\cfrac{\begin{matrix}(\varphi')\\ \Gamma \vdash \Delta, (A\{x \leftarrow t\})_{\nu'}\end{matrix}}{\Gamma \vdash \Delta, ((\exists x)A)_\nu}\ \exists{:}\, r$$

By definition of q we have $q(\varphi, \nu) = q(\varphi', \nu')$, and so $H^\star(\varphi) = H^\star(\varphi')$. That $H^\star(\varphi)$ is a Herbrand sequent thus follows immediately from the induction hypothesis.

- The last rule is a binary logical rule. We only consider $\wedge{:}\, r$, the cases $\vee{:}\, l$ and $\rightarrow{:}\, l$ are analogous. So φ is of the form

$$\cfrac{\begin{matrix}(\varphi_1)\\ \Gamma \vdash \Delta, (A)_{\nu_1}\end{matrix} \quad \begin{matrix}(\varphi_2)\\ \Gamma \vdash \Delta, (B)_{\nu_2}\end{matrix}}{\Gamma \vdash \Delta, (A \wedge B)_\nu}\ \wedge{:}\, r$$

As the end-sequent is prenex the formulas A and B do not contain quantifiers. So $q(\varphi_1, \nu_1) = A$, $q(\varphi_2, \nu_2) = B$ and $q(\varphi, \nu) = A \wedge B$. Now consider the sequents $H^\star(\varphi_1)$ and $H^\star(\varphi_2)$; $H^\star(\varphi_1)$ is a permutation variant of a sequent $\Gamma_1 \vdash \Delta_1, A$, $H^\star(\varphi_1)$ is a permutation variant of a sequent $\Gamma_2 \vdash \Delta_2, B$ (note that, in general, Γ_1, Γ_2 and Δ_1, Δ_2 are different from each other). Therefore $H^\star(\varphi)$ can be obtained by the following derivation:

$$\cfrac{\cfrac{\cfrac{H^\star(\varphi_1)}{\Gamma_1, \Gamma_2 \vdash \Delta_1, \Delta_2, A}\ s^* \quad \cfrac{H^\star(\varphi_2)}{\Gamma_1, \Gamma_2 \vdash \Delta_1, \Delta_2, B}\ s^*}{\Gamma_1, \Gamma_2 \vdash \Delta_1, \Delta_2, A \wedge B}\ \wedge{:}\, r}{H^\star(\varphi)}\ s^*$$

By (IH) $H^\star(\varphi_1)$ and $H^\star(\varphi_2)$ are valid in $\mathcal{A}$. As all rules in the derivation above are sound $H^\star(\varphi)$ is valid in $\mathcal{A}$.

- The last rule of φ is an atomic cut and φ is of the form

$$\dfrac{\begin{matrix}(\varphi_1) & (\varphi_2) \\ \Gamma \vdash \Delta, A & A, \Pi \vdash \Lambda\end{matrix}}{\Gamma, \Pi \vdash \Delta, \Lambda}\ cut$$

As A is atomic and thus quantifier-free $H^\star(\varphi_1)$ is a permutation variant of a sequent $\Gamma^* \vdash \Delta^*, A$ and $H^\star(\varphi_2)$ a permutation variant of a sequent $A, \Pi^* \vdash \Lambda^*$. By definition of $H^\star$, $H^\star(\varphi)$ is a structural variant of $\Gamma^*, \Pi^* \vdash \Delta^*, \Lambda^*$. Therefore, $H^\star(\varphi)$ can be obtained from $H^\star(\varphi_1)$ and $H^\star(\varphi_2)$ as follows:

$$\dfrac{\dfrac{\dfrac{H^\star(\varphi_1)}{\Gamma^* \vdash \Delta^*, A}\ s^* \quad \dfrac{H^\star(\varphi_2)}{A, \Pi^* \vdash \Lambda^*}\ s^*}{\Gamma^*, \Pi^* \vdash \Delta^*, \Lambda^*}\ cut}{H^\star(\varphi)}\ s^*$$

By (IH) $H^\star(\varphi_1), H^\star(\varphi_2)$ are valid in $\mathcal{A}$; as all rules in the derivation above are sound $H^\star(\varphi)$ is valid in $\mathcal{A}$.

This concludes the proof of (1).

(2) is easy to show: by definition $H^\star(\varphi)$ contains only formulas which also appear in φ; for each occurrence of a formula A in $H^\star$ there are one or more occurrences of this formula in φ. Therefore $\|H^\star(\varphi)\| \leq \|\varphi\|$. □

Definition 4.2.7 (Herbrand complexity) Let S be a weakly quantified sequent which is valid in $\mathcal{A}$. Then we define

$$\mathrm{HC}_{\mathcal{A}}(S) = \min\{\|S^*\| \mid S^* \text{ is an } \mathcal{A}\text{-Herbrand sequent of } S\}.$$

If S is not valid in $\mathcal{A}$ then $\mathrm{HC}_{\mathcal{A}}(S)$ is undefined. $\mathrm{HC}_{\mathcal{A}}(S)$ is called the *$\mathcal{A}$-Herbrand complexity* of S. If $\mathcal{A}$ is the standard axiom set then we write HC instead of $\mathrm{HC}_{\mathcal{A}}$ and call $\mathrm{HC}(S)$ the Herbrand complexity of S. ◇

Definition 4.2.8 (proof complexity) Let S be an arbitrary sequent and $\mathcal{A}$ be an axiom set. We define

$$\mathrm{PC}^{\mathcal{A}}(S) = \min\{\|\varphi\| \mid \varphi \in \Phi^{\mathcal{A}} \text{ and } \varphi \text{ proves } S\}.$$

$\mathrm{PC}^{\mathcal{A}}(S)$ is called the *proof complexity* of S w.r.t. $\mathcal{A}$.

Let

$$\begin{aligned} \mathrm{PC}_0^{\mathcal{A}}(S) &= \min\{\|\varphi\| \mid \varphi \in \Phi_0^{\mathcal{A}} \text{ and } \varphi \text{ proves } S\}, \\ \mathrm{PC}_\emptyset^{\mathcal{A}}(S) &= \min\{\|\varphi\| \mid \varphi \in \Phi_\emptyset^{\mathcal{A}} \text{ and } \varphi \text{ proves } S\}. \end{aligned}$$

Then $\mathrm{PC}_0^{\mathcal{A}}(S)$ ($\mathrm{PC}_\emptyset^{\mathcal{A}}$) denotes the proof compiexity of S w.r.t. $\mathcal{A}$ if only atomic cuts (no cuts at all) are admitted. If $\mathcal{A}$ is the standard axiom set we write $\mathrm{PC}, \mathrm{PC}_0$ and $\mathrm{PC}_\emptyset$. ◇

Note that the proof complexity does not just depend on the number of nodes in the proof tree but on the symbol occurrences.

Proposition 4.2.1 *For every sequent S and every axiom set we have* $\mathrm{PC}^{\mathcal{A}}(S) \leq \mathrm{PC}_0^{\mathcal{A}}(S) \leq \mathrm{PC}_\emptyset^{\mathcal{A}}$.

Proof: trivial. □

Herbrand complexity defines a lower bound for proof complexity if only cut-free proofs or proofs with at most atomic cuts are considered:

Theorem 4.2.2 *Let S be a prenex weakly quantified sequent and $\mathcal{A}$ be an axiom set. Then* $\mathrm{HC}_{\mathcal{A}}(S) \leq \mathrm{PC}_0^{\mathcal{A}}(S)$.

Proof: Let $\varphi \in \Phi_0^{\mathcal{A}}$ be a proof of S. By Theorem 4.2.1 (2) we know

$$\|H^\star(\varphi)\| \leq \|\varphi\|.$$

But

$$\mathrm{HC}_{\mathcal{A}}(S) \leq \min\{\|H^\star(\varphi)\| \mid \varphi \in \Phi_0^{\mathcal{A}}\} \leq \min\{\|\varphi\| \mid \varphi \in \Phi_0^{\mathcal{A}}, \varphi \text{ proves } S\}.$$

By definition of $\mathrm{PC}_0^{\mathcal{A}}$ we thus obtain $\mathrm{HC}_{\mathcal{A}}(S) \leq \mathrm{PC}_0^{\mathcal{A}}(S)$. □

4.3 The Proof Sequence of R. Statman

Definition 4.3.1 Let $e : \mathbb{N}^2 \to \mathbb{N}$ be the following function

$$\begin{aligned} e(0, m) &= m \\ e(n+1, m) &= 2^{e(n,m)}. \end{aligned}$$

A function $f : \mathbb{N}^k \to \mathbb{N}^m$ for $k, m \geq 1$ is called *elementary* if there exists an $n \in \mathbb{N}$ and a Turing machine T computing f s.t. the computing time of T on input $(l_1, \ldots, l_k)$ is less than or equal to $e(n, |(l_1, \ldots, l_k)|)$ where $|\ |$ denotes the maximum norm on $\mathbb{N}^k$ (see also [28]).
The function $s : \mathbb{N} \to \mathbb{N}$ is defined as $s(n) = e(n, 1)$ for $n \in \mathbb{N}$.
A function which is not elementary is called *nonelementary*. ◇

Remark: Note that the functions s and e are nonelementary. In general, any function f which grows "too fast", i.e. for which there exists no number k s.t.

$$f(n) \leq e(k, n),$$

is nonelementary. ◇

In [72] R. Statman proved that there exists a sequence of short proofs γ_n of sentences A_n s.t. the Herbrand complexity of A_n is inherently nonelementary in the length of the γ_n; in fact Statman did not explicitly address a specific formal calculus, leaving the formalization of the proof sequence to the reader. Independently V. Orevkov [67] proved the nonelementary complexity of cut-elimination for function-free predicate logic without equality. The proof sequences of Statman and Orevkov are different, but both encode the principle of iterated exponentiation best described by P. Pudlak in [68]. We decided to choose Statman's sequence for our complexity analysis of cut-elimination methods, though Orevkov's or Pudlak's sequence would be equally appropriate.

We first give a "mathematical" (or informal) description of Statman's proof sequence:

We have two basic axioms:

$$\begin{aligned}
&(\mathrm{Ax})\colon\ (\forall x)px = p(qx),\\
&(\mathrm{Ax}_T)\colon\ (\forall x)(\forall y)\mathbf{T}xy = x(xy).
\end{aligned}$$

where $\mathbf{T}$ is a constant symbol defining the exponential combinator and p, q are arbitrary constant symbols. As usual in combinatory logic we write stw for $(st)w$. Formally we need a binary function symbol g and $g(g(\mathbf{T}, x), y)$ to denote the term $\mathbf{T}xy$. As g is not associative terms must be denoted with care. For all terms s, t we define

$$s^1 t = st,\ \ s^{n+1}t = s(s^n t).$$

So, e.g., $x^3 y = x(x(xy))$.

We first show that $\mathbf{T}$ is indeed the exponential combinator:

Proposition 4.3.1 *Let x, y be variables. Then, for all $i \geq 1$, $\mathbf{T}^i xy = x^{2^i} y$.*

Proof: By induction on i.

$i = 1$: immediately by definition of $\mathbf{T}$ and by $x^2 y = x(xy)$.

(IH) Assume that $\mathbf{T}^i xy = x^{2^i} y$ for some $i \geq 1$.

Then

$$\mathbf{T}^{i+1}xy = \mathbf{T}(\mathbf{T}^i x)y =_{(\mathrm{Ax}_T)} (\mathbf{T}^i x)(\mathbf{T}^i xy) =_{(\mathrm{IH})} x^{2^i}(x^{2^i}y) = x^{2^{i+1}}y. \ \square$$

The iteration of the exponential operator $\mathbf{T}$ is defined as

$$\mathbf{T}_1 = \mathbf{T},\ \mathbf{T}_{n+1} = \mathbf{T}_n\mathbf{T}.$$

Proposition 4.3.2 *Let x, y be variables. Then $\mathbf{T}_n xy = x^{s(n)}y$ for s in Definition 4.3.1.*

Proof: By induction on n.

$$(\mathrm{IB})\mathbf{T}_1 xy = \mathbf{T}xy = x^2y = x^{s(1)}y.$$

(IH) Assume $\mathbf{T}_n xy = x^{s(n)}y$. Then

$$\mathbf{T}_{n+1}xy = \mathbf{T}_n\mathbf{T}xy =_{(\mathrm{IH})} \mathbf{T}^{s(n)}xy.$$

By Proposition 4.3.1 we have

$$\mathbf{T}^{s(n)}xy = x^{2^{s(n)}}y = x^{s(n+1)}y. \ \square$$

Now we prove the equation $pq = p(\mathbf{T}_n q)$ for some constants p, q. It is easy to show that, for all k, $pq = p(q^k q)$. For $k = 1$ this follows directly from (Ax). Assume that $pq = p(q^k q)$ is already derived; from (Ax) we obtain

$$p(q^k q) = p(q(q^k q)),\ p(q(q^k q)) = p(q^{k+1}q),$$

and, by transitivity of $=$, $pq = p(q^{k+1}q)$. This way we can derive $pq = p(q^{s(n)}q)$ (in $s(n)$ steps). By Proposition 4.3.2 we know that $\mathbf{T}_n qq = q^{s(n)}q$, and so we obtain

$$E_n\colon\ pq = \mathbf{T}_n qq$$

by using equational inference on the axioms Ax and Ax_T. This proof is very simple, but also very long; indeed, we need more than $s(n)$ equational inferences to obtain the result. The following short proof of E_n from the axioms Ax, Ax_T using lemmas is described in [72]. We define the sets

$$\begin{aligned} H_1 &= \{y \mid \text{ for all } x\colon\ px = p(yx)\}, \\ H_{i+1} &= \{y \mid \text{ for all } x \text{ in } H_i\colon\ yx \in H_i\}. \end{aligned}$$

We show first that $\mathbf{T} \in H_i$ for $i \geq 2$.

Assume that $y \in H_1$. Then $px = p(yx)$ and $p(yx) = p(y(yx))$. So, by transitivity and (Ax_T), $px = p(\mathbf{T}yx)$ for all x. In particular we obtain: for all $y \in H_1$ also $\mathbf{T}y \in H_1$. By definition of H_2 we obtain $\mathbf{T} \in H_2$.

Now let $i \geq 2$ and assume $z \in H_i$. Then, by definition of H_i, for all $x \in H_{i-1}$: $zx \in H_{i-1}$. So also $z(zx) \in H_{i-1}$, or $\mathbf{T}zx \in H_{i-1}$. This holds for all x and therefore $\mathbf{T}z \in H_i$. We obtain

For all $z \in H_i$: $\mathbf{T}z \in H_i$.

By definition of H_{i+1} we obtain $\mathbf{T} \in H_{i+1}$.

So we have shown $\mathbf{T} \in H_i$ for $i \geq 2$. In particular we get $\mathbf{T} \in H_{n+1}$. Now we prove that

$$(+)\ \mathbf{T}_i \in H_{n+2-i} \text{ for } 1 \leq i \leq n.$$

For $i = 1$ we already obtained the desired result. Assume now that $\mathbf{T}_i \in H_{n+2-i}$ for $i < n$. By definition of H_{n+2-i} this means

For all $x \in H_{n+1-i}$: $\mathbf{T}_i x \in H_{n+1-i}$.

But as $\mathbf{T} \in H_{n+1-i}$ we also get $\mathbf{T}_{i+1} \in H_{n+1-i}$, or $\mathbf{T}_{i+1} \in H_{n+2-(i+1)}$. This proves (+).

In particular (+) yields $\mathbf{T}_n \in H_2$. But $\mathbf{T}_n \in H_2$ means that, for all $x \in H_1$,

$$(\star)\ \mathbf{T}_n x \in H_1.$$

By (Ax) we get $q \in H_1$, and $(\star)$ yields $\mathbf{T}_n q \in H_1$, which – by definition of H_1 – means

$$(\forall x) px = p((\mathbf{T}_n q)x.$$

We obtain E_n: $pq = p(\mathbf{T}_n q)q$ just by instantiation. This proof is much shorter than the former one, but at the same time more complex. In fact it uses the sets H_i and properties of the H_i as lemmas. Indeed, the first one corresponds to a cut-free **LK**-proof (strictly speaking an **LK**-proof with only atomic cuts), while the latter one heavily uses cut. The proof with cuts is also less "explicit" as the term $\mathbf{T}_n q$ is not evaluated in the proof.

Definition 4.3.2 We formalize the informal proof sequence defined above by the following sequence of **LK**-proofs γ_n from the axiom set $\mathcal{A}_e$:
The end-sequents of γ_n (for $n \geq 1$) are of the form

$$S_n\colon\ \mathrm{Ax}_T, (\forall x_1)px_1 = p(qx_1) \vdash pq = p((\mathbf{T}_n q)q),$$

where

$$\mathbf{T}_1 \equiv \mathbf{T},\ \mathbf{T}_{n+1} \equiv \mathbf{T}_n\mathbf{T},$$
$$\mathrm{Ax}_T = (\forall y)(\forall x)\mathbf{T}yx = y(yx).$$

Note that $\mathbf{T}yx$ stands for $(\mathbf{T}y)x$.

For the cut formulas we need representations of the sets H_i defined above. For $i \geq 1$ we define

$$\begin{array}{rcl} H_1(y_1) & \equiv & (\forall x_1)px_1 = p(y_1x_1), \\ H_{i+1}(y_{i+1}) & \equiv & (\forall x_{i+1})(H_i(x_{i+1}) \to H_i(y_{i+1}x_{i+1})) \text{ for } i \geq 1. \end{array}$$

We define γ_n for $n \geq 1$ as

$$\dfrac{\dfrac{\overset{\delta_n}{\mathrm{Ax}_T \vdash H_2(\mathbf{T}_n)} \quad \overset{(\varphi_1)}{H_2(\mathbf{T}_n), H_1(q) \vdash H_1(\mathbf{T}_n q)}}{\mathrm{Ax}_T, H_1(q) \vdash H_1(\mathbf{T}_n q)}\,cut \qquad \dfrac{pq = p((\mathbf{T}_n q)q) \vdash pq = p((\mathbf{T}_n q)q)}{H_1(\mathbf{T}_n q) \vdash pq = p((\mathbf{T}_n q)q)}\,\forall{:}\,l}{\mathrm{Ax}_T, H_1(q) \vdash pq = p((\mathbf{T}_n q)q)}\,cut$$

Note that, by definition of H_1, $H_1(q) \equiv (\forall x_1)px_1 = p(qx_1)$.

We have to define the proof sequences δ_i and φ_i:

$\delta_1 = \psi_{n+1}$.

For $1 \leq i < n$ we define $\delta_{i+1} =$

$$\dfrac{\dfrac{\overset{(\delta_i)}{\mathrm{Ax}_T \vdash H_{n-i+2}(\mathbf{T}_i)} \qquad \dfrac{\overset{(\psi_{n-i+1})}{\mathrm{Ax}_T \vdash H_{n-i+1}(\mathbf{T})} \quad \overset{(\varphi_{i+1})}{H_{n-i+1}(\mathbf{T}), H_{n-i+2}(\mathbf{T}_i) \vdash H_{n-i+1}(\mathbf{T}_{i+1})}}{\mathrm{Ax}_T, H_{n-i+2}(\mathbf{T}_i) \vdash H_{n-i+1}(\mathbf{T}_{i+1})}\,cut}{\mathrm{Ax}_T, \mathrm{Ax}_T \vdash H_{n-i+1}(\mathbf{T}_{i+1})}\,p + cut}{\mathrm{Ax}_T \vdash H_{n-i+1}(\mathbf{T}_{i+1})}\,c{:}\,l$$

where the proofs ψ_j and φ_j will be defined below.

$\varphi_1 =$

$$\dfrac{\dfrac{\dfrac{\overset{(\pi(H_1(q)))}{H_1(q) \vdash H_1(q)} \quad \overset{(\pi(H_1(\mathbf{T}_n q)))}{H_1(\mathbf{T}_n q) \vdash H_1(\mathbf{T}_n q)}}{H_1(q) \to H_1(\mathbf{T}_n q), H_1(q) \vdash H_1(\mathbf{T}_n q)}\,\to{:}\,l}{(\forall x_2)(H_1(x_2) \to H_1(\mathbf{T}_n x_2)), H_1(q) \vdash H_1(\mathbf{T}_n q)}\,\forall{:}\,l}{H_2(\mathbf{T}_n), H_1(q) \vdash H_1(\mathbf{T}_n q)}\,(\mathrm{def})$$

where the $\pi(A)$ are defined in Lemma 4.1.1

For $1 \leq i < n$ we define

$\varphi_{i+1} =$

$$
\dfrac{\dfrac{\dfrac{\dfrac{\dfrac{\begin{matrix}(\pi(H_{n-i+1}(\mathbf{T})))\\ H_{n-i+1}(\mathbf{T}) \vdash H_{n-i+1}(\mathbf{T})\end{matrix} \quad \begin{matrix}(\pi(H_{n-i+1}(\mathbf{T}_{i+1})))\\ H_{n-i+1}(\mathbf{T}_{i+1}) \vdash H_{n-i+1}(\mathbf{T}_{i+1})\end{matrix}}{H_{n-i+1}(\mathbf{T}) \rightarrow H_{n-i+1}(\mathbf{T}_{i+1}), H_{n-i+1}(\mathbf{T}) \vdash H_{n-i+1}(\mathbf{T}_{i+1})}\ \rightarrow: l}{(\forall x_{n-i+2})(H_{n-i+1}(x_{n-i+2}) \rightarrow H_{n-i+1}(\mathbf{T}_i x_{n-i+2})), H_{n-i+1}(\mathbf{T}) \vdash H_{n-i+1}(\mathbf{T}_{i+1})}\ \forall: l}{H_{n-i+2}(\mathbf{T}_i), H_{n-i+1}(\mathbf{T}) \vdash H_{n-i+1}(\mathbf{T}_{i+1})}\ (\mathrm{def})}{H_{n-i+1}(\mathbf{T}), H_{n-i+2}(\mathbf{T}_i) \vdash H_{n-i+1}(\mathbf{T}_{i+1})}\ p: l}{}
$$

It remains to define the proofs ψ_i for $2 \leq i \leq n$:

$\psi_2 =$

$$
\dfrac{\dfrac{\dfrac{\dfrac{\begin{matrix}(\psi_2')\\ \mathrm{Ax}_T, (\forall x_1) p x_1 = p(\beta x_1) \vdash (\forall x_1) p x_1 = p((\mathbf{T}\beta) x_1)\end{matrix}}{\mathrm{Ax}_T, H_1(\beta) \vdash H_1(\mathbf{T}\beta)}\ (\mathrm{def})}{\mathrm{Ax}_T \vdash H_1(\beta) \rightarrow H_1(\mathbf{T}\beta)}\ \rightarrow: r}{\mathrm{Ax}_T \vdash (\forall x_2)(H_1(x_2) \rightarrow H_1(\mathbf{T} x_2))}\ \forall: r}{\mathrm{Ax}_T \vdash H_2(\mathbf{T})}\ (\mathrm{def})
$$

where $\psi_2' =$

$$
\dfrac{\dfrac{\dfrac{\dfrac{\dfrac{\dfrac{\dfrac{\begin{matrix}(\psi_2'')\\ \mathbf{T}\beta\alpha = \beta(\beta\alpha), p\alpha = p(\beta\alpha), p(\beta\alpha) = p(\beta(\beta\alpha)) \vdash p\alpha = p((\mathbf{T}\beta)\alpha)\end{matrix}}{\mathbf{T}\beta\alpha = \beta(\beta\alpha), p\alpha = p(\beta\alpha), (\forall x_1) p x_1 = p(\beta x_1) \vdash p\alpha = p((\mathbf{T}\beta)\alpha)}\ \forall: l + p}{\mathbf{T}\beta\alpha = \beta(\beta\alpha), (\forall x_1) p x_1 = p(\beta x_1), (\forall x_1) p x_1 = p(\beta x_1) \vdash p\alpha = p((\mathbf{T}\beta)\alpha)}\ \forall: l + p}{\mathbf{T}\beta\alpha = \beta(\beta\alpha), (\forall x_1) p x_1 = p(\beta x_1) \vdash p\alpha = p((\mathbf{T}\beta)\alpha)}\ c: l + p}{(\forall x)\mathbf{T}\beta x = \beta(\beta x), (\forall x_1) p x_1 = p(\beta x_1) \vdash p\alpha = p((\mathbf{T}\beta)\alpha)}\ \forall: l}{(\forall y)(\forall x)\mathbf{T} y x = y(y x), (\forall x_1) p x_1 = p(\beta x_1) \vdash p\alpha = p((\mathbf{T}\beta)\alpha)}\ \forall: l}{(\forall y)(\forall x)\mathbf{T} y x = y(y x), (\forall x_1) p x_1 = p(\beta x_1) \vdash (\forall x_1) p x_1 = p((\mathbf{T}\beta) x_1)}\ \forall: r}{\mathrm{Ax}_T, (\forall x_1) p x_1 = p(\beta x_1) \vdash (\forall x_1) p x_1 = p((\mathbf{T}\beta) x_1)}\ (\mathrm{def})
$$

and $\psi_2'' =$

$$
\dfrac{\begin{matrix}((\mathrm{trans}))\\ p\alpha = p(\beta\alpha), p(\beta\alpha) = p(\beta(\beta\alpha)) \vdash p\alpha = p(\beta(\beta\alpha))\end{matrix} \quad \psi_2^{(3)}}{\mathbf{T}\beta\alpha = \beta(\beta\alpha), p\alpha = p(\beta\alpha), p(\beta\alpha) = p(\beta(\beta\alpha)) \vdash p\alpha = p((\mathbf{T}\beta)\alpha)}\ cut + p
$$

and $\psi_2^{(3)} =$

$$
\dfrac{\begin{matrix}(\mathrm{symm})\\ \mathbf{T}\beta\alpha = \beta(\beta\alpha) \vdash \beta(\beta\alpha) = \mathbf{T}\beta\alpha\end{matrix} \quad \begin{matrix}(\psi_2^{(4)})\\ \beta(\beta\alpha) = \mathbf{T}\beta\alpha, p\alpha = p(\beta(\beta\alpha)) \vdash p\alpha = p((\mathbf{T}\beta)\alpha)\end{matrix}}{p\alpha = p(\beta(\beta\alpha)), \mathbf{T}\beta\alpha = \beta(\beta\alpha) \vdash p\alpha = p(\mathbf{T}\beta\alpha)}\ cut
$$

$\psi_2^{(4)} =$

$$\dfrac{\begin{matrix}\rho(\beta(\beta\alpha),\mathbf{T}\beta\alpha,p)\\ \beta(\beta\alpha)=\mathbf{T}\beta\alpha\vdash p(\beta(\beta\alpha))=p((\mathbf{T}\beta)\alpha)\end{matrix}\quad \psi_2^{(5)}}{\beta(\beta\alpha)=\mathbf{T}\beta\alpha, p\alpha=p(\beta(\beta\alpha))\vdash p\alpha=p((\mathbf{T}\beta)\alpha)}\ cut$$

where $\rho(s,t,p) =$

$$\dfrac{\begin{matrix}\text{ref}\\ \vdash ps=ps\end{matrix}\quad \dfrac{\text{subst}}{ps=ps, s=t\vdash ps=pt}\ p\colon l}{s=t\vdash ps=pt}\ cut$$

and $\psi_2^{(5)} =$

$$\dfrac{\text{trans}}{p(\beta(\beta\alpha))=p((\mathbf{T}\beta)\alpha), p\alpha=p(\beta(\beta\alpha))\vdash p\alpha=p((\mathbf{T}\beta)\alpha)}\ p\colon l$$

Now let us assume that ψ_i is already defined. Then we set
$\psi_{i+1} =$

$$\dfrac{\dfrac{\dfrac{\begin{matrix}(\psi'_{i+1})\\ \mathrm{Ax}_T, H_i(\beta)\vdash H_i(\mathbf{T}\beta)\end{matrix}}{\mathrm{Ax}_T\vdash H_i(\beta)\to H_i(\mathbf{T}\beta)}\ \to\colon r}{\mathrm{Ax}_T\vdash(\forall x_{i+1})(H_i(x_{i+1})\to H_i(\mathbf{T}x_{i+1}))}\ \forall\colon r}{\mathrm{Ax}_T\vdash H_{i+1}(\mathbf{T})}\ (\mathrm{def})$$

where $\psi'_{i+1} =$

$$\dfrac{\dfrac{\dfrac{\dfrac{\dfrac{\dfrac{\dfrac{\dfrac{\begin{matrix}\tau(H_{i-1},(\mathbf{T}\beta)\alpha,\beta(\beta\alpha),\alpha,\beta\alpha)\\ H_{i-1}(\alpha),\mathbf{T}\alpha\beta=\beta(\beta\alpha),H_{i-1}(\alpha)\to H_{i-1}(\beta\alpha),H_{i-1}(\beta\alpha)\to H_{i-1}(\beta(\beta\alpha))\vdash H_{i-1}((\mathbf{T}\beta)\alpha)\end{matrix}}{H_{i-1}(\alpha),\mathbf{T}\alpha\beta=\beta(\beta\alpha),H_{i-1}(\alpha)\to H_{i-1}(\beta\alpha),H_i(\beta)\vdash H_{i-1}((\mathbf{T}\beta)\alpha)}\ *}{H_{i-1}(\alpha),\mathbf{T}\alpha\beta=\beta(\beta\alpha),H_i(\beta),H_i(\beta)\vdash H_{i-1}((\mathbf{T}\beta)\alpha)}\ \forall\colon l+p^*}{H_{i-1}(\alpha),\mathbf{T}\alpha\beta=\beta(\beta\alpha),H_i(\beta)\vdash H_{i-1}((\mathbf{T}\beta)\alpha)}\ c\colon l+p^*}{\mathbf{T}\alpha\beta=\beta(\beta\alpha),H_i(\beta)\vdash H_{i-1}(\alpha)\to H_{i-1}((\mathbf{T}\beta)\alpha)}\ \to\colon r}{(\forall y)(\forall x)\mathbf{T}yx=y(yx),H_i(\beta)\vdash H_{i-1}(\alpha)\to H_{i-1}((\mathbf{T}\beta)\alpha)}\ \forall\colon l\ 2\times}{(\forall y)(\forall x)\mathbf{T}yx=y(yx),H_i(\beta)\vdash(\forall x_i)(H_{i-1}(x_i)\to H_{i-1}((\mathbf{T}\beta)x_i))}\ \forall\colon r}{\mathrm{Ax}_T,H_i(\beta)\vdash H_i(\mathbf{T}\beta)}\ (\mathrm{def})$$

for $\tau(A,t,w,s_1,s_2) =$

$$\dfrac{\dfrac{\begin{matrix}\pi(A(s_1))\\ A(s_1)\vdash A(s_1)\end{matrix}\quad \dfrac{\begin{matrix}\alpha(A(s_2))\\ A(s_2)\vdash A(s_2)\end{matrix}\quad\begin{matrix}\pi(A(w))\\ A(w)\vdash A(w)\end{matrix}}{A(s_2),A(s_2)\to A(w)\vdash A(w)}\ \to\colon l+p^*}{A(s_1),A(s_1)\to A(s_2),A(s_2)\to A(w)\vdash A(w)}\ \to\colon l+p^*\quad \eta}{t=w,A(s_1),A(s_1)\to A(s_2),A(s_2)\to A(w)\vdash A(t)}\ cut+p^*$$

for $\eta =$

$$\frac{\dfrac{(\text{symm})}{t = w \vdash w = t} \quad \dfrac{\lambda(A, w, t)}{w = t, A(w) \vdash A(t)}}{t = w, A(w) \vdash A(t)}\ cut$$

where the proofs $\pi(\ )$ are defined in Lemma 4.1.1 and $\lambda(\ )$ in Lemma 4.1.2. $\diamond$

Proposition 4.3.3 *Let $(\gamma_n)_{n \in \mathbb{N}}$ be the sequence of proofs defined in Definition 4.3.2. Then there exists a constant m s.t. $\|\gamma_n\| \leq 2^{2*n+m}$ for all $n \geq 1$.*

Proof: We first show that there exists a constant k s.t. for all $n \geq 1$

$$l(\gamma_n) \leq 2^{n+k}.$$

The lengths of the proofs γ_n depend on the complexity of the formulas H_i which appear as cuts in the proofs (note that the axioms are atomic). By definition of the formulas H_i we have:

$$\begin{array}{rcl} comp(H_1) & = & 1, \\ comp(H_{i+1}) & = & 2 * comp(H_i) + 2 \text{ for } i \geq 1. \end{array}$$

In particular we get $comp(H_i) < 2^{i+1}$ for all $i \geq 1$
For the length of the sequence γ_n we get

$$l(\gamma_n) = l(\delta_n) + l(\varphi_n) + 4.$$

where (by the Lemmas 4.1.1 and 4.1.2)

$$\begin{array}{rcl} l(\delta_1) & = & l(\psi_{n+1}), \\ l(\delta_{i+1}) & = & l(\delta_i) + l(\psi_{n-i+1}) + l(\varphi_{i+1}) + 4, \\ l(\varphi_n) & = & 3 + 2 * l(\pi(H_1(q))) \leq 3 + 2 * (4 * comp(H_1(q)) + 1) = 13, \\ l(\varphi_{i+1}) & = & 4 + 2 * \pi(H_{n-i+1}(\mathbf{T})) \leq \\ & & 4 + 2 * (4 * comp(H_{n-i+1}(\mathbf{T}) + 1)) < 2^{n-i+8}. \\ l(\psi_2) & = & s \text{ for some constant } s, \\ l(\psi_{i+1}) & = & 2 + l(\psi'_{i+1}) = 12 + l(\tau(H_{i-1}, \mathbf{T}z\alpha, z(z\alpha), \alpha, z\alpha)) \leq \\ & & k_1 + 2 * l(\pi(H_{i-1}(\alpha)) + l(\lambda(H_{i-1}, w, t)) \leq \\ & & k_1 + 2 * (4 * comp(H_{i-1}(\alpha)) + 1) + 11 * comp(H_{i-1}) + 1 \leq \\ & & 2^{i+k_2} \text{ for constants } k_1, k_2. \end{array}$$

Therefore

$$l(\delta_{i+1}) \leq l(\delta_i) + 2^{n-i+k_2} + 2^{n-i+8} + 4.$$

A solution of this recursive inequality is $l(\delta_n) \leq 2^{n+k_3}$ for a constant k_3. Putting things together we can define a constant k s.t.

$$l(\gamma_n) < 2^{n+k}.$$

Moreover all sequents in the proofs γ_n contain at most 5 formulas; the logical complexity of all formulas in γ_n is $< 2^{n+2}$, and the maximal number of term occurrences in atoms is $\leq r * n$ for some constant r. Therefore there exists a constant m s.t.

$$\|\gamma_n\| \leq 2^{n+k} * 5 * 2^{n+2} * r * n \leq 2^{2*n+m}. \ \square$$

The following theorem shows that there exists no elementary bound on the Herbrand complexity (w.r.t. $\mathcal{A}_e$) of the S_n in terms of $\|\gamma_n\|$. Therefore, the Herbrand complexity of the S_n grows *nonelementarily* in the lengths of the shortest proofs of the S_n.

Proposition 4.3.4 *Let $(S_n)_{n \in \mathbb{N}}$ be the sequence of end-sequents of the proofs γ_n defined in Definition 4.3.2. Then there exists a constant k s.t. for all $n \geq 1$: $\mathrm{HC}_{\mathcal{A}_e}(S_n) > \frac{1}{k}s(n)$.*

Proof: In [72] R. Statman proved that the Herbrand complexity of the proof sequence is greater than $\frac{1}{2}s(n)$. But our presentation differs from that in [72]; we introduced $\mathbf{T}$ as a new constant, while $\mathbf{T}$ is defined in [72] as $\mathbf{T} \equiv (\mathbf{SB})(\mathbf{CBI})$, where the standard combinators $\mathbf{S}, \mathbf{B}, \mathbf{C}$ are defined by the formulas:

$$\begin{array}{rcl} (\mathbf{S})\ (\forall x)(\forall y)(\forall z)\mathbf{S}xyz & = & xz(yz), \\ (\mathbf{B})\ (\forall x)(\forall y)(\forall z)\mathbf{B}xyx & = & x(yz), \\ (\mathbf{C})\ (\forall x)(\forall y)(\forall z)\mathbf{C}xyz & = & xzy. \end{array}$$

Now let S_n^* be the sequent

$$(\mathbf{S}), (\mathbf{B}), (\mathbf{C}), \mathrm{Ax} \vdash pq = p((((\mathbf{SB})(\mathbf{CBI}))_n q)q).$$

From Statman's result we know that $\mathrm{HC}_{\mathcal{A}_e}(S_n^*) > \frac{1}{2}s(n)$. Let $\mathcal{S}$ be an $\mathcal{A}_e$-Herbrand sequent of S_n. We construct an $\mathcal{A}_e$-Herbrand sequent $\mathcal{S}^*$ of S_n^* s.t.

$$\|\mathcal{S}^*\| \leq m\|\mathcal{S}\|.$$

for a constant m independent of n.
To this aim we replace all occurrences of terms $\mathbf{T}rs$ in $\mathcal{S}$ by the terms $(\mathbf{SB})((\mathbf{CB})\mathbf{I})rs$. Afterwards, we add the equations

$$\begin{aligned}
(\mathbf{SB})(\mathbf{CBI})r &= (\mathbf{B}r)(\mathbf{CBI}r),\\
(\mathbf{B}r)(\mathbf{CBI}r)s &= r(\mathbf{CBI}rs),\\
\mathbf{CBI}r &= \mathbf{B}r\mathbf{I}\\
\mathbf{B}r\mathbf{I}s &= r(\mathbf{I}s),\\
\mathbf{I}s &= s
\end{aligned}$$

to the left hand side of the sequent. We thus obtain a sequent $\mathcal{S}^*$ which is indeed an instance sequent of S_n^*. As $\mathcal{S}$ is equationally valid, $\mathcal{S}^*$ is too, thus $\mathcal{S}^*$ is an $\mathcal{A}_e$-Herbrand sequent of S_n^*. Moreover the size of the whole sequent is multiplied at most by a constant k, and so

$$\|\mathcal{S}^*\| \leq k\|\mathcal{S}\|.$$

But then

$$\frac{1}{2}s(n) < \mathrm{HC}_{\mathcal{A}_e}(S_n^*) \leq \|\mathcal{S}^*\| \leq k\|\mathcal{S}\| \text{ and}$$

By choosing $m = 2 * k$ we obtain

$$\frac{1}{m}s(n) < \mathrm{HC}_{\mathcal{A}_e}(S_n). \ \square$$

Definition 4.3.3 Let $\bar{x}\colon (x_n)_{n\in\mathbb{N}}$ and $\bar{y}\colon (y_n)_{n\in\mathbb{N}}$ be sequences of natural numbers. We call $\bar{x}$ *elementary* in $\bar{y}$ if there exists a $k \in \mathbb{N}$ s.t.

$$x_n \leq e(k, y_n)$$

for all $n \in \mathbb{N}$. Otherwise we call $\bar{x}$ *nonelementary* in $\bar{y}$. $\diamond$

Theorem 4.3.1 *The sequence* $(\mathrm{HC}_{\mathcal{A}_e}(S_n))_{n\in\mathbb{N}}$ *is nonelementary in* $(\mathrm{PC}^{\mathcal{A}_e}(S_n))_{n\in\mathbb{N}}$.

Proof: By Proposition 4.3.4 there exists a constant m s.t.

$$\frac{s(n)}{m} < \mathrm{HC}_{\mathcal{A}_e}(S_n)$$

for $n \geq 1$. Proposition 4.3.3 gives us

$$\mathrm{PC}^{\mathcal{A}_e}(S_n) \leq 2^{2*n+k}$$

for a constant k independent of n.
Now let $k \in \mathbb{N}$. Then there exists a number r s.t.

$$e(k, \mathrm{PC}^{\mathcal{A}_e}(S_n)) < e(k+r, n) \text{ for all } n.$$

But $e(k+r, n) < \frac{s(n)}{m}$ almost everywhere. Putting things together we find: for all $k \in \mathbb{N}$ there exists a constant M s.t. for all $n \geq M$:

$$e(k, \mathrm{PC}^{\mathcal{A}_e}(S_n)) < \mathrm{HC}_{\mathcal{A}_e}(S_n);$$

but this means that $(\mathrm{HC}_{\mathcal{A}_e}(S_n))_{n \in \mathbb{N}}$ is nonelementary in $(\mathrm{PC}^{\mathcal{A}_e}(S_n))_{n \in \mathbb{N}}$.

□

Corollary 4.3.1 *The elimination of cuts on the sequence $(S_n)_{n \in \mathbb{N}}$ is nonelementary, i.e. $(\mathrm{PC}_0^{\mathcal{A}_e}(S_n))_{n \in \mathbb{N}}$ is nonelementary in $(\mathrm{PC}^{\mathcal{A}_e}(S_n))_{n \in \mathbb{N}}$.*

Proof: By Theorem 4.3.1 and $\mathrm{HC}_{\mathcal{A}_e}(S_n) \leq \mathrm{PC}_0^{\mathcal{A}_e}(S_n)$, which follows from Theorem 4.2.2. □

Chapter 5

Reduction and Elimination

5.1 Proof Reduction

In Gentzen's famous paper cut-elimination is a constructive method for proving the "Hauptsatz" which is used to constitute such important principles as the existence of a mid-sequent and the decidability of propositional intuitionistic logic. The idea of the Hauptsatz is connected to the elimination of ideal objects in mathematical proofs according to Hilbert's program. In this sense **LK** (with the standard axiom set) is consistent because the empty sequent is not cut-free derivable; any proof of a contradiction would need ideal (indirect) arguments. By shift of emphasis mathematicians began to focus on the proof transformation by cut-elimination itself. In fact, cut-elimination is an essential tool for making implicit contents of proofs explicit. It also allows the construction of Herbrand disjunctions and interpolants for real mathematical proofs. Furthermore elementary proofs can be obtained from abstract ones; one of the most important examples from literature is the transformation of the Fürstenberg–Weiss proof into the original (van der Waerden's) proof [40]. This transformation, however, is informal and therefore it is not evident that van der Waerden's proof is the only elementary proof corresponding to that of Fürstenberg and Weiss. In extremis any elementary proof could be a possible target proof of informal cut-elimination.

In this book we consider cut-elimination from a formal point of view. Therefore it is necessary to formulate the cut-reduction and elimination methods under consideration (Gentzen's, Tait's method, CERES) in a broad sense to provide an adequate spectrum of target proofs and allow the formulation of negative results, e.g. a specific cut-free proof (Herbrand sequent, interpolant) cannot be obtained from a given proof. Furthermore the formal

M. Baaz, A. Leitsch, *Methods of Cut-Elimination*, Trends in Logic 34, 63
DOI 10.1007/978-94-007-0320-9_5, © Springer Science+Business Media B.V. 2011

specification of cut-elimination is useful for computer implementations to obtain additional information from proof by faithful experiments.
Let $\Phi = \Phi^{\mathcal{A}}$ (see Definition 3.2.8) for an arbitrary but fixed axiom set $\mathcal{A}$.

Definition 5.1.1 (proof reduction relation) Any binary relation on Φ is called a *proof reduction relation.* $\diamond$

Remark: Let $>$ be a proof reduction relation and $\varphi > \psi$. As $\Phi = \Phi^{\mathcal{A}}$, φ and ψ are both **LK**-proofs from the same axiom set $\mathcal{A}$. $\diamond$

Definition 5.1.2 (reduction sequence) Let $>$ be a proof reduction relation. A sequence $\gamma\colon \varphi_1, \ldots \varphi_n$ is called a $>$-sequence if $\varphi_i > \varphi_{i+1}$ for all $i \in \{1, \ldots n-1\}$. γ is also called a *$>$-derivation* of φ_n from φ_1. $\diamond$

Definitions 5.1.1 and 5.1.2 are very general. Clearly we are interested in specific reduction relations, particularly in transformations of proofs to Φ_0.

Definition 5.1.3 (cut-elimination sequence) Let $>$ be a proof reduction relation. Let $\gamma\colon \varphi_1, \ldots, \varphi_n$ be a $>$-derivation s.t.

1. $\varphi_1, \ldots, \varphi_n$ have all the same end-sequent,
2. $\varphi_n \in \Phi_0$.

Then γ is called a *cut-elimination sequence* of φ_1 w.r.t. $>$. $\diamond$

Definition 5.1.4 (cut-elimination relation) The reduction relation $>$ is called a *cut-elimination relation* if on every $\varphi \in \Phi$ there exists a cut-elimination sequence w.r.t. $>$. $\diamond$

Definition 5.1.5 (ACNF) Let $\varphi = \varphi_1$ and $\varphi_1, \ldots, \varphi_n$ be a cut-elimination sequence of φ_1 w.r.t. $>$. Then φ_n is called an ACNF (atomic cut normal form) of φ. $\diamond$

Note that we did not require $>$ to be terminating or confluent. However, all cut-elimination relations we are investigating in this book are terminating, but in general not confluent (so ACNFs are not unique).
Below we will define a set of rules defining a cut-elimination relation $>$ which can be extracted from Gentzen's famous proof of cut-elimination in **LK**.
Gentzen's method of cut-elimination is based on the transformation of uppermost mix-derivations ψ in φ (with a single final cut only) into other **LK**-proofs ψ'. The subproof ψ then is replaced by ψ' in φ, i.e. $\varphi[\psi]_\lambda > \varphi[\psi']_\lambda$,

where $\varphi.\lambda = \psi$. We define the rules $\mathcal{R}$ below without the restriction of ψ being an uppermost cut-derivation in φ, because we use $\mathcal{R}$ also for the definition of other cut-elimination methods. On the other hand we require the final cut in the cut-derivation to be a mix. We have seen in Chapter 3 that the restriction of cuts to mixes is inessential as every cut can easily be transformed into a mix + some structural rules.

Definition 5.1.6 (the cut reduction rules $\mathcal{R}$) Let ψ be an essential mix-derivation. We define a set of rules $\mathcal{R}$ transforming ψ into an **LK**-proof ψ'. For the sake of simplicity we assume that all cuts below are in fact mixes; in particular we do not write "mix" but "cut" in all derivations. For the names we assume that $\psi =$

$$\dfrac{\begin{matrix}(\rho) \\ \Gamma \vdash \Delta\end{matrix} \quad \begin{matrix}(\sigma) \\ \Pi \vdash \Lambda\end{matrix}}{\Gamma, \Pi^* \vdash \Delta^*, \Lambda}\ cut$$

Moreover we assume that the proofs are regular.

The cases below are labelled by the numbers also used in Gentzen's proof. Basically we distinguish the cases $\mathrm{rank}(\psi) = 2$ and $\mathrm{rank}(\psi) > 2$ (see Definition 3.2.17).

3.11. $\mathrm{rank}(\psi) = 2$.

3.113.1. the last inference in ρ is $w : r$:

$$\dfrac{\dfrac{\begin{matrix}(\rho') \\ \Gamma \vdash \Delta\end{matrix}}{\Gamma \vdash \Delta, A}\ w:r \qquad \begin{matrix}(\sigma) \\ \Pi \vdash \Lambda\end{matrix}}{\Gamma, \Pi^* \vdash \Delta, \Lambda}\ cut(A)$$

transforms to

$$\dfrac{\begin{matrix}(\rho') \\ \Gamma \vdash \Delta\end{matrix}}{\Gamma, \Pi^* \vdash \Delta, \Lambda}\ s^*$$

3.113.2. the last inference in ψ_2 is $w : l$: symmetric to 3.113.1.

The last inferences in ρ, σ are logical ones and the cut-formula is the principal formula of these inferences:

3.113.31.

$$\dfrac{\dfrac{\overset{(\rho_1)}{\Gamma \vdash \Delta, A} \quad \overset{(\rho_2)}{\Gamma \vdash \Delta, B}}{\Gamma \vdash \Delta, A \wedge B}\ \wedge{:}\,r \qquad \dfrac{\overset{(\sigma')}{A, \Pi \vdash \Lambda}}{A \wedge B, \Pi \vdash \Lambda}\ \wedge{:}\,l_1}{\Gamma, \Pi \vdash \Delta, \Lambda}\ cut(A \wedge B)$$

transforms to

$$\dfrac{\dfrac{\overset{(\rho_1)}{\Gamma \vdash \Delta, A} \quad \overset{(\sigma')}{A, \Pi \vdash \Lambda}}{\Gamma, \Pi^* \vdash \Delta^*, \Lambda}\ cut(A)}{\Gamma, \Pi \vdash \Delta, \Lambda}\ s^*$$

For $\wedge{:}\,l_2$ the transformation is analogous.

3.113.32. The last inferences of ρ, σ are $\vee{:}\,r_1$ ($\vee{:}\,r_2$) and $\vee{:}\,l$: symmetric to 3.113.31.

3.113.33.

$$\dfrac{\dfrac{\overset{(\rho'\{x \leftarrow \alpha\})}{\Gamma \vdash \Delta, B\{x \leftarrow \alpha\}}}{\Gamma \vdash \Delta, (\forall x)B}\ \forall{:}\,r \qquad \dfrac{\overset{(\sigma')}{B\{x \leftarrow t\}, \Pi \vdash \Lambda}}{(\forall x), \Pi \vdash \Lambda}\ \forall{:}\,l}{\Gamma, \Pi \vdash \Delta, \Lambda}\ cut((\forall x)B)$$

transforms to

$$\dfrac{\dfrac{\overset{(\rho'\{x \leftarrow t\})}{\Gamma \vdash \Delta, B\{x \leftarrow t\}} \quad \overset{(\sigma')}{B(x/t), \Pi \vdash \Lambda}}{\Gamma, \Pi^* \vdash \Delta^*, \Lambda}\ cut(B\{x \leftarrow t\})}{\Gamma, \Pi \vdash \Delta, \Lambda}\ s^*$$

3.113.34. The last inferences in ρ, σ are $\exists{:}\,r, \exists{:}\,l$: symmetric to 3.113.33.

3.113.35.

$$\dfrac{\dfrac{\overset{(\rho')}{A, \Gamma \vdash \Delta}}{\Gamma \vdash \Delta, \neg A}\ \neg{:}\,r \qquad \dfrac{\overset{(\sigma')}{\Pi \vdash \Lambda, A}}{\neg A, \Pi \vdash \Lambda}\ \neg{:}\,l}{\Gamma, \Pi \vdash \Delta, \Lambda}\ cut(\neg A)$$

reduces to

$$\dfrac{\dfrac{\overset{(\sigma')}{\Pi \vdash \Lambda, A} \quad \overset{(\rho')}{A, \Gamma \vdash \Delta}}{\Pi, \Gamma^* \vdash \Lambda^*, \Delta}\ cut(A)}{\Gamma, \Pi \vdash \Delta, \Lambda}\ s^*$$

3.113.36.

$$\dfrac{\dfrac{\begin{array}{c}(\rho')\\ A,\Gamma\vdash\Delta,B\end{array}}{\Gamma\vdash\Delta,A\to B}\to:r \qquad \dfrac{\begin{array}{c}(\sigma_1)\\ \Pi_1\vdash\Lambda_1,A\end{array}\quad\begin{array}{c}(\sigma_2)\\ B,\Pi_2\vdash\Lambda_2\end{array}}{A\to B,\Pi_1,\Pi_2\vdash\Lambda_1,\Lambda_2}\to:l}{\Gamma,\Pi_1,\Pi_2\vdash\Delta,\Lambda_1,\Lambda_2}\,cut(A\to B)$$

reduces to

$$\dfrac{\dfrac{\begin{array}{c}(\sigma_1)\\ \Pi_1\vdash\Lambda_1,A\end{array}\qquad\dfrac{\begin{array}{c}(\rho')\\ A,\Gamma\vdash\Delta,B\end{array}\quad\begin{array}{c}(\sigma_2)\\ B,\Pi_2\vdash\Lambda_2\end{array}}{A,\Gamma,\Pi_2^*\vdash\Delta^*,\Lambda_2}\,cut(B)}{\Pi_1,\Gamma^+\Pi_2^{*+}\vdash\Delta^*,\Lambda_1^+,\Lambda_2}\,cut(A)}{\Gamma,\Pi_1,\Pi_2\vdash\Delta,\Lambda_1,\Lambda_2}\,s^*$$

3.12. rank$(\psi) > 2$:

3.121. rank$_r(\psi) > 1$:

3.121.1. The cut formula occurs in Γ.

$$\dfrac{\begin{array}{c}(\rho)\\ \Gamma\vdash\Delta\end{array}\quad\begin{array}{c}(\sigma)\\ \Pi\vdash\Lambda\end{array}}{\Gamma,\Pi^*\vdash\Delta^*,\Lambda}\,cut(A)$$

transforms to

$$\dfrac{\begin{array}{c}(\sigma)\\ \Pi\vdash\Lambda\end{array}}{\Gamma,\Pi^*\vdash\Delta^*,\Lambda}\,s^*$$

3.121.2. The cut formula does not occur in Γ.

3.121.21. Let ξ be one of the rules $w\colon l$, $c\colon l$ or $\pi\colon l$; then

$$\dfrac{\begin{array}{c}(\rho)\\ \Gamma\vdash\Delta\end{array}\quad\dfrac{\begin{array}{c}(\sigma')\\ \Sigma\vdash\Lambda\end{array}}{\Pi\vdash\Lambda}\,\xi}{\Gamma,\Pi^*\vdash\Delta^*,\Lambda}\,cut(A)$$

transforms to

$$\dfrac{\dfrac{\begin{array}{c}(\rho)\\ \Gamma\vdash\Delta\end{array}\quad\begin{array}{c}(\sigma')\\ \Sigma\vdash\Lambda\end{array}}{\Gamma,\Sigma^*\vdash\Delta^*,\Lambda}\,cut(A)}{\Gamma,\Pi^*\vdash\Delta^*,\Lambda}\,s^*$$

Note that the sequence of structural rules s^* may be empty, i.e. it can be skipped if the sequent does not change.

3.121.22. Let ξ be an arbitrary unary rule (different from $c : l, w : l, p\colon l$)

and let C^* be empty if $C = A$ and C otherwise. The formulas B and C may be equal or different or simply nonexisting (in case ξ is a right rule). Let us assume that ψ is of the form

$$\dfrac{\begin{matrix}(\rho)\\ \Gamma \vdash \Delta\end{matrix} \quad \dfrac{\begin{matrix}(\sigma')\\ B, \Pi \vdash \Sigma\end{matrix}}{C, \Pi \vdash \Lambda}\,\xi}{\Gamma, C^*, \Pi^* \vdash \Delta^*, \Lambda}\,cut(A)$$

Let τ be the proof

$$\dfrac{\dfrac{\dfrac{\begin{matrix}(\rho)\\ \Gamma \vdash \Delta\end{matrix} \quad \begin{matrix}(\sigma')\\ B, \Pi \vdash \Sigma\end{matrix}}{\Gamma, B^*, \Pi^* \vdash \Delta^*, \Sigma}\,cut(A)}{\Gamma, B, \Pi^* \vdash \Delta^*, \Sigma}\,s^*}{\Gamma, C, \Pi^* \vdash \Delta^*, \Lambda}\,\xi + s^*$$

3.121.221. $A \neq C$, including the case that ξ is a right rule and B, C do not exist at all: then ψ transforms to τ.

3.121.222. $A = C$ and $A \neq B$: in this case C is the principal formula of ξ. Then ψ transforms to

$$\dfrac{\dfrac{\begin{matrix}(\rho)\\ \Gamma \vdash \Delta\end{matrix} \quad \begin{matrix}(\tau)\\ \Gamma, A, \Pi^* \vdash \Delta^*, \Lambda\end{matrix}}{\Gamma, \Gamma^*, \Pi^* \vdash \Delta^*, \Delta^*, \Lambda}\,cut(A)}{\Gamma, \Pi^* \vdash \Delta^*, \Lambda}\,s^*$$

3.121.223. $A = B = C$. Then $\Sigma \neq \Lambda$ and ψ transforms to

$$\dfrac{\dfrac{\begin{matrix}(\rho)\\ \Gamma \vdash \Delta\end{matrix} \quad \begin{matrix}(\sigma')\\ A, \Pi \vdash \Sigma\end{matrix}}{\Gamma, \Pi^* \vdash \Delta^*, \Sigma}\,cut(A)}{\Gamma, \Pi^* \vdash \Delta^*, \Lambda}\,\xi$$

3.121.23. The last inference in σ is binary:

3.121.231. The case $\wedge\colon r$. Here

$$\dfrac{\begin{matrix}(\rho)\\ \Gamma \vdash \Delta\end{matrix} \quad \dfrac{\begin{matrix}(\sigma_1)\\ \Pi \vdash \Lambda, B\end{matrix} \quad \begin{matrix}(\sigma_2)\\ \Pi \vdash \Lambda, C\end{matrix}}{\Pi, \vdash \Lambda, B \wedge C}\,\wedge\colon r}{\Gamma, \Pi^* \vdash \Delta^*, \Lambda, B \wedge C}\,cut(A)$$

transforms to

$$\dfrac{\dfrac{\overset{(\rho)}{\Gamma \vdash \Delta} \quad \overset{(\sigma_1)}{\Pi \vdash \Lambda, B}}{\Gamma, \Pi^* \vdash \Delta^*, \Lambda, B}\ cut(A) \qquad \dfrac{\overset{(\rho)}{\Gamma \vdash \Delta} \quad \overset{(\sigma_2)}{\Pi \vdash \Lambda, C}}{\Gamma, \Pi^* \vdash \Delta^*, \Lambda, C}\ cut(A)}{\Gamma, \Pi^* \vdash \Delta^*, \Lambda, B \wedge C}\ \wedge\colon r$$

3.121.232. The case $\vee\colon l$. Then ψ is of the form

$$\dfrac{\overset{(\rho)}{\Gamma \vdash \Delta} \qquad \dfrac{\overset{(\sigma_1)}{B, \Pi \vdash \Lambda} \quad \overset{(\sigma_2)}{C, \Pi \vdash \Lambda}}{B \vee C, \Pi \vdash \Lambda}\ \vee\colon l}{\Gamma, (B \vee C)^*, \Pi^* \vdash \Delta^*, \Lambda}\ cut(A)$$

Again $(B \vee C)^*$ is empty if $A = B \vee C$ and $B \vee C$ otherwise.
We first define the τ as the regularization of the proof:

$$\dfrac{\dfrac{\dfrac{\overset{(\rho)}{\Gamma \vdash \Delta} \quad \overset{(\sigma_1)}{B, \Pi \vdash \Lambda}}{B^*, \Gamma, \Pi^* \vdash \Delta^*, \Lambda}\ cut(A)}{B, \Gamma, \Pi^* \vdash \Delta^*, \Lambda}\ \xi \qquad \dfrac{\dfrac{\overset{(\rho)}{\Gamma \vdash \Delta} \quad \overset{(\sigma_2)}{C, \Pi \vdash \Lambda}}{C^*, \Gamma, \Pi^* \vdash \Delta^*, \Lambda}\ cut(A)}{C, \Gamma, \Pi^* \vdash \Delta^*, \Lambda}\ \xi}{B \vee C, \Gamma, \Pi^* \vdash \Delta^*, \Lambda}\ \vee\colon l$$

Note that, in case $A = B$ or $A = C$, the inference ξ is $w : l$; otherwise ξ is the identical transformation and can be dropped.
If $(B \vee C)^* = B \vee C$ then ψ transforms to τ.
If, on the other hand, $(B \vee C)^*$ is empty (i.e. $B \vee C = A$) then we transform ψ to

$$\dfrac{\dfrac{\overset{(\rho)}{\Gamma \vdash \Delta} \quad \tau}{\Gamma, \Gamma, \Pi^* \vdash \Delta^*, \Delta^*, \Lambda}\ cut(A)}{\Gamma, \Pi^* \vdash \Delta^*, \Lambda}\ s^*$$

3.121.233. The last inference in σ is $\rightarrow\colon l$. Then ψ is of the form:

$$\dfrac{\overset{(\rho)}{\Gamma \vdash \Delta} \qquad \dfrac{\overset{(\sigma_1)}{\Pi_1 \vdash \Lambda_1, B} \quad \overset{(\sigma_2)}{C, \Pi_2 \vdash \Lambda_2}}{B \rightarrow C, \Pi_1, \Pi_2 \vdash \Lambda_1, \Lambda_2}\ \rightarrow\colon l}{\Gamma, (B \rightarrow C)^*, \Pi_1^*, \Pi_2^* \vdash \Delta^*, \Lambda_1, \Lambda_2}\ cut(A)$$

As in 3.121.232 $(B \rightarrow C)^* = B \rightarrow C$ for $B \rightarrow C \neq A$ and $(B \rightarrow C)^*$ empty otherwise.

3.121.233.1. A occurs in Π_1 and in Π_2. Again we define a proof τ:

$$
\dfrac{\dfrac{\overset{(\rho)}{\Gamma \vdash \Delta} \quad \overset{(\sigma_1)}{\Pi_1 \vdash \Lambda_1, B}}{\Gamma, \Pi_1^* \vdash \Delta^*, \Lambda_1, B}\, cut(A) \quad \dfrac{\dfrac{\overset{(\rho)}{\Gamma \vdash \Delta} \quad \overset{(\sigma_2)}{C, \Pi_2 \vdash \Lambda_2}}{C^*, \Gamma, \Pi_2^* \vdash \Delta^*, \Lambda_2}\, cut(A)}{C, \Gamma, \Pi_2^* \vdash \Delta^*, \Lambda_2}\, \xi}{B \rightarrow C, \Gamma, \Pi_1^*, \Gamma, \Pi_2^* \vdash \Delta^*, \Lambda_1, \Delta^*, \Lambda_2}\, \rightarrow: l
$$

ξ is either weakening or the inference can be dropped. If $(B \rightarrow C)^* = B \rightarrow C$ then, as in 3.121.232, ψ is transformed to τ + some unary structural rule applications.
If $(B \rightarrow C)^*$ is empty then we transform ψ to

$$
\dfrac{\dfrac{\overset{(\rho)}{\Gamma \vdash \Delta} \quad \tau}{\Gamma, \Gamma, \Pi_1^*, \Gamma, \Pi_2^* \vdash \Delta, \Delta^*, \Lambda_1, \Delta^*, \Lambda_2}\, cut(A)}{\Gamma, \Pi_1^*, \Pi_2^* \vdash \Delta^*, \Lambda_1, \Lambda_2}\, s^*
$$

3.121.233.2. A occurs in Π_2, but not in Π_1. As in 3.121.233.1 we define a proof τ:

$$
\dfrac{\overset{(\sigma_1)}{\Pi_1 \vdash \Lambda_1, B} \quad \dfrac{\dfrac{\overset{(\rho)}{\Gamma \vdash \Delta} \quad \overset{(\sigma_2)}{C, \Pi_2 \vdash \Lambda_2}}{C^*, \Gamma, \Pi_2^* \vdash \Delta^*, \Lambda_2}\, cut(A)}{C, \Gamma, \Pi_2^* \vdash \Delta^*, \Lambda_2}\, \xi}{B \rightarrow C, \Pi_1, \Gamma, \Pi_2^* \vdash \Lambda_1, \Delta^*, \Lambda_2}\, \rightarrow: l
$$

Again we distinguish the cases $B \rightarrow C = A$ and $B \rightarrow C \neq A$ and define the transformation of ψ exactly like in 3.121.233.1.

3.121.233.3. A occurs in Π_1, but not in Π_2: analogous to 3.121.233.2.

3.121.234. The last inference in σ is $cut(B)$ for some formula B. Then ψ is of the form

$$
\dfrac{\overset{(\rho)}{\Gamma \vdash \Delta} \quad \dfrac{\overset{(\sigma_1)}{\Pi_1 \vdash \Lambda_1} \quad \overset{(\sigma_2)}{\Pi_2 \vdash \Lambda_2}}{\Pi_1, {\Pi_2}^+ \vdash {\Lambda_1}^+, \Lambda_2}\, cut(B)}{\Gamma, {\Pi_1}^*, {\Pi_2}^{+*} \vdash \Delta^*, {\Lambda_1}^+, \Lambda_2}\, cut(A)
$$

3.121.234.1. A occurs in Π_1 and in Π_2. Then ψ transforms to the regularization of

$$\dfrac{\dfrac{\dfrac{\overset{(\rho)}{\Gamma\vdash\Delta}\quad\overset{(\sigma_1)}{\Pi_1\vdash\Lambda_1}}{\Gamma,{\Pi_1}^*\vdash\Delta^*,\Lambda_1}\,cut(A)\qquad \dfrac{\overset{(\rho)}{\Gamma\vdash\Delta}\quad\overset{(\sigma_2)}{\Pi_2\vdash\Lambda_2}}{\Gamma,{\Pi_2}^*\vdash\Delta^*,\Lambda_2}\,cut(A)}{\Gamma,{\Pi_1}^*,\Gamma^*,{\Pi_2}^{+*}\vdash\Delta^{*+},{\Lambda_1}^+,\Delta^*,\Lambda_2}\,cut(B)}{\Gamma,{\Pi_1}^*,{\Pi_2}^{+*}\vdash\Delta^*,{\Lambda_1}^+,\Lambda_2}\,s^*$$

Note that, for $A = B$, we have ${\Pi_2}^{*+} = \Pi^*$ and $\Delta^{*+} = \Delta^*$; ${\Pi_2}^{*+} = {\Pi_2}^{+*}$ holds in all cases.

3.121.234.2. A occurs in Π_1, but not in Π_2. In this case we have ${\Pi_2}^{+*} = {\Pi_2}^+$ and we transform ψ to

$$\dfrac{\dfrac{\dfrac{\overset{(\rho)}{\Gamma\vdash\Delta}\quad\overset{(\sigma_1)}{\Pi_1\vdash\Lambda_1}}{\Gamma,{\Pi_1}^*\vdash\Delta^*,\Lambda_1}\,cut(A)\qquad \overset{(\sigma_2)}{\Pi_2\vdash\Lambda_2}}{\Gamma,{\Pi_1}^*,{\Pi_2}^+\vdash\Delta^{*+},{\Lambda_1}^+,\Lambda_2}\,cut(B)}{\Gamma,{\Pi_1}^*,{\Pi_2}^+\vdash\Delta^*,{\Lambda_1}^+,\Lambda_2}\,s^*$$

3.121.234.3. A is in Π_2, but not in Π_1: symmetric to 3.121.234.2.

3.122. $\text{rank}_r(\psi) = 1$ and $\text{rank}_l(\psi) > 1$: symmetric to 3.121.

◇

Remark: You might have observed that there are missing numbers in the rules between 3.11 and 3.113.1; this can be explained by the fact that we do not eliminate atomic cuts and that our axioms need not be of the form $A \vdash A$. On the other hand, case 3.121.234 does not occur in Gentzen's proof; but this case is necessary when reductions of cut-derivations with several cuts have to be considered. ◇

The rules in Definition 5.1.6 define a proof reduction relation $>_{\mathcal{R}}$ in an obvious way. Every subproof ψ of a proof φ can be replaced by a proof ψ' if there exists a rule transforming ψ into ψ'. Strictly speaking we define the compatible closure of $>_{\mathcal{R}}$.

Definition 5.1.7 Let $\varphi \in \Phi$ and ν be a node in φ with $\varphi.\nu = \psi$. Now assume that there exists a rule in $\mathcal{R}$ transforming ψ into ψ'. Then

$$\varphi \;>_{\mathcal{R}}\; \varphi[\psi']_\nu.$$

Proposition 5.1.1 $>_{\mathcal{R}}$ *is a proof reduction relation on* $\Phi^{\mathcal{A}}$.

Proof: We have to ensure that $\varphi \in \Phi^{\mathcal{A}}$ and $\varphi >_{\mathcal{R}} \chi$ implies $\chi \in \Phi_{\mathcal{A}}$. Only in cases 3.113.33 and 3.113.34 of Definition 5.1.6 the initial sequents may be changed. Indeed, initial sequents S may change to $S\theta$ for $\theta = \{x \leftarrow t\}$. But, as $\mathcal{A}$ is an axiom set, $S \in \mathcal{A}$ implies $S\theta \in \mathcal{A}$. Thus $\chi \in \Phi_{\mathcal{A}}$. □

The relation $>_{\mathcal{R}}$ is very liberal; indeed every cut-derivation on any position can be replaced. It is easy to see that this flexibility makes the relation $>_{\mathcal{R}}$ nonterminating. This can be easily seen in looking at the rule 3.121.234. Here $cut(A)$ and $cut(B)$ can be interchanged infinitely often leading to non-termination. Note that there are other forms of nontermination as well. Though we could show that $>_{\mathcal{R}}$ is a cut-elimination relation we intend to prove a stronger result, namely the existence of a *terminating* subrelation $>$ of $>_{\mathcal{R}}$ which is also a cut-elimination relation. It is obvious that the existence of such a $>$ implies that also $>_{\mathcal{R}}$ is a cut-elimination relation. Gentzen's proof of the "Hauptsatz" yields such a terminating relation $>$. The main principle is to select an uppermost essential cut and to reduce it by one of the rules in $\mathcal{R}$. In mathematical practice Gentzen type reductions appear in the transformation of non-elementary proofs in elementary ones starting from the simplest nonelementary proof parts.

Definition 5.1.8 Let φ be an essential cut-derivation

$$\frac{\psi_1 \quad \psi_2}{S}\ cut$$

with $\psi_1, \psi_2 \in \Phi_0$. Then φ is called a *simple* cut-derivation. ◇

Definition 5.1.9 Let $\varphi \in \Phi$ and ψ be a subproof of φ which is a simple cut-derivation. Then ψ is called an *uppermost cut-derivation* in φ. ◇

Definition 5.1.10 Let $\varphi \in \Phi$ and ν be a position in φ s.t. $\varphi.\nu$ is an uppermost cut-derivation in φ. Suppose there exists a rule in $\mathcal{R}$ rewriting $\varphi.\nu$ to ψ. Then $\varphi >_G \varphi[\psi]_\nu$. ◇

Gentzen's proof is based on a specific use of a relation quite similar to $>_G$ (note that we do not eliminate atomic cuts – except possibly in the weakening rules) and on a double induction on rank and grade. Gentzen did not define the corresponding relation explicitly but rather defined the proof reductions in cases within the proof. This is not surprising, as the main aim of the cut-elimination theorem was to give a *constructive proof* of the existence of cut-free **LK**-proofs. The abstraction to proof reduction rules and their computational use make sense only in the more modern perspective of computational proof theory and computer-aided proof transformation.

5.2 The Hauptsatz

Our first aim is to prove that $>_G$ is terminating. That $>_G$ is also a cut-elimination relation then follows quite easily. In order to prove termination we need some technical definitions and lemmas.

Proposition 5.2.1 *Let φ be an **LK**-proof which is irreducible under $>_G$. Then $\varphi \in \Phi_0$.*

Proof: Let us assume that $\varphi \notin \Phi_0$. Then there exist cut-derivations in φ. Among those cut-derivations we select an uppermost one. But then one of the cases in $\mathcal{R}$ apply and φ can be reduced by $>_G$. □

Definition 5.2.1 Let $\rhd$ be an arbitrary binary relation on a set M and $m \geq 1$. We define a relation $\rhd_m$ of type $M^m \times M^m$ by
$(x_1, \ldots, x_m) \rhd_m (y_1, \ldots, y_m)$ iff
there exists an $i \in \{1, \ldots, m\}$ s.t. $x_i \rhd y_i$, but for all $j \in \{1, \ldots, m\}$ with $j \neq i$: $y_j = x_j$. ◇

$\rhd_m$ represents the principle of parallel reduction on M^m w.r.t. $\rhd$. Indeed, $x_1 \rhd^* y_1, \ldots, x_m \rhd^* y_m$ iff $(x_1, \ldots, x_m) \rhd_m^* (y_1, \ldots, y_m)$.

Lemma 5.2.1 *Let $\rhd$ be a terminating relation on a set M. Then, for every $m \geq 1$, $\rhd_m$ terminates on M^m.*

Proof: Let $x_1, \ldots, x_r$ be a $\rhd_m$-derivation from x_1 for $x_1 \in M^m$. Let $x_1 = (z_1, \ldots, z_m)$ for $z_j \in M$. By assumption $\rhd$ terminates on all $z_1, \ldots, z_m$. Let

$$k_i = \max\{l(\delta) \mid \delta \text{ is a } \rhd-\text{derivation from } z_i\}$$

and $p = \max\{k_1, \ldots, k_m\}$.
Then, by definition of $\rhd_m$, the maximal length of a derivation from x_1 is $\leq m * p$. In particular $r \leq m * p$. Thus $\rhd_m$ is terminating. □

Lemma 5.2.2 *Let $\varphi \in \Phi$ s.t. φ is of the form*

$$\frac{\psi_1 \quad \psi_2}{S}\ \xi$$

where ξ is a binary logical rule (or an atomic cut). Let us assume that $>_G$ terminates on ψ_1 and on ψ_2. Then $>_G$ terminates on φ.

Proof: By Definition 5.1.6 rules in $\mathcal{R}$ are only applicable to cut-derivations. Now let $\varphi >^*_G \varphi'$; as the last inference ξ is not an essential cut, φ' is of the form

$$\frac{\psi_1' \quad \psi_2'}{S}\ \xi$$

with $\psi_1 >^*_G \psi_1'$ and $\psi_2 >^*_G \psi_2'$. Clearly, by Definition 5.2.1

$$\varphi >^* \varphi' \ \text{ iff } \ (\psi_1, \psi_2) \ >^*_{G2} \ (\psi_1', \psi_2').$$

By assumption $>_G$ terminates on ψ_1, ψ_2. Therefore, by Lemma 5.2.1, $>_{G2}$ terminates on (ψ_1, ψ_2). Thus $>_G$ terminates on φ. □

The following proposition gives us the main key for proving termination of $>_G$ on Φ.

Proposition 5.2.2 *Let φ be a simple cut-derivation. Then $>_G$ terminates on φ.*

Proof: Like Gentzen's proof of the Hauptsatz also this one proceeds by induction on rank and grade. But note that only cuts of logical complexity > 0 are eliminated.
The proof is based on induction using the ordering

$$\mathit{order}(\varphi) = (\mathrm{grade}(\varphi), \mathrm{rank}(\varphi)),$$

where $(k, l) < (k', l')$ if either $k < k'$ or $k = k'$ and $l < l'$.
(IB): $\mathit{order}(\varphi) = (0, m)$ for arbitrary m.
Then the last cut is not essential and no $>_G$-reduction is possible; this contradicts the assumption that φ is simple.

(IH):
Let us assume that $>_G$ terminates on φ for all φ with $\mathit{order}(\varphi) < (n+1, m)$.

We distinguish two cases (a) $m = 2$ and (b) $m > 2$.

case (a) $m = 2$:

Now grade$(\varphi) = n + 1$ for $n \geq 2$ and rank$(\varphi) = 2$.
Now the cut-derivation splits up into one or more cut-derivations of lower grade and (IH) can be applied. In cases 3.113.1, 3.113.2 the proof is transformed to a Φ_0-proof directly. We only show two typical cases, 3.113.33 and 3.113.36; the other cases are analogous.

3.113.33. The proof φ:

$$\dfrac{\dfrac{\begin{array}{c}(\rho'\{x \leftarrow \alpha\})\\ \Gamma \vdash \Delta, B\{x \leftarrow \alpha\}\end{array}}{\Gamma \vdash \Delta, (\forall x)B}\ \forall{:}\,r \qquad \dfrac{\begin{array}{c}(\sigma')\\ B\{x \leftarrow t\}, \Pi \vdash \Lambda\end{array}}{(\forall x)B, \Pi \vdash \Lambda}\ \forall{:}\,l}{\Gamma, \Pi \vdash \Delta, \Lambda}\ cut((\forall x)B)$$

transforms to φ':

$$\dfrac{\dfrac{\begin{array}{c}(\rho'\{x \leftarrow t\})\\ \Gamma \vdash \Delta, B\{x \leftarrow t\}\end{array} \quad \begin{array}{c}(\sigma')\\ B\{x \leftarrow t\}, \Pi \vdash \Lambda\end{array}}{\Gamma, \Pi^* \vdash \Delta^*, \Lambda}\ cut(B\{x \leftarrow t\})}{\Gamma, \Pi \vdash \Delta, \Lambda}\ s^*$$

Now φ' contains the cut-derivation ψ:

$$\dfrac{\begin{array}{c}(\rho'(x/t))\\ \Gamma \vdash \Delta, B(x/t)\end{array} \quad \begin{array}{c}(\sigma')\\ B(x/t), \Pi \vdash \Lambda\end{array}}{\Gamma, \Pi^* \vdash \Delta^*, \Lambda}\ cut(B(x/t))$$

with grade$(\psi) = n$. By (IH) $>_G$ terminates on ψ. But then $>_G$ also terminates on φ' as all reductions on φ' apply to the subproof ψ only (by definition of $>_G$ only essential cut-derivations are reduced within φ'). As φ only reduces to φ' (in one step) $>_G$ terminates also on φ.

3.113.36. The proof φ:

$$\dfrac{\dfrac{\begin{array}{c}(\rho')\\ A, \Gamma \vdash \Delta, B\end{array}}{\Gamma \vdash \Delta, A \rightarrow B}\ \rightarrow{:}\,r \qquad \dfrac{\begin{array}{c}(\sigma_1)\\ \Pi_1 \vdash \Lambda_1, A\end{array} \quad \begin{array}{c}(\sigma_2)\\ B, \Pi_2 \vdash \Lambda_2\end{array}}{A \rightarrow B, \Pi_1, \Pi_2 \vdash \Lambda_1, \Lambda_2}\ \rightarrow{:}\,l}{\Gamma, \Pi_1, \Pi_2 \vdash \Delta, \Lambda_1, \Lambda_2}\ cut(A \rightarrow B)$$

reduces to φ':

$$\dfrac{\dfrac{\begin{array}{c}(\sigma_1)\\ \Pi_1 \vdash \Lambda_1, A\end{array} \quad \dfrac{\begin{array}{c}(\rho')\\ A, \Gamma \vdash \Delta, B\end{array} \quad \begin{array}{c}(\sigma_2)\\ B, \Pi_2 \vdash \Lambda_2\end{array}}{A, \Gamma, \Pi_2^* \vdash \Delta^*, \Lambda_2}\ cut(B)}{\Pi_1, \Gamma^+, \Pi_2^{*+} \vdash \Lambda_1^+, \Delta^*, \Lambda_2}\ cut(A)}{\Gamma, \Pi_1, \Pi_2 \vdash \Delta, \Lambda_1, \Lambda_2}\ s^*$$

The (only) uppermost cut-derivation in φ' is ψ:

$$\frac{\overset{(\rho')}{A, \Gamma \vdash \Delta, B} \quad \overset{(\sigma_2)}{B, \Pi_2 \vdash \Lambda_2}}{A, \Gamma, \Pi_2^* \vdash \Delta^*, \Lambda_2} \; cut(B)$$

Clearly grade$(\psi) \leq n$ and by (IH) $>_G$ terminates on ψ with an LK-proof χ. By Proposition 5.2.1 $\chi \in \Phi_0$. Let μ be the position of ψ in φ'. Then termination on ψ in turn yields the proof $\varphi'[\chi]_\mu$. Note that before termination of $>_G$ on the subproof ψ no other reduction is possible! $\varphi'[\chi]_\mu$ also contains a single essential cut-derivation, namely ψ':

$$\frac{\overset{(\sigma_1)}{\Pi_1 \vdash \Lambda_1, A} \quad \overset{\chi}{A, \Gamma, \Pi_2^* \vdash \Delta^*, \Lambda_2}}{\Gamma^+, \Pi_1, \Pi_2^{*+} \vdash \Delta^*, \Lambda_1^+, \Lambda_2} \; cut(A)$$

Again grade$(\psi') \leq n$ and by (IH) $>_G$ terminates on ψ'. But then $>_G$ also terminates on $\varphi'[\chi]_\mu$. Putting things together we obtain the termination of $>_G$ on φ'. Note that, again, φ only reduces to φ' and thus $>_G$ terminates on φ.

case (b): $m > 2$:

As rank$(\varphi) = m$ for $m > 2$ we have rank$_l(\varphi) > 1$ or rank$_r(\varphi) > 1$. We consider only the case where rank$_r(\varphi) > 1$. The other case is symmetric. Now we have to use the rules which do not reduce the grade, but the rank of cut-derivations. Among the different cases in the definition of $\mathcal{R}$ we select 3.121.222, 3.121.232 and 3.121.234. These cases are typical and, in some sense, the most complicated ones (requiring a maximal number of additional cuts). The other cases are either similar (e.g. 3.121.233) or simpler (e.g., for binary rules, 3.121.231).

3.121.222. Let ξ be an arbitrary unary rule (different from $c : l, w : l$) and let $A \neq B$. Let us assume that φ is of the form

$$\frac{\overset{(\rho)}{\Gamma \vdash \Delta} \quad \dfrac{\overset{(\sigma')}{B, \Pi \vdash \Sigma}}{A, \Pi \vdash \Lambda} \; \xi}{\Gamma, C^*, \Pi^* \vdash \Delta^*, \Lambda} \; cut(A)$$

Let τ be the proof

$$\dfrac{\dfrac{\overset{(\rho)}{\Gamma \vdash \Delta} \quad \overset{(\sigma')}{B, \Pi \vdash \Lambda}}{\Gamma, B, \Pi^* \vdash \Delta^*, \Lambda} \; cut(A)}{\Gamma, A, \Pi^* \vdash \Delta^*, \Lambda} \; \xi + s^*$$

Then φ transforms to φ':

$$\cfrac{\cfrac{\overset{(\rho)}{\Gamma \vdash \Delta} \quad \overset{(\tau)}{\Gamma, A, \Pi^* \vdash \Delta^*, \Lambda}}{\Gamma, \Gamma^*, \Pi^* \vdash \Delta^*, \Delta^*, \Lambda}\ cut(A)}{\Gamma, \Pi^* \vdash \Delta^*, \Lambda}\ s^*$$

By construction the uppermost essential cut-derivation within φ' lies in τ. Let us call it ψ:

$$\cfrac{\overset{(\rho)}{\Gamma \vdash \Delta} \quad \overset{(\sigma')}{B, \Pi \vdash \Lambda}}{\Gamma, B, \Pi^* \vdash \Delta^*, \Lambda}\ cut(A)$$

As the cut has been shifted towards σ' we have $\text{rank}_r(\psi) < \text{rank}_r(\varphi)$, $\text{rank}_l(\psi) = \text{rank}_l(\varphi)$, so $\text{rank}(\psi) < m$ and $\text{grade}(\psi) = n + 1$. Therefore, according to (IH), $>_G$ terminates on ψ. As ψ is the only essential uppermost cut-derivation in φ' all reductions have to apply to ψ until it is normalized to a proof χ. Thus after termination of $>_G$ on ψ we obtain the proof $\varphi'[\chi]_\mu$ (where $\varphi'.\mu = \psi$). By Proposition 5.2.1 $\chi \in \Phi_0$. Therefore the only essential cut-derivation in $\varphi'[\chi]$ is ψ':

$$\cfrac{\overset{(\rho)}{\Gamma \vdash \Delta} \quad \cfrac{\overset{(\chi)}{\Gamma, B, \Pi^* \vdash \Delta^*, \Lambda}}{\Gamma, A, \Pi^* \vdash \Delta^*, \Lambda}\ \xi + s^*}{\Gamma, \Gamma^*, \Pi^* \vdash \Delta^*, \Delta^*, \Lambda}\ cut(A)$$

By construction $\text{rank}_r(\psi') = 1$, $\text{rank}_l(\psi') = \text{rank}_l(\varphi)$ and thus $\text{rank}(\psi') < m$, $\text{grade}(\psi') = n + 1$. By (IH), $>_G$ terminates on ψ' and therefore terminates on φ'.
By definition of $\mathcal{R}$ only the reduction of φ to φ' is possible if $\text{rank}_r(\varphi) > 1$ (even if also $\text{rank}_l(\varphi) > 1$). Therefore $>_G$ also terminates on φ.

3.121.232. The case $\vee\colon l$. Then φ is of the form

$$\cfrac{\overset{(\rho)}{\Gamma \vdash \Delta} \quad \cfrac{\overset{(\sigma_1)}{B, \Pi \vdash \Lambda} \quad \overset{(\sigma_2)}{C, \Pi \vdash \Lambda}}{B \vee C, \Pi \vdash \Lambda}\ \vee\colon l}{\Gamma, (B \vee C)^*, \Pi^* \vdash \Delta^*, \Lambda}\ cut(A)$$

Again we consider the most interesting case where $A = B \vee C$. Like above we define a proof τ:

$$\cfrac{\cfrac{\overset{(\rho)}{\Gamma \vdash \Delta} \quad \overset{(\sigma_1)}{B, \Pi \vdash \Lambda}}{B, \Gamma, \Pi^* \vdash \Delta^*, \Lambda}\ cut(B \vee C) \quad \cfrac{\overset{(\rho)}{\Gamma \vdash \Delta} \quad \overset{(\sigma_2)}{C, \Pi \vdash \Lambda}}{C, \Gamma, \Pi^* \vdash \Delta^*, \Lambda}\ cut(B \vee C)}{B \vee C, \Gamma, \Pi^* \vdash \Delta^*, \Lambda}\ \vee\colon l$$

Then φ transforms to φ':

$$\cfrac{\cfrac{\overset{(\rho)}{\Gamma \vdash \Delta} \quad \tau}{\Gamma, \Gamma, \Pi^* \vdash \Delta^*, \Delta^*, \Lambda}\ cut(B \vee C)}{\Gamma, \Pi^* \vdash \Delta^*, \Lambda}\ s^*$$

There are two uppermost essential cuts in φ', both of them lying in τ. We consider the corresponding cut-derivations ψ_1:

$$\cfrac{\overset{(\rho)}{\Gamma \vdash \Delta} \quad \overset{(\sigma_1)}{B, \Pi \vdash \Lambda}}{B, \Gamma, \Pi^* \vdash \Delta^*, \Lambda}\ cut(B \vee C)$$

and ψ_2:

$$\cfrac{\overset{(\rho)}{\Gamma \vdash \Delta} \quad \overset{(\sigma_2)}{C, \Pi \vdash \Lambda}}{C, \Gamma, \Pi^* \vdash \Delta^*, \Lambda}\ cut(B \vee C)$$

Let ψ be one of ψ_1, ψ_2. Then $\mathrm{rank}_r(\psi) < \mathrm{rank}_r(\varphi)$ and $\mathrm{rank}_l(\psi) = \mathrm{rank}_l(\varphi)$, and therefore $\mathrm{rank}(\psi_1), \mathrm{rank}(\psi_2) < m$. So, by (IH), $>_G$ terminates on ψ_1 (with χ_1) and on ψ_2 (with χ_2). By Lemma 5.2.2 $>_G$ also terminates on τ itself giving the result τ':

$$\cfrac{\overset{(\chi_1)}{B, \Gamma, \Pi^* \vdash \Delta^*, \Lambda} \quad \overset{(\chi_2)}{C, \Gamma, \Pi^* \vdash \Delta^*, \Lambda}}{B \vee C, \Gamma, \Pi^* \vdash \Delta^*, \Lambda}\ \vee{:}\,l$$

By Proposition 5.2.1 $\tau' \in \Phi_0$. Note that all reductions on φ' have to act on τ till all cuts of logical complexity > 0 are eliminated there. Now let μ be the node with $\varphi'.\mu = \tau$. Then after termination on τ we obtain the proof $\varphi'[\tau']_\mu$. The only essential cut-derivation in $\varphi'[\tau']_\mu$ is ψ':

$$\cfrac{\overset{(\rho)}{\Gamma \vdash \Delta} \quad \tau'}{\Gamma, \Gamma, \Pi^* \vdash \Delta^*, \Delta^*, \Lambda}\ cut(B \vee C)$$

In ψ' we have $\mathrm{rank}_r(\psi') = 1$ and therefore $\mathrm{rank}(\psi') < m$. By (IH) $>_G$ terminates on ψ' and thus also on φ'. As, by definition of $\mathcal{R}$, φ only reduces to φ', $>_G$ terminates on φ.

3.121.234. The case of cut. Here φ is of the form

$$\cfrac{\overset{(\rho)}{\Gamma \vdash \Delta} \quad \cfrac{\overset{(\sigma_1)}{\Pi_1 \vdash \Lambda_1} \quad \overset{(\sigma_2)}{\Pi_2 \vdash \Lambda_2}}{\Pi_1, {\Pi_2}^+ \vdash {\Lambda_1}^+, \Lambda_2}\ cut(B)}{\Gamma, {\Pi_1}^*, {\Pi_2}^{+*} \vdash \Delta^*, {\Lambda_1}^+, \Lambda_2}\ cut(A)$$

As φ is a simple (and essential) cut-derivation, the formula B is an atom and A contains logical operators (in particular we have $A \neq B$). We consider the most interesting case (3.121.234.1) where A occurs in Π_1 and in Π_2. Then φ transforms to φ' for $\varphi' =$

$$\dfrac{\dfrac{\overset{(\rho)}{\Gamma \vdash \Delta} \quad \overset{(\sigma_1)}{\Pi_1 \vdash \Lambda_1}}{\Gamma, {\Pi_1}^* \vdash \Delta^*, \Lambda_1}\ cut(A) \quad \dfrac{\overset{(\rho)}{\Gamma \vdash \Delta} \quad \overset{(\sigma_2)}{\Pi_2 \vdash \Lambda_2}}{\Gamma, {\Pi_2}^* \vdash \Delta^*, \Lambda_2}\ cut(A)}{\dfrac{\Gamma, \Gamma^+, {\Pi_1}^*, {\Pi_2}^{+*} \vdash \Delta^{*+}, \Delta^*, {\Lambda_1}^+, \Lambda_2}{\Gamma, {\Pi_1}^*, {\Pi_2}^{+*} \vdash \Delta^*, {\Lambda_1}^+, \Lambda_2}\ s^*}\ cut(B)$$

Above we write $*$ for the cut on A and $+$ for the cut on B. There are two uppermost (nonatomic) cuts in φ'; the corresponding cut-derivations are ψ_1:

$$\dfrac{\overset{(\rho)}{\Gamma \vdash \Delta} \quad \overset{(\sigma_1)}{\Pi_1 \vdash \Lambda_1}}{\Gamma, {\Pi_1}^* \vdash \Delta^*, \Lambda_1}\ cut(A)$$

and ψ_2:

$$\dfrac{\overset{(\rho)}{\Gamma \vdash \Delta} \quad \overset{(\sigma_2)}{\Pi_2 \vdash \Lambda_2}}{\Gamma, {\Pi_2}^* \vdash \Delta^*, \Lambda_2}\ cut(A)$$

Let ψ be one of ψ_1, ψ_2. Then $\text{rank}_r(\psi) < \text{rank}_r(\varphi)$ and $\text{rank}_l(\psi) = \text{rank}_l(\varphi)$, and therefore $\text{rank}(\psi_1), \text{rank}(\psi_2) < m$. So, by (IH), $>_G$ terminates on ψ_1 (with χ_1) and on ψ_2 (with χ_2). By Lemma 5.2.1 $>_{G2}$ terminates on (ψ_1, ψ_2). But (as the atomic cut with B is irreducible under $>_G$) then $>_G$ terminates on φ' with the result φ'':

$$\dfrac{\dfrac{\overset{(\chi_1)}{\Gamma, {\Pi_1}^* \vdash \Delta^*, \Lambda_1} \quad \overset{(\chi_2)}{\Gamma, {\Pi_2}^* \vdash \Delta^*, \Lambda_2}}{\Gamma, \Gamma^+, {\Pi_1}^*, {\Pi_2}^{+*} \vdash \Delta^{*+}, \Delta^*, {\Lambda_1}^+, \Lambda_2}\ cut(B)}{\Gamma, {\Pi_1}^*, {\Pi_2}^{+*} \vdash \Delta^*, {\Lambda_1}^+, \Lambda_2}\ s^*$$

But φ only reduces to φ', thus $>_G$ terminates on φ. □

Theorem 5.2.1 (termination of $>_G$) *$>_G$ terminates on all* **LK***-proofs.*

Proof: By induction on the number *cutnr* of nonatomic cuts in an **LK**-proof φ.
If $cutnr(\varphi) = 0$ (i.e. $\varphi \in \Phi_0$) then there are no cut-derivations in φ and thus no reductions under $>_G$; thus $>_G$ trivially terminates on φ.

(IH):
Let us assume that $>_G$ terminates on all $\varphi \in \Phi$ with $cutnr(\varphi) \leq k$.

Now let φ be an **LK**-proof with $cutnr(\varphi) = k+1$. We distinguish two cases:

(a) There exists a subproof ψ of φ which is a cut-derivation and contains all the cuts in φ.

In this case a unique lowermost cut exists which is the last inference of ψ. In particular ψ is of the form

$$\frac{\begin{array}{cc}(\psi_1) & (\psi_2)\\ S_1 & S_2\end{array}}{S}\ cut$$

where $cutnr(\psi_1) \leq k$ and $cutnr(\psi_2) \leq k$. By (IH) $>_G$ terminates on ψ_1 and on ψ_2. By Lemma 5.2.1 $>_{G2}$ terminates on (ψ_1, ψ_2). According to the definition of $>_G$ the lowermost cut in ψ can only be reduced if there are no nonatomic cuts above. In particular this cut can only be reduced if ψ_1 and ψ_2 are normalized. Let χ_1 be a normal form of ψ_1 and χ_2 of ψ_2 w.r.t. $>_G$ and let r be the total number of steps in the normalization of (ψ_1, ψ_2). Then $\psi >_G^r \psi'$ for $\psi' =$

$$\frac{\begin{array}{cc}(\chi_1) & (\chi_2)\\ S_1 & S_2\end{array}}{S}\ cut$$

But ψ' is a simple cut-derivation and, by Proposition 5.2.2, $>_G$ terminates on ψ'. Therefore $>_G$ terminates on ψ. According to the definition of ψ all $>_G$-reductions in φ take place within ψ; thus $>_G$ terminates on φ.

(b) φ does not contain a unique lowermost cut.

Then let $\psi_1, \ldots, \psi_m$ be all maximal cut-derivations in φ (i.e. cut-derivations which are no proper subproofs of other cut-derivations in φ). Then φ is of the form $\varphi[\psi_1, \ldots, \psi_m]_{\bar{\mu}}$ where $\bar{\mu}$ is the vector of positions of the ψ_j.

Clearly $cutnr(\psi_j) \leq k$ for all $j \in \{1, \ldots, m\}$. Thus, by (IH), $>_G$ terminates on $\psi_1, \ldots, \psi_m$. By Lemma 5.2.1 $>_{Gm}$ terminates on $(\psi_1, \ldots, \psi_m)$. It is obvious that the number of possible $>_G$-reductions on φ coincides with that on $(\psi_1, \ldots, \psi_m)$. Therefore $>_G$ terminates on φ. To any normal form $(\psi_1^*, \ldots, \psi_m^*)$ of $(\psi_1, \ldots, \psi_n)$ we obtain a normal form of φ of the form $\varphi[\psi_1^*, \ldots, \psi_m^*]_{\bar{\mu}}$, and vice versa.

□

Theorem 5.2.2 *$>_G$ is a cut-elimination relation.*

Proof: Let φ be an **LK**-proof of a sequent S. Then, by Theorem 5.2.1, $>_G$ terminates on φ. Let γ: $\varphi, \varphi_1, \ldots, \varphi_n$ be a corresponding $>_G$-derivation (s.t. φ_n is irreducible under $>_G$) . By definition of $>_G$ all φ_j have the same end-sequent S. By Proposition 5.2.1 $\varphi_n \in \Phi_0$. Therefore γ is a cut-elimination sequence on φ. □

Theorem 5.2.3 (the Hauptsatz) *Let φ be an* **LK***-proof of a sequent S. Then there exists an* **LK***-proof ψ of S s.t. ψ does not contain nonatomic cuts.*

Proof: Immediate by Theorem 5.2.2. □

The Hauptsatz formulated in Theorem 5.2.3 does not fully coincide with the original form in Gentzen's paper [38]. In fact Gentzen used a slightly different version of **LK** which made it possible to eliminate all (including the atomic) cuts. The initial sequents in Gentzen's original version of **LK** are $A \vdash A$ for arbitrary formulas A, instead of the form $A_1, \ldots, A_n \vdash B_1, \ldots, B_m$ for atoms A_i, B_j; as a consequence cuts with axioms are absorbed. Gentzen's form of the initial sequents made the various simulations of calculi (**LK**, natural deduction, Hilbert type) easier. In this book we do not consider such simulations, but concentrate on cut-elimination in **LK**; to this aim it is more appropriate to use *logic-free* initial sequents. But restricting the axiom sets to the standard one (see Definition 3.2.2) would complicate mathematical applications, where theory axioms can be used as initial sequents (e.g. take the transitivity axiom $P(x, y), P(y, z) \vdash P(x, z)$). However we will show that, for **LK**-proofs from the standard axiom set, all cuts can be eliminated.

Theorem 5.2.4 *Let $\varphi \in \Phi^{\mathcal{A}_T}$ for the standard axiom set $\mathcal{A}_T$ and φ be a proof of S. Then there exists a cut-free* **LK***-proof of S.*

Proof: By Theorem 5.2.3 there exists an **LK**-proof ψ of S with at most atomic cuts. We eliminate these cuts by the rank reduction rules in $\mathcal{R}$ applied to uppermost derivations of atomic cuts. So let χ be a subproof of the form

$$\frac{\begin{matrix}(\chi_1) & & (\chi_2) \\ \Gamma \vdash \Delta, A & & A, \Pi \vdash \Lambda\end{matrix}}{\Gamma, \Pi \vdash \Delta, \Lambda}\ cut(A)$$

where A is an atom and χ_1, χ_2 are cut-free. We proceed by induction on rank(χ).

rank(χ) = 2:
Then either χ_1 or χ_2 or both are of the form $A \vdash A$ or A is generated by weakening. We consider two typical cases, the other ones are symmetric.
(a): $\chi =$

$$\dfrac{\dfrac{\begin{matrix}(\chi_1')\\ \Gamma \vdash \Delta\end{matrix}}{\Gamma \vdash \Delta, A}\, w:r \qquad \begin{matrix}(\chi_2)\\ A, \Pi \vdash \Lambda\end{matrix}}{\Gamma, \Pi \vdash \Delta, \Lambda}\, cut(A)$$

Then χ reduces to τ:

$$\dfrac{\begin{matrix}(\chi_1')\\ \Gamma \vdash \Delta\end{matrix}}{\Gamma, \Pi \vdash \Delta, \Lambda}\, s^*$$

Clearly τ is cut-free.

(b): $\chi =$

$$\dfrac{A \vdash A \qquad \begin{matrix}(\chi_2)\\ A, \Pi \vdash \Lambda\end{matrix}}{A, \Pi \vdash \Lambda}\, cut(A)$$

In this case χ reduces to χ_2; χ_2 is cut-free by assumption.

rank(χ) > 2:
We assume $\mathrm{rank}_r(\chi) > 1$ like in the definition of $\mathcal{R}$. Now the rank reduction proceeds exactly as in Proposition 5.2.2 (note that the complexity of the cut-formulas does not play a role in the arguments). □

Remark: Theorem 5.2.4 can be generalized to atomic axiom systems which are closed under cut. The elimination of atomic cuts is (due to the rank reduction rules) exponential. ◇

Below we give an example of a cut-elimination sequence w.r.t. $>_G$.

Example 5.2.1 Let ψ be the proof

$$\dfrac{\dfrac{\dfrac{P(y) \vdash P(y)}{(\forall x)P(x) \vdash P(y)}\, \forall{:}\,l}{(\forall x)P(x) \vdash (\forall x)P(x)}\, \forall{:}\,r \qquad \dfrac{\dfrac{Q(b) \vdash}{\vdash \neg Q(b)}\, \neg:r}{(\forall x)P(x) \vdash \neg Q(b)}\, w:l}{(\forall x)P(x) \vdash (\forall x)P(x) \land \neg Q(b)}\, \land{:}\,r$$

and $\varphi =$

$$\dfrac{\psi \qquad \dfrac{\dfrac{\dfrac{P(b) \vdash P(a)}{(\forall x)P(x) \vdash P(a)}\ \forall\!: l}{(\forall x)P(x) \wedge \neg Q(b) \vdash P(a)}\ \wedge\!: l \qquad \dfrac{\dfrac{\dfrac{Q(b) \vdash}{\vdash \neg Q(b)}\ \neg : r}{\vdash (\exists x)\neg Q(x)}\ \exists\!: r}{(\forall x)P(x) \wedge \neg Q(b) \vdash (\exists x)\neg Q(x)}\ w : l}{(\forall x)P(x) \wedge \neg Q(b) \vdash P(a) \wedge (\exists x)\neg Q(x)}\ \wedge\!: r}{(\forall x)P(x) \vdash P(a) \wedge (\exists x)\neg Q(x)}\ cut$$

Now φ itself is a simple cut-derivation with grade$(\varphi) = 3$, rank$_l(\varphi) = 1$, rank$_r(\varphi) = 2$ and so rank$(\varphi) = 3$. There is only one rule in $\mathcal{R}$ which is applicable, the rank-reduction rule 3.121.231 (inducing so-called cross cuts). The resulting **LK**-proof is φ_1:

$$\dfrac{\dfrac{\psi \qquad \dfrac{\dfrac{P(b) \vdash P(a)}{(\forall x)P(x) \vdash P(a)}\ \forall\!: l}{(\forall x)P(x) \wedge \neg Q(b) \vdash P(a)}\ \wedge\!: l_1}{(\forall x)P(x) \vdash P(a)}\ cut \qquad \dfrac{\psi \qquad \dfrac{\dfrac{\dfrac{Q(b) \vdash}{\vdash \neg Q(b)}\ \neg : r}{\vdash (\exists x)\neg Q(x)}\ \exists\!: r}{(\forall x)P(x) \wedge \neg Q(b) \vdash (\exists x)\neg Q(x)}\ w : l}{(\forall x)P(x) \vdash (\exists x)\neg Q(x)}\ cut}{(\forall x)P(x) \vdash P(a) \wedge (\exists x)\neg Q(x)}\ \wedge\!: r$$

In φ_1 there are two (simple) cut-derivations. We select the left one and call it ψ_1:

$$\dfrac{\dfrac{\dfrac{\dfrac{P(y) \vdash P(y)}{(\forall x)P(x) \vdash P(y)}\ \forall\!: l}{(\forall x)P(x) \vdash (\forall x)P(x)}\ \forall\!: r \qquad \dfrac{\dfrac{Q(b) \vdash}{\vdash \neg Q(b)}\ \neg : r}{(\forall x)P(x) \vdash \neg Q(b)}\ w : l}{(\forall x)P(x) \vdash (\forall x)P(x) \wedge \neg Q(b)}\ \wedge\!: r \qquad \dfrac{\dfrac{P(b) \vdash P(a)}{(\forall x)P(x) \vdash P(a)}\ \forall\!: l}{(\forall x)P(x) \wedge \neg Q(b) \vdash P(a)}\ \wedge\!: l_1}{(\forall x)P(x) \vdash P(a)}\ cut$$

For ψ_1 we have grade$(\psi_1) = 3$, rank$_l(\psi_1) =$ rank$_r(\psi_1) = 1$. Therefore case 3.113.31 applies and we obtain a proof ψ_2:

$$\dfrac{\dfrac{\dfrac{P(y) \vdash P(y)}{(\forall x)P(x) \vdash P(y)}\ \forall\!: l}{(\forall x)P(x) \vdash (\forall x)P(x)}\ \forall\!: r \qquad \dfrac{P(b) \vdash P(a)}{(\forall x)P(x) \vdash P(a)}\ \forall\!: l}{(\forall x)P(x) \vdash P(a)}\ cut$$

By substituting ψ_2 for ψ_1 in φ_1 we obtain the proof φ_2:

$$\dfrac{\psi_2 \qquad \dfrac{\psi \quad \dfrac{\dfrac{\dfrac{Q(b)\vdash}{\vdash \neg Q(b)}\ \neg:r}{\vdash (\exists x)\neg Q(x)}\ \exists:r}{(\forall x)P(x)\wedge\neg Q(b)\vdash(\exists x)\neg Q(x)}\ w:l}{(\forall x)P(x)\vdash(\exists x)\neg Q(x)}\ cut}{(\forall x)P(x)\vdash P(a)\wedge(\exists x)\neg Q(x)}\ \wedge:r$$

Again there are two simple cut-derivations in φ_2; this time we select the right one and call it ψ_3:

$$\dfrac{\dfrac{\dfrac{\dfrac{P(y)\vdash P(y)}{(\forall x)P(x)\vdash P(y)}\ \forall:l}{(\forall x)P(x)\vdash(\forall x)P(x)}\ \forall:r \qquad \dfrac{\dfrac{Q(b)\vdash}{\vdash\neg Q(b)}\ \neg:r}{(\forall x)P(x)\vdash\neg Q(b)}\ w:l}{(\forall x)P(x)\vdash(\forall x)P(x)\wedge\neg Q(b)}\ \wedge:r \qquad \psi_3'}{(\forall x)P(x)\vdash(\exists x)\neg Q(x)}\ cut$$

where $\psi_3' =$

$$\dfrac{\dfrac{\dfrac{Q(b)\vdash}{\vdash\neg Q(b)}\ \neg:r}{\vdash(\exists x)\neg Q(x)}\ \exists:r}{(\forall x)P(x)\wedge\neg Q(b)\vdash(\exists x)\neg Q(x)}\ w:l$$

For ψ_3 we have $\mathrm{grade}(\psi_3) = 3$ and $\mathrm{rank}_l(\psi_3) = \mathrm{rank}_r(\psi_3) = 1$. So case 3.113.2 applies and we obtain a proof ψ_4:

$$\dfrac{\dfrac{\dfrac{Q(b)\vdash}{\vdash\neg Q(b)}\ \neg:r}{\vdash(\exists x)\neg Q(x)}\ \exists:r}{(\forall x)P(x)\vdash(\exists x)\neg Q(x)}\ w:l$$

By substituting ψ_4 for ψ_3 in φ_2 we obtain the proof φ_3:

$$\dfrac{\dfrac{\dfrac{\dfrac{P(y)\vdash P(y)}{(\forall x)P(x)\vdash P(y)}\ \forall:l}{(\forall x)P(x)\vdash(\forall x)P(x)}\ \forall:r \qquad \dfrac{P(b)\vdash P(a)}{(\forall x)P(x)\vdash P(a)}\ \forall:l}{(\forall x)P(x)\vdash P(a)}\ cut \qquad \dfrac{\dfrac{\dfrac{Q(b)\vdash}{\vdash\neg Q(b)}\ \neg:r}{\vdash(\exists x)\neg Q(x)}\ \exists:r}{(\forall x)P(x)\vdash(\exists x)\neg Q(x)}\ w:l}{(\forall x)P(x)\vdash P(a)\wedge(\exists x)\neg Q(x)}\ \wedge:r$$

In φ_3 there is only a single cut-derivation which we call ψ_5:

$$\dfrac{\dfrac{\dfrac{P(y) \vdash P(y)}{(\forall x)P(x) \vdash P(y)}\ \forall: l}{(\forall x)P(x) \vdash (\forall x)P(x)}\ \forall: r \qquad \dfrac{P(b) \vdash P(a)}{(\forall x)P(x) \vdash P(a)}\ \forall: l}{(\forall x)P(x) \vdash P(a)}\ cut$$

Now $\mathrm{grade}(\psi_5) = 1$, $\mathrm{rank}_l(\psi_5) = \mathrm{rank}_r(\psi_5) = 1$. Therefore case 3.113.33 applies and ψ_5 reduces to ψ_6:

$$\dfrac{\dfrac{P(b) \vdash P(b)}{(\forall x)P(x) \vdash P(b)}\ \forall: l \qquad P(b) \vdash P(a)}{(\forall x)P(x) \vdash P(a)}\ cut$$

Now we substitute ψ_5 by ψ_6 in φ_3 and obtain φ_4:

$$\dfrac{\dfrac{\dfrac{P(b) \vdash P(b)}{(\forall x)P(x) \vdash P(b)}\ \forall: l \qquad P(b) \vdash P(a)}{(\forall x)P(x) \vdash P(a)}\ cut \qquad \dfrac{\dfrac{\dfrac{Q(b) \vdash}{\vdash \neg Q(b)}\ \neg : r}{\vdash (\exists x)\neg Q(x)}\ \exists: r}{(\forall x)P(x) \vdash (\exists x)\neg Q(x)}\ w : l}{(\forall x)P(x) \vdash P(a) \wedge (\exists x)\neg Q(x)}\ \wedge: r$$

φ_4 is irreducible under $>_G$ because $\varphi_4 \in \Phi_0$ (thus φ_4 does not contain cut-derivations). So we have obtained a "Gentzen"-normal form of φ. The sequence $\varphi, \varphi_1, \varphi_2, \varphi_3, \varphi_4$ is a cut-elimination sequence on φ w.r.t. $>_G$.

5.3 The Method of Tait and Schütte

The relation $>_G$ extracted from Gentzen's original proof of cut-elimination is characterized by selections of uppermost cuts in **LK**-proofs. Another way to show the eliminability of cuts is to select a cut of maximal complexity; this way was chosen by W. Tait [73] and K. Schütte [70] in the context of infinitary proofs where Gentzen's method fails. Again Tait and Schütte did not define a computational method directly, but rather a method of proof. Tait's proof of cut-elimination does not even deal with usual variants of **LK**. Thus we have to adapt this method of proof to our rewriting system $\mathcal{R}$. Another problem in formalizing this method within $\mathcal{R}$ is that rank-reduction is not used in the proofs of Tait and Schütte (the reduction of cuts is achieved by immediate pruning). But we will illustrate below that

our reduction method (based on the relation $>_T$) is more fine-grained and in fact simulates the methods directly extracted from the proofs of Tait and Schütte.
To facilitate the arguments and definitions we introduce the following concept:

Definition 5.3.1 Let ψ be a cut-derivation with A being the cut-formula of the last inference. We call ψ *strict* if for all non-final cuts in ψ with cut formulas B we have $comp(B) < comp(A)$. ◇

Definition 5.3.2 (cut-complexity) Let φ be a proof and A be a cut formula in φ for which $comp(A)$ is maximal. Then we say that the cut-complexity of φ (denoted by $cutcomp(A)$) is $comp(A)$. ◇

Definition 5.3.3 Let ψ, ψ' be cut-derivations in **LK** and $\psi >_{\mathcal{R}} \psi'$. Let φ be an **LK**-proof and $\varphi.\nu = \psi$ for a node ν in φ. Then $\varphi >_T \varphi[\psi']_\nu$ if the following conditions are fulfilled:

(a) The final cut in ψ has maximal complexity in φ (i.e. its grade is the cut-complexity of φ).

(b) ψ is strict.

◇

We show below that every **LK**-proof which is irreducible under $>_T$ is in Φ_0. Thus every terminating $>_T$-reduction chain leads to normalized proofs.

Proposition 5.3.1 *Let φ be an* **LK***-proof which is irreducible under $>_T$. Then $\varphi \in \Phi_0$.*

Proof: Let φ be an **LK**-proof with $\varphi \notin \Phi_0$. We show that there exists a proof φ' s.t. $\varphi >_T \varphi'$. Let k be the cut-complexity of φ; clearly $k > 0$ as $\varphi \notin \Phi_0$. Then φ must contain a cut-derivation ψ s.t. the final cut is of complexity k. If ψ is not strict then it contains a proper cut-derivation ψ' which is strict and of cut-complexity k. As $k > 0$ there exists a proof ρ with $\psi' >_{\mathcal{R}} \rho$ (this follows directly from the definition of $\mathcal{R}$). Let $\varphi.\nu = \psi'$. Then, by definition of $>_T$, $\varphi >_T \varphi[\rho]_\nu$, i.e. φ is reducible under $>_T$. □

It is intuitively clear that the role of simple cut-derivations in $>_G$ is analogous to that of strict cut-derivations in $>_T$. However the structure of the reduction method $>_T$ requires another form of termination proof. This time it is not the grade of a cut-derivation which is relevant to the induction argument but another measure we are going to define below.

Definition 5.3.4 (weight) Let φ be an **LK**-proof. The number of occurrences of maximal cuts A in φ (i.e. $comp(A) = cutcomp(\varphi)$) is denoted by $nmc(\varphi)$. The *weight* of φ (denoted by $weight(\varphi)$) is defined by

$$weight(\varphi) \quad = \quad (cutcomp(\varphi), nmc(\varphi)).$$

weights are compared w.r.t. the usual tuple-ordering defined by:
$(n, m) < (l, k)$ if either $n < l$, or if $n = l$ and $m < k$. $\diamond$

Remark: Let φ be a strict cut-derivation. Then $weight(\varphi) = (k, 1)$ for some $k \in \mathbb{N}$. $\diamond$

Theorem 5.3.1 (termination of $>_T$) *$>_T$ terminates on Φ.*

Proof: We proceed by (double) induction on the weight of an **LK**-proof φ with an inner induction on the rank (for strict proofs).
(IB-1):
Let $weight(\varphi) = (0, m)$ for some number m.
Then there are only atomic cuts in φ and, according to the definition of $\mathcal{R}$, φ is irreducible under $>_T$; thus we obtain immediate termination.

(IH-1):
Let us assume that $>_T$ terminates on φ for all φ with $weight(\varphi) < (n+1, 1)$.

Now let $weight(\varphi) = (n + 1, 1)$.
Then $\varphi = \varphi[\psi]_\nu$ (for some node ν) where ψ is the (single) cut-derivation in φ with $cutcomp(\psi) = n + 1$.
We first show that $>_T$ terminates on ψ. Like in the case of $>_G$ we consider the rank of ψ.

(IB-2) Let rank$(\psi) = 2$.
In this case the arguments are similar to the case of $>_G$ as the cut-derivation splits up into one or more cut-derivations of lower weight and (IH-1) can be applied. In cases 3.113.1, 3.113.2 the proof is transformed to a Φ_0-proof directly and thus to (IB-1). Like for $>_G$ we only show two typical cases, 3.113.33 and 3.113.36; the other cases are analogous.

3.113.33. The proof ψ:

$$\cfrac{\cfrac{\cfrac{(\rho'(x/y))}{\Gamma \vdash \Delta, B(x/y)}}{\Gamma \vdash \Delta, (\forall x)B(x)}\,\forall{:}\,r \quad \cfrac{\cfrac{(\sigma')}{B(x/t), \Pi \vdash \Lambda}}{(\forall x)B(x), \Pi \vdash \Lambda}\,\forall{:}\,l}{\Gamma, \Pi \vdash \Delta, \Lambda}\,cut((\forall x)B)$$

transforms to ψ':

$$\cfrac{\cfrac{\begin{array}{cc}(\rho'(x/t)) & (\sigma') \\ \Gamma \vdash \Delta, B(x/t) & B(x/t), \Pi \vdash \Lambda\end{array}}{\Gamma, \Pi^* \vdash \Delta^*, \Lambda}\ cut(B(x/t))}{\Gamma, \Pi \vdash \Delta, \Lambda}\ s^*$$

Now $weight(\psi') = (n, l)$ for some $l \in \mathbb{N}$. By (IH-1) $>_T$ terminates on ψ'. As ψ, by strictness, only reduces to ψ' (in one step) $>_T$ terminates also on ψ.

3.113.36. The proof ψ:

$$\cfrac{\cfrac{\begin{array}{c}(\rho') \\ A, \Gamma \vdash \Delta, B\end{array}}{\Gamma \vdash \Delta, A \to B}\ \to: r \qquad \cfrac{\begin{array}{cc}(\sigma_1) & (\sigma_2) \\ \Pi_1 \vdash \Lambda_1, A & B, \Pi_2 \vdash \Lambda_2\end{array}}{A \to B, \Pi_1, \Pi_2 \vdash \Lambda_1, \Lambda_2}\ \to: l}{\Gamma, \Pi_1, \Pi_2 \vdash \Delta, \Lambda_1, \Lambda_2}\ cut(A \to B)$$

reduces to ψ':

$$\cfrac{\cfrac{\begin{array}{c}(\sigma_1) \\ \Pi_1 \vdash \Lambda_1, A\end{array} \qquad \cfrac{\begin{array}{cc}(\rho') & (\sigma_2) \\ A, \Gamma \vdash \Delta, B & B, \Pi_2 \vdash \Lambda_2\end{array}}{A, \Gamma, \Pi_2^* \vdash \Delta^*, \Lambda_2}\ cut(B)}{\Gamma^+, \Pi_1, \Pi_2^{*+} \vdash \Delta^*, \Lambda_1^+, \Lambda_2}\ cut(A)}{\Gamma, \Pi_1, \Pi_2 \vdash \Delta, \Lambda_1, \Lambda_2}\ s^*$$

Again $weight(\psi') = (k, l)$ for some $k \leq n$ and arbitrary l. By (IH-1) $>_T$ terminates on ψ'. As ψ only reduces to ψ' $>_T$ terminates on ψ.

To complete the proof of the termination on ψ we have to show that $>_T$ terminates on strict cut-derivations ψ with $weight(\psi) = (n + 1, 1)$ with arbitrary rank. So we define

(IH-2):
Let us assume that $>_T$ terminates on all strict cut-derivations ψ with $weight(\psi) = (n + 1, 1)$ and $\text{rank}(\psi) \leq m$, for some $m \geq 2$.

Now let ψ be a proof with $weight(\psi) = (n + 1, 1)$ and $\text{rank}(\psi) = m + 1$. Then either $\text{rank}_l(\psi) > 1$ or $\text{rank}_r(\psi) > 1$ (of course both may be > 1). We consider only the case where $\text{rank}_r(\psi) > 1$.
Now we have to use the rules which do not reduce the grade, but rather the rank of cut-derivations. Among the different cases in the definition of $\mathcal{R}$ we (again) select 3.121.222, 3.121.232 and 3.121.234. Like for $>_G$ the other cases are either similar (e.g. 3.121.233) or simpler (e.g., for binary rules,

3.121.231).

3.121.222. Let ξ be an arbitrary unary rule (different from $c:l, w:l$) and let $A \neq B$. Let us assume that ψ is of the form

$$\cfrac{\begin{array}{c}(\rho)\\ \Gamma \vdash \Delta\end{array} \quad \cfrac{\begin{array}{c}(\sigma')\\ B, \Pi \vdash \Sigma\end{array}}{A, \Pi \vdash \Lambda}\ \xi}{\Gamma, \Pi^* \vdash \Delta^*, \Lambda}\ cut(A)$$

Let τ be the proof

$$\cfrac{\cfrac{\begin{array}{c}(\rho)\\ \Gamma \vdash \Delta\end{array} \quad \begin{array}{c}(\sigma')\\ B, \Pi \vdash \Sigma\end{array}}{\Gamma, B, \Pi^* \vdash \Delta^*, \Sigma}\ cut(A)}{\Gamma, A, \Pi^* \vdash \Delta^*, \Sigma}\ \xi + s^*$$

Then ψ transforms to ψ':

$$\cfrac{\cfrac{\begin{array}{c}(\rho)\\ \Gamma \vdash \Delta\end{array} \quad \begin{array}{c}(\tau)\\ \Gamma, A, \Pi^* \vdash \Delta^*, \Lambda\end{array}}{\Gamma, \Gamma^*, \Pi^* \vdash \Delta^*, \Delta^*, \Lambda}\ cut(A)}{\Gamma, \Pi^* \vdash \Delta^*, \Lambda}\ s^*$$

The only strict subderivation in ψ' lies in τ. Note that the second A-cut derivation below contains another cut-derivation with the same formula A. Let us call the strict subderivation χ:

$$\cfrac{\begin{array}{c}(\rho)\\ \Gamma \vdash \Delta\end{array} \quad \begin{array}{c}(\sigma')\\ B, \Pi \vdash \Sigma\end{array}}{\Gamma, B, \Pi^* \vdash \Delta^*, \Sigma}\ cut(A)$$

As the cut has been shifted upwards in σ' we have $\mathrm{rank}_r(\chi) < \mathrm{rank}_r(\psi)$, $\mathrm{rank}_l(\chi) = \mathrm{rank}_l(\psi)$, so $\mathrm{rank}(\chi) \leq m$ and $weight(\chi) = (n+1, 1)$. Therefore, according to (IH-2), $>_T$ terminates on χ. As χ is the uppermost maximal cut-derivation derivation in ψ' all $>_T$-reductions have to apply to χ' for $\chi >^*_{\mathcal{R}} \chi'$ until we obtain a proof χ' with $weight(\chi') = (k, l)$ for some $k \leq n$ and arbitrary l. After this transformation of χ to χ' the only strict and maximal cut-derivation in $\psi'[\chi']$ is ψ'':

$$\cfrac{\begin{array}{c}(\rho)\\ \Gamma \vdash \Delta\end{array} \quad \cfrac{\begin{array}{c}(\chi)\\ \Gamma, B, \Pi^* \vdash \Delta^*, \Sigma\end{array}}{\Gamma, A, \Pi^* \vdash \Delta^*, \Sigma}\ \xi + s^*}{\Gamma, \Gamma^*, \Pi^* \vdash \Delta^*, \Delta^*, \Sigma}\ cut(A)$$

By construction $\text{rank}_r(\psi'') = 1$, $\text{rank}_l(\psi'') = \text{rank}_l(\psi)$ and thus $\text{rank}(\psi'') \leq m$, $weight(\psi'') = (n+1, 1)$. By (IH-2), $>_T$ terminates on ψ'' and therefore terminates on ψ'.
By definition of $\mathcal{R}$ only the reduction of ψ to ψ' is possible if $\text{rank}_r(\psi) > 1$ (even if also $\text{rank}_l(\psi) > 1$). Therefore $>_T$ also terminates on ψ.

3.121.232. The case $\vee\colon l$. Then ψ is of the form

$$\dfrac{\begin{matrix}(\rho)\\ \Gamma \vdash \Delta\end{matrix} \quad \dfrac{\begin{matrix}(\sigma_1)\\ B, \Pi \vdash \Lambda\end{matrix} \quad \begin{matrix}(\sigma_2)\\ C, \Pi \vdash \Lambda\end{matrix}}{B \vee C, \Pi \vdash \Lambda}\ \vee\colon l}{\Gamma, (B \vee C)^*, \Pi^* \vdash \Delta^*, \Lambda}\ cut(A)$$

Again we consider the most interesting case where $A = B \vee C$. Like above we define a proof τ:

$$\dfrac{\dfrac{\begin{matrix}(\rho)\\ \Gamma \vdash \Delta\end{matrix} \quad \begin{matrix}(\sigma_1)\\ B, \Pi \vdash \Lambda\end{matrix}}{B, \Gamma, \Pi^* \vdash \Delta^*, \Lambda}\ cut(B \vee C) \quad \dfrac{\begin{matrix}(\rho)\\ \Gamma \vdash \Delta\end{matrix} \quad \begin{matrix}(\sigma_2)\\ C, \Pi \vdash \Lambda\end{matrix}}{C, \Gamma, \Pi^* \vdash \Delta^*, \Lambda}\ cut(B \vee C)}{B \vee C, \Gamma, \Pi^* \vdash \Delta^*, \Lambda}\ \vee\colon l$$

Then ψ transforms to ψ':

$$\dfrac{\dfrac{\begin{matrix}(\rho)\\ \Gamma \vdash \Delta\end{matrix} \quad \tau}{\Gamma, \Gamma, \Pi^* \vdash \Delta^*, \Delta^*, \Lambda}\ cut(B \vee C)}{\Gamma, \Pi^* \vdash \Delta^*, \Lambda}\ s^*$$

There are three maximal cut-derivations in ψ', two of them being strict. The strict cut-derivations, let us denote them by χ_1 and χ_2, both lie in τ. In particular we have $\chi_1 =$

$$\dfrac{\begin{matrix}(\rho)\\ \Gamma \vdash \Delta\end{matrix} \quad \begin{matrix}(\sigma_1)\\ B, \Pi \vdash \Lambda\end{matrix}}{B, \Gamma, \Pi^* \vdash \Delta^*, \Lambda}\ cut(B \vee C)$$

and $\chi_2 =$

$$\dfrac{\begin{matrix}(\rho)\\ \Gamma \vdash \Delta\end{matrix} \quad \begin{matrix}(\sigma_2)\\ C, \Pi \vdash \Lambda\end{matrix}}{C, \Gamma, \Pi^* \vdash \Delta^*, \Lambda}\ cut(B \vee C)$$

Let χ be one of χ_1, χ_2. Then $\text{rank}_r(\chi) < \text{rank}_r(\psi)$ and $\text{rank}_l(\chi) = \text{rank}_l(\psi)$, hence $\text{rank}(\chi_1), \text{rank}(\chi_2) \leq m$. So, by (IH-2), $>_T$ terminates on χ_1 and on χ_2. (with χ_2). By Lemma 5.2.1 the parallel relation $>_{T2}$ also terminates on (χ_1, χ_2). Let $\psi'.\nu_1 = \chi_1$ and $\psi'.\nu_2 = \chi_2$. As long as the reductions in ψ' apply to subproofs χ'_1, χ'_2 which can be obtained by reduction from χ_1, χ_2 we obtain a proof $\psi'[\chi'_1, \chi'_2]_{(\nu_1,\nu_2)}$. Only if $cutcomp(\chi'_1) < n+1$ and $cutcomp(\chi'_2) < n+1$ the lowermost maximal cut in ψ' may be reduced. Thus, by termination of $>_{T2}$ on (χ_1, χ_2), ψ' reduces to a proof $\psi'[\chi'_1, \chi'_2]_{(\nu_1,\nu_2)}$ with

$$cutcomp(\chi'_1) < n+1 \text{ and } cutcomp(\chi'_2) < n+1.$$

At this stage of reduction the only strict cut-derivation in $\psi'[\chi']$ is $\xi =$

$$\frac{\begin{matrix}(\rho) \\ \Gamma \vdash \Delta\end{matrix} \quad \tau'}{\Gamma, \Gamma, \Pi^* \vdash \Delta^*, \Delta^*, \Lambda}\ cut(B \vee C)$$

for $\tau' = \tau[\chi'_1, \chi'_2]_{(\mu_1,\mu_2)}$ (μ_1, μ_2 being the root nodes of χ_1, χ_2 in τ). By construction we have $\text{rank}_r(\xi) = 1$ and therefore $\text{rank}(\xi) \leq m$. So we obtain $weight(\xi) = (n+1, 1)$ and $\text{rank}(\xi) \leq m$. By (IH-2) $>_T$ terminates on ξ and thus also on ψ'. As, by definition of $\mathcal{R}$, ψ only reduces to ψ' and ψ' reduces only to $\psi'[\chi']$, $>_T$ terminates on ψ.

3.121.234. The case of cut. Here ψ is of the form

$$\frac{\begin{matrix}(\rho) \\ \Gamma \vdash \Delta\end{matrix} \quad \dfrac{\begin{matrix}(\sigma_1) \\ \Pi_1 \vdash \Lambda_1\end{matrix} \quad \begin{matrix}(\sigma_2) \\ \Pi_2 \vdash \Lambda_2\end{matrix}}{\Pi_1, {\Pi_2}^+ \vdash {\Lambda_1}^+, \Lambda_2}\ cut(B)}{\Gamma, {\Pi_1}^*, {\Pi_2}^{+*} \vdash \Delta^*, {\Lambda_1}^+, \Lambda_2}\ cut(A)$$

As ψ is a strict cut-derivation we have $comp(B) < comp(A)$ (in particular $A \neq B$). We consider the most interesting case (3.121.234.1) where A occurs in Π_1 and in Π_2. Then ψ transforms to ψ' for $\psi' =$

$$\frac{\dfrac{\dfrac{\begin{matrix}(\rho) \\ \Gamma \vdash \Delta\end{matrix} \quad \begin{matrix}(\sigma_1) \\ \Pi_1 \vdash \Lambda_1\end{matrix}}{\Gamma, {\Pi_1}^* \vdash \Delta^*, \Lambda_1}\ cut(A) \quad \dfrac{\begin{matrix}(\rho) \\ \Gamma \vdash \Delta\end{matrix} \quad \begin{matrix}(\sigma_2) \\ \Pi_2 \vdash \Lambda_2\end{matrix}}{\Gamma, {\Pi_2}^* \vdash \Delta^*, \Lambda_2}\ cut(A)}{\Gamma, {\Pi_1}^*, \Gamma^+, {\Pi_2}^{+*} \vdash \Delta^{*+}, {\Lambda_1}^+, \Delta^*, \Lambda_2}\ cut(B)}{\Gamma, {\Pi_1}^*, {\Pi_2}^{+*} \vdash \Delta^*, {\Lambda_1}^+, \Lambda_2}\ s^*$$

There are two strict cut-derivations in ψ', the derivations $\chi_1 =$

$$\frac{\begin{matrix}(\rho) \\ \Gamma \vdash \Delta\end{matrix} \quad \begin{matrix}(\sigma_1) \\ \Pi_1 \vdash \Lambda_1\end{matrix}}{\Gamma, {\Pi_1}^* \vdash \Delta^*, \Lambda_1}\ cut(A)$$

and $\chi_2 =$

$$\frac{\begin{array}{cc}(\rho) & (\sigma_2)\\ \Gamma \vdash \Delta & \Pi_2 \vdash \Lambda_2\end{array}}{\Gamma, \Pi_2{}^* \vdash \Delta^*, \Lambda_2}\ cut(A)$$

Let χ be one of χ_1, χ_2. Then $\text{rank}_r(\chi) < \text{rank}_r(\psi)$ and $\text{rank}_l(\chi) = \text{rank}_l(\psi)$, and therefore $\text{rank}(\chi_1), \text{rank}(\chi_2) \leq m$. Moreover $weight(\chi_1) = weight(\chi_2) = (n+1, 1)$. So, by (IH-2), $>_T$ terminates on χ_1 and on χ_2. By Lemma 5.2.1 $>_{T2}$ terminates on (χ_1, χ_2). Let $\psi'.\nu_1 = \chi_1$ and $\psi'.\nu_2 = \chi_2$. Then the reduction proceeds on proofs χ_1', χ_2' reachable by $>_{\mathcal{R}}$ from of χ_1, χ_2 until we obtain a derivation $\psi'' \colon \psi'[\chi_1', \chi_2']_{(\nu_1,\nu_2)}$ with

$$cutcomp(\chi_1') \leq comp(B) \text{ and } cutcomp(\chi_2') \leq comp(B).$$

Then the only strict cut-derivation ξ in ψ'' (ending with the cut on B) fulfils $weight(\xi) \leq (n, l)$ for some $l \in \mathbb{N}$. In case $comp(B) = 0$ there is no strict cut-derivation anymore. Therefore (IH-1) applies and $>_T$ terminates on ξ, and thus on ψ'. But ψ only reduces to ψ', thus $>_T$ terminates on ψ.

Remember that $\varphi = \varphi[\psi]_\nu$ for $weight(\varphi) = (n+1, 1)$, where ψ is the only strict cut-derivation in φ with $weight(\psi) = (n+1, 1)$. We have shown above that $>_T$ terminates on ψ. By definition of $>_T$, reductions on φ must lead to a proof $\varphi' \colon \varphi[\psi']_\nu$ with $weight(\psi') < (n+1, 1)$ (note that such a ψ' is actually obtained by termination of $>_T$ on ψ). But then also $weight(\varphi') < (n+1, 1)$ and, by (IH-1), $>_T$ terminates on φ'. As arbitrary reductions on φ lead to such a proof φ', $>_T$ terminates on φ. So we have shown:

(IB-3) $>_T$ terminates on all φ with $weight(\varphi) = (n+1, 1)$.

(IH-3) Let us assume that $>_T$ terminates on all φ with $weight(\varphi) \leq (n+1, k)$.

So let us assume that $weight(\varphi) = (n+1, k+1)$. Then $\varphi = \varphi[\psi_1, \ldots, \psi_l]_{\bar{\nu}}$, where $\psi_1, \ldots \psi_l$ (for $l \leq k+1$) are the strict cut-derivations in φ with $cutcomp(\psi_j) = n+1$ and $\bar{\nu}$ is the vector $(\nu_1, \ldots, \nu_l)$ with $\varphi.\nu_j = \psi_j$.
By (IB3) $>_T$ terminates on ψ_j for $j = 1, \ldots k+1$; indeed, as the ψ_j are strict we have $weight(\psi_j) = (n+1, 1)$. By Lemma 5.2.1 $>_{Tk+1}$ terminates on the vector $(\psi_1, \ldots, \psi_l)$. Thus every sequence of reductions must lead to a proof

$$\varphi' \colon\ \varphi[\psi_1', \ldots, \psi_l']_{\bar{\nu}}$$

where for some ψ_j' $weight(\psi_j') < (n+1, 1)$. But then

$$weight(\varphi') = (n+1, k) < weight(\varphi).$$

By (IH-3) $>_T$ terminates on φ'. As every reduction sequence on φ leads to such a proof φ', $>_T$ terminates on φ.

□

Theorem 5.3.2 *$>_T$ is a cut-elimination relation.*

Proof: Let $\varphi \in \Phi$ be a proof of a S. Then, by Theorem 5.3.1, $>_T$ terminates on φ. Let γ: $\varphi, \varphi_1, \ldots, \varphi_n$ be a corresponding $>_T$-derivation (s.t. φ_n is irreducible under $>_T$) . By definition of $>_T$ all φ_i have the end-sequent S. By Proposition 5.3.1 $\varphi_n \in \Phi_0$. Therefore γ is a cut-elimination sequence on φ. □

In usual mathematics Tait–Schütte reductions occur when most complex defined properties are replaced by their explicit definitions in order to provide a better understanding of the proof. For example, in the concept hierarchy integral – Riemann sum – limes, the integrals are eliminated first.

5.4 Complexity of Cut-Elimination Methods

In Chapter 4 we have shown that the problem of cut-elimination in **LK**-proofs is nonelementary. That means there are sequences $(\varphi_n)_{n\in\mathbb{N}}$ of **LK**-proofs where *all* cut-elimination sequences w.r.t. *all* cut-elimination relations are of nonelementary size. Thus, clearly, both $>_G$ and $>_T$ only define nonelementary cut-elimination sequences on $(\varphi_n)_{n\in\mathbb{N}}$. So is there a point in further analyzing the complexity of methods? The answer is yes! Indeed, though all methods define only long cut-elimination sequences on worst-case examples they may strongly differ on other sequences. Moreover it may be the case that one method is always at least as good as the other one. In particular two methods differ essentially if their difference on specific sequences is of the complexity of cut-elimination itself. Below we develop a formal framework for a mathematical comparison of cut-elimination relations.

Definition 5.4.1 (NE-improvement) Let η be a cut-elimination sequence. We denote by $\|\eta\|$ the number of all symbol occurrences in η (i.e. the symbolic length of η). Let $>_x$ and $>_y$ be two cut-elimination relations (e.g. $>_T$ and $>_G$). We say that $>_x$ NE-improves $>_y$ (NE stands for nonelementarily) if there exists a sequence of **LK**-proofs $(\varphi_n)_{n\in\mathbb{N}}$ with the following properties:

1. There exists a $k \in \mathbb{N}$ s.t. for all n there exists a cut-elimination sequence η_n on φ_n w.r.t. $>_x$ with $\|\eta_n\| < e(k, \|\varphi_n\|)$,

2. For all $k \in \mathbb{N}$ there exists an $m \in \mathbb{N}$ s.t. for all n with $n > m$ and for all cut-elimination sequences θ on φ_n w.r.t. $>_y$: $\|\theta\| > e(k, \|\varphi_n\|)$.

◇

Definition 5.4.2 Let $>_x$ and $>_y$ be two cut-elimination relations s.t. $>_x$ NE-improves $>_y$ and $>_y$ NE-improves $>_x$. Then $>_x$ and $>_y$ are called *incomparable.* ◇

Our aim is to prove that the method of Gentzen and the method of Tait–Schütte NE-improve each other and thus are not comparable.
For the first speed-up theorem we need some auxiliary definitions and constructions.

Definition 5.4.3 Let A be an atom, $A_0 = A$ and $A_{m+1} = \neg A_m$ for all $m \geq 0$. Let π_0 be the **LK**-proof $A_0 \vdash A_0$ and $\pi_{m+1} =$

$$\cfrac{\cfrac{\cfrac{\begin{matrix}(\pi_m)\\ A_m \vdash A_m\end{matrix}}{A_{m+1}, A_m \vdash}\ \neg : l}{A_m, A_{m+1} \vdash}\ p : l}{A_{m+1} \vdash A_{m+1}}\ \neg : r$$

for all $m \geq 0$. Furthermore, for all $m \geq 0$ let τ_m be

$$\cfrac{\begin{matrix}(\pi_m)\\ A_m \vdash A_m\end{matrix} \quad \begin{matrix}(\pi_m)\\ A_m \vdash A_m\end{matrix}}{A_m \vdash A_m}\ cut(A_m)$$

◇

Lemma 5.4.1 *Let τ_m be the sequence in Definition 5.4.3. Then there exists a cut-elimination sequence ξ_m on τ_m w.r.t. $>_T$ and constants c, k independent of m s.t.*

$$\|\xi_m\| \leq c + k * m^3$$

for all m.

Proof: We define a cut-elimination sequence ξ_0 and give a definition of ξ_{m+1} in terms of ξ_m.
If $m = 0$ then τ_0 has only one atomic cut and we define $\xi_0 = \tau_0$. Now let us assume inductively that we have a cut elimination sequence ξ_m for the proof τ_m.

By Definition 5.4.3 τ_{m+1} is of the form

$$\dfrac{\dfrac{\dfrac{\dfrac{\begin{matrix}(\pi_m)\\ A_m \vdash A_m\end{matrix}}{A_m, A_{m+1} \vdash}\,\neg l}{A_m, A_{m+1} \vdash}\,p:l}{A_{m+1} \vdash A_{m+1}}\,\neg r \qquad \dfrac{\dfrac{\dfrac{\begin{matrix}(\pi_m)\\ A_m \vdash A_m\end{matrix}}{A_{m+1}, A_m \vdash}\,\neg l}{A_m, A_{m+1} \vdash}\,p:l}{A_{m+1} \vdash A_{m+1}}\,\neg r}{A_{m+1} \vdash A_{m+1}}\,cut(A_{m+1})$$

τ_{m+1} is a cut-derivation with a single cut where $\mathrm{rank}_l(\tau_{m+1}) = 1$ and $\mathrm{rank}_r(\tau_{m+1}) = 3$. Therefore we have to apply the rank reduction rules in Definition 5.1.6; in particular the rule 3.121.22 applies and the result is the proof τ^1_{m+1}:

$$\dfrac{\dfrac{\dfrac{\begin{matrix}(\pi_{m+1})\\ A_{m+1} \vdash A_{m+1}\end{matrix} \qquad \dfrac{\dfrac{\begin{matrix}(\pi_m)\\ A_m \vdash A_m\end{matrix}}{A_{m+1}, A_m \vdash}\,\neg l}{A_m, A_{m+1} \vdash}\,p:l}{A_{m+1}, A_m \vdash}\,cut(A_{m+1})}{A_m, A_{m+1} \vdash}\,p:l}{A_{m+1} \vdash A_{m+1}}\,\neg r$$

For the (single) cut-derivation τ^1_{m+1} in χ we have $\mathrm{rank}_r(\chi) = 2$ and we may apply the rank reduction rule 3.121.21 in Definition 5.1.6; the result is a proof τ^2_{m+1}:

$$\dfrac{\dfrac{\dfrac{\begin{matrix}(\pi_{m+1})\\ A_{m+1} \vdash A_{m+1}\end{matrix} \qquad \dfrac{\begin{matrix}(\pi_m)\\ A_m \vdash A_m\end{matrix}}{A_{m+1}, A_m \vdash}\,\neg l}{A_{m+1}, A_m \vdash}\,cut(A_{m+1})}{A_m, A_{m+1} \vdash}\,p:l}{A_{m+1} \vdash A_{m+1}}\,\neg r$$

Now the rank of the cut-derivation in τ^2_{m+1} is 2 and we may apply the grade reduction rule 3.113.35. The result is the proof τ^3_{m+1} with cut formula A_m:

$$\dfrac{\dfrac{\dfrac{\dfrac{\begin{matrix}(\pi_m)\\ A_m \vdash A_m\end{matrix} \qquad \dfrac{\dfrac{\begin{matrix}(\pi_m)\\ A_m \vdash A_m\end{matrix}}{A_{m+1}, A_m \vdash}\,\neg:l}{A_m, A_{m+1} \vdash}\,p:l}{A_m, A_{m+1} \vdash}\,cut(A_m)}{A_{m+1}, A_m \vdash}\,p:l}{A_m, A_{m+1} \vdash}\,p:l}{A_{m+1} \vdash A_{m+1}}\,\neg:r$$

The new cut-derivation in τ^3_{m+1} has a right rank > 1 and, again, we apply the rule 3.121.21 and obtain a proof τ^4_{m+1}:

$$\cfrac{\cfrac{\cfrac{\cfrac{\begin{matrix}(\pi_m)\\ A_m \vdash A_m\end{matrix} \quad \cfrac{\begin{matrix}(\pi_m)\\ A_m \vdash A_m\end{matrix}}{A_{m+1}, A_m \vdash}\ \neg : l}{A_m, A_{m+1} \vdash}\ cut(A_m)}{A_{m+1}, A_m \vdash}\ p : l}{A_m, A_{m+1} \vdash}\ p : l}{A_{m+1} \vdash A_{m+1}}\ \neg : r$$

In τ^4_{m+1} the rank reduction rule 3.121.22 applies and we obtain τ^5_{m+1}:

$$\cfrac{\cfrac{\cfrac{\cfrac{\cfrac{\cfrac{\begin{matrix}(\pi_m)\\ A_m \vdash A_m\end{matrix} \quad \begin{matrix}(\pi_m)\\ A_m \vdash A_m\end{matrix}}{A_m \vdash A_m}\ cut(A_m)}{A_{m+1}, A_m \vdash}\ \neg l}{A_m, A_{m+1} \vdash}\ p : l}{A_{m+1}, A_m \vdash}\ p : l}{A_m, A_{m+1} \vdash}\ p : l}{A_{m+1} \vdash A_{m+1}}\ \neg : r$$

But τ^5_{m+1} is just the proof

$$\cfrac{\cfrac{\cfrac{\cfrac{\cfrac{\begin{matrix}(\tau_m)\\ A_m \vdash A_m\end{matrix}}{A_{m+1}, A_m \vdash}\ \neg l}{A_m, A_{m+1} \vdash}\ p : l}{A_{m+1}, A_m \vdash}\ p : l}{A_m, A_{m+1} \vdash}\ p : l}{A_{m+1} \vdash A_{m+1}}\ \neg : r$$

By induction we have a cut-elimination sequence ξ_m on τ_m. The sequence ξ_m can be transformed into a cut-elimination sequence ξ'_m on τ^5_{m+1} in an obvious manner. But then the sequence ξ_{m+1}:

$$\tau_{m+1}, \tau^1_{m+1}, \ldots \tau^4_{m+1}, \xi'_m$$

is a cut-elimination sequence on τ_{m+1} w.r.t. $>_T$.
It is obvious that the length of the sequence ξ_m is $5 * m + 1$ for all m.

Let us investigate the size of ξ_m. In the sequence $\tau_{m+1}, \tau^1_{m+1}, \ldots, \tau^5_{m+1}$ we have $\|\tau^1_{m+1}\| > \|\tau_{m+1}\|$, but $\|\tau^i_{m+1}\| \leq \|\tau_{m+1}\|$ for $i = 2, 3, 4, 5$ and even

$$\|\tau_{m+1}, \tau^1_{m+1}, \ldots \tau^4_{m+1}\| \leq 5 * \|\tau_{m+1}\|$$

and

$$\|\tau^5_{m+1}\| < \|\tau_{m+1}\|.$$

Therefore $\|\xi_m\| \leq 5 * m * \|\tau_m\| + \|\tau_m\|$.
By Definition 5.4.3 we have (for some constant c with $\|A_0\| = c$):

$$\begin{array}{rcl} \|\tau_m\| & = & 2 * (m + c) + 2 * \|\pi_m\|, \\ \|\pi_{m+1}\| & \leq & 6 * (m + c + 1) + \|\pi_m\|, \\ \|\pi_0\| & = & 2 * c. \end{array}$$

Therefore there exists constants d_1, d_2, c, k s.t.

$$\begin{array}{rcl} \|\tau_m\| & \leq & d_1 + d_2 * m^2 \text{ and} \\ \|\xi_m\| & \leq & c + k * m^3. \end{array}$$

□

Theorem 5.4.1 $>_T$ *NE-improves* $>_G$.

Proof: Let γ_n be Statman's sequence defined in Chapter 4. We know that the maximal complexity of cut formulas in γ_n is less than 2^{n+3}. Let $g(n) = 2^{n+3}$ and the formulas A and A_i be as in Definition 5.4.3. Then clearly $comp(A_{g(n)}) = g(n)$ and thus $comp(A_{g(n)})$ is greater than the cut-complexity of γ_n. We will integrate $A_{g(n)}$ into a more complex formula, making this formula the principal formula of a cut. For every $n \in \mathbb{N}$ let ρ_n be the **LK**-proof:

$$\dfrac{\dfrac{\begin{matrix}(\pi_{g(n)})\\ A_{g(n)} \vdash A_{g(n)}\end{matrix} \quad \begin{matrix}(\gamma_n)\\ \Delta \vdash D_n\end{matrix}}{A_{g(n)}, \Delta \vdash A_{g(n)} \wedge D_n}\wedge\!: r \qquad \dfrac{\dfrac{\dfrac{\begin{matrix}(\pi_{g(n)})\\ A_{g(n)} \vdash A_{g(n)}\end{matrix} \quad A \vdash A}{A_{g(n)} \to A, A_{g(n)} \vdash A}\to\!: l}{A_{g(n)}, A_{g(n)} \to A \vdash A}\,p : l}{A_{g(n)} \wedge D_n, A_{g(n)} \to A \vdash A}\wedge\!: l}{A_{g(n)}, \Delta, A_{g(n)} \to A \vdash A}\,cut$$

where the π_m are the proofs from Definition 5.4.3 and $\mathcal{D}_n = p((\mathbf{T}_n q)q)$, $\Delta = \mathrm{Ax}, \mathrm{Ax}_T$ as defined in Section 4.3. From the proof of Lemma 5.4.1 we know that $\|\pi_m\| \leq c_1 + c_2 * m^2$ for constants c_1, c_2 and, by definition of $g(n)$:

$$\|\pi_{g(n)}\| \leq c_1 + c_2 * 2^{2*n+6}.$$

By definition of γ_n the proofs γ_n and thus also the ρ_n contain (only) an exponential number of sequents; thus there exist constants d_1, d_2 with

$$l(\rho_n) \leq 2^{d_1 + d_2 * n}$$

where the size of each sequent is less or equal than $2^{d_3 * n}$ for some constant d_3 independent of n. Consequently there exists a constant d s.t.

$$\|\rho_n\| \leq 2^{d(n+1)}$$

for all n.
We now construct a cut-elimination sequence on ρ_n based on $>_T$. As $comp(A_{g(n)})$ is greater than the cut-complexity of γ_n, and $comp(A_{g(n)} \wedge D_n) > comp(A_{g(n)})$, the most complex cut formula in ρ_n is $A_{g(n)} \wedge D_n$. This formula is selected by $>_T$ and we obtain $\rho_n >_T \rho'_n$ (via rule 3.113.31 in Definition 5.1.6) for the proof ρ^1_n:

$$\dfrac{\dfrac{\dfrac{(\pi_{g(n)})}{A_{g(n)} \vdash A_{g(n)}} \qquad \dfrac{\dfrac{\dfrac{(\pi_{g(n)})}{A_{g(n)} \vdash A_{g(n)}} \quad A \vdash A}{A_{g(n)} \rightarrow A, A_{g(n)} \vdash A}\ {\rightarrow}{:}\,l}{A_{g(n)}, A_{g(n)} \rightarrow A \vdash A}\ p:l}{A_{g(n)}, A_{g(n)} \rightarrow A \vdash A}\ cut(A_{g(n)})}{A_{g(n)}, \Delta, A_{g(n)} \rightarrow A \vdash A}\ s^*$$

ρ^1_n contains only one single cut with cut formula $A_{g(n)}$ and $\|\rho^1_n\| < \|\rho_n\|$. The right-rank of the corresponding cut-derivation is greater than 1 and we have to apply the rule 3.121.31. The result is the proof ρ^2_n:

$$\dfrac{\dfrac{\dfrac{(\pi_{g(n)})}{A_{g(n)} \vdash A_{g(n)}} \qquad \dfrac{\dfrac{(\pi_{g(n)})}{A_{g(n)} \vdash A_{g(n)}} \quad A \vdash A}{A_{g(n)} \rightarrow A, A_{g(n)} \vdash A}\ {\rightarrow}{:}\,l}{A_{g(n)}, A_{g(n)} \rightarrow A \vdash A}\ cut(A_{g(n)})}{A_{g(n)}, \Delta, A_{g(n)} \rightarrow A \vdash A}\ s^*$$

Clearly $\|\rho^2_n\| < \|\rho_n\|$. Still the right rank of the (single) cut-derivation in ρ^2_n is greater than 1 and we apply 3.121.233.3 in Definition 5.1.6. The corresponding result is the proof ρ^3_n:

$$\dfrac{\dfrac{\dfrac{\dfrac{(\pi_{g(n)})}{A_{g(n)} \vdash A_{g(n)}} \quad \dfrac{(\pi_{g(n)})}{A_{g(n)} \vdash A_{g(n)}}}{A_{g(n)} \vdash A_{g(n)}}\ cut(A_{g(n)}) \qquad A \vdash A}{A_{g(n)} \rightarrow A, A_{g(n)} \vdash A}\ {\rightarrow}{:}\,l}{A_{g(n)}, \Delta, A_{g(n)} \rightarrow A \vdash A}\ s^*$$

The (only) cut-derivation in ρ_n^3 is just the proof $\tau_{g(n)}$ defined in Definition 5.4.3. By Lemma 5.4.1 we know that there exists a cut-elimination sequence $\xi_{g(n)}$ based on $>_T$ with

$$\|\xi_{g(n)}\| \leq c + k * g(n)^3.$$

By definition of g there exists a constant d with

$$\|\xi_{g(n)}\| \leq 2^{d*(n+1)}$$

for all n.

$\xi_{g(n)}$ immediately defines a cut-elimination sequence $\xi'_{g(n)}$ on ρ_n^3 (with the last two sequents unchanged) and

$$\|\xi'_{g(n)}\| \leq 2^{d'*(n+1)}.$$

for a constant d'. Putting things together we obtain a constant r and a cut-elimination sequence ζ_n on ρ_n s.t.

$$\|\zeta_n\| \leq 2^{r*(n+1)} \quad \text{for all } n \geq 1.$$

In the second part of the proof we show that *every* cut-elimination sequence on ρ_n based on the relation $>_G$ is of nonelementary length in n and thus also in terms of $\|\rho_n\|$.

Note that every cut in γ_n lies *above* the cut with cut formula $D_n \wedge E_n$. Therefore, in Gentzen's method, we have to eliminate all non-atomic cuts in γ_n before eliminating the cut with $D_n \wedge E_n$. So every cut-elimination sequence on ρ_n based on $>_G$ must contain a proof of the form

$$\dfrac{\dfrac{\overset{(\pi_{g(n)})}{A_{g(n)} \vdash A_{g(n)}} \qquad \overset{(\gamma_n^*)}{\Delta \vdash D_n}}{A_{g(n)}, \Delta \vdash A_{g(n)} \wedge D_n}\ \wedge : r \qquad \dfrac{\dfrac{\dfrac{\overset{(\pi_{g(n)})}{A_{g(n)} \vdash A_{g(n)}} \quad A \vdash A}{A_{g(n)} \to A, A_{g(n)} \vdash A}\ \to : l}{A_{g(n)}, A_{g(n)} \to A \vdash A}\ p : l}{A_{g(n)} \wedge D_n, A_{g(n)} \to A \vdash A}\ \wedge : l}{A_{g(n)}, \Delta, A_{g(n)} \to A \vdash A}\ cut$$

where $\gamma_n^* \in \Phi_0$. But according to Statman's result we have $l(\gamma_n^*) > \frac{s(n)}{2}$. Clearly the length of γ_n^* is a lower bound on the length of every cut-elimination sequence on ρ_n based on $>_G$. Thus for all cut-elimination sequences θ on ρ_n w.r.t. $>_G$ we obtain

$$\|\theta\| > \frac{s(n)}{2}.$$

□

A nonelementary speed-up is possible also the other way around. In this case it is an advantage to select the cuts from upwards instead by formula complexity.

Theorem 5.4.2 *$>_G$ NE-improves $>_T$.*

Proof: Consider Statman's sequence γ_n defined in Chapter 4. Locate the uppermost proof δ_1 in γ_n; note that δ_1 is identical to ψ_{n+1}. In γ_n we first replace the proof δ_1 (or ψ_{n+1}) of $\Gamma \vdash H_{n+1}(\mathbf{T})$ by the proof $\hat{\delta}_1$ below (where Q is an arbitrary atom):

$$\dfrac{\dfrac{\begin{matrix}(\omega)\\ P \wedge \neg P \vdash\end{matrix}}{P \wedge \neg P \vdash \neg Q}\; w\colon r \qquad \dfrac{\begin{matrix}(\psi_{n+1})\\ \mathrm{Ax}_T \vdash H_{n+1}(\mathbf{T})\end{matrix}}{\neg Q, \mathrm{Ax}_T \vdash H_{n+1}(\mathbf{T})}\; w\colon l}{P \wedge \neg P, \mathrm{Ax}_T \vdash H_{n+1}(\mathbf{T})}\; cut$$

The subproof ω is a proof of $P \wedge \neg P \vdash$ of constant length. Furthermore we use the same inductive definition in defining $\hat{\delta}_k$ as that of δ_k in Chapter 4. Finally we obtain a proof φ_n in place of γ_n. Note that φ_n differs from γ_n only by an additional cut with cut-formula $\neg Q$ for an atom Q and the formula $P \wedge \neg P$ in the antecedents of sequents. We obtain

$$(+) \quad \|\varphi_n\| \leq c + 2 * \|\gamma_n\| \leq 2^{dn+r}$$

for appropriate constants c, d and r.

Our aim is to define a cut-elimination sequence on φ_n w.r.t. $>_G$ which is of elementary complexity. Let S_k be the end sequent of the proof $\hat{\delta}_k$. We first investigate cut-elimination on the proof $\hat{\delta}_k$; the remaining two cuts are eliminated in a similar way. To this aim we prove by induction on k:

(*) There exists a cut-elimination sequence $\hat{\delta}_{k,1}, \ldots, \hat{\delta}_{k,m}$ of $\hat{\delta}_k$ w.r.t. $>_G$ with the following properties:

(1) $m \leq l(\hat{\delta}_k)$,

(2) $\|\hat{\delta}_{k,i}\| \leq \|\hat{\delta}_k\|$ for $i = 1, \ldots, m$,

(3) $\hat{\delta}_{k,m}$ is of the form

$$\dfrac{\begin{matrix}(\omega)\\ P \wedge \neg P \vdash\end{matrix}}{S_k}\; w^* + p$$

Induction basis $k = 1$:
In $\hat{\delta}_1$ there is only one nonatomic cut (with the formula $\neg Q$) where the cut formula is introduced by weakening. Thus by definition of $>_G$, using the rule 3.113.1, we get $\hat{\delta}_1 >_G \hat{\delta}_{1,2}$ where $\hat{\delta}_{1,2}$ is the proof

$$\frac{\begin{matrix}(\omega)\\ P \wedge \neg P \vdash\end{matrix}}{P \wedge \neg P, \mathrm{Ax}_T \vdash H_{n+1}(\mathbf{T})}\ w^* + p$$

Clearly $2 \leq l(\hat{\delta}_1)$ and $\|\hat{\delta}_{1,2}\| \leq \|\hat{\delta}_1\|$. Moreover $\hat{\delta}_{1,2}$ is of the form (3). This gives $(*)$ for $k = 1$.

(IH) Assume that $(*)$ holds for k.

By definition, $\hat{\delta}_{k+1}$ is of the form

$$\frac{\begin{matrix}(\psi_{n-k+1})\\ \mathrm{Ax}_T \vdash H_{n-k+1}(\mathbf{T})\end{matrix} \quad \rho_k}{P \wedge \neg P, \mathrm{Ax}_T \vdash H_{n-k+1}(\mathbf{T}_{k+1})}\ cut$$

for $\rho_k =$

$$\frac{\begin{matrix}(\hat{\delta}_k)\\ P \wedge \neg P, \mathrm{Ax}_T \vdash H_{n-k+2}(\mathbf{T}_k)\end{matrix} \quad \begin{matrix}(\varphi_{k+1})\\ H_{n-k+2}(\mathbf{T}_k), H_{n-k+1}(\mathbf{T}) \vdash H_{n-k+1}(\mathbf{T}_{k+1})\end{matrix}}{P \wedge \neg P, \mathrm{Ax}_T, H_{n-k+1}(\mathbf{T}) \vdash H_{n-k+1}(\mathbf{T}_{k+1})}\ cut$$

By (IH) there exists a cut-elimination sequence $\hat{\delta}_{k,1}, \ldots, \hat{\delta}_{k,m}$ on $\hat{\delta}_k$ w.r.t. $>_G$ fulfilling (1), (2) and (3). In particular we have $\|\hat{\delta}_{k,m}\| \leq \|\hat{\delta}_k\|$ and $\hat{\delta}_{k,m}$ is of the form

$$\frac{\begin{matrix}(\omega)\\ P \wedge \neg P \vdash\end{matrix}}{S_k}\ w^* + p$$

All formulas in S_k, except $P \wedge \neg P$, are introduced by weakening in $\hat{\delta}_{k,m}$. In particular this holds for the formula $H_{n-k+2}(\mathbf{T}_k)$ which is a cut formula in $\hat{\delta}_{k+1}$. After cut-elimination on $\hat{\delta}_k$ the proof ρ_k is transformed (via $>_G$) into a proof $\hat{\rho}_k$:

$$\frac{\begin{matrix}(\hat{\delta}_{k,m})\\ P \wedge \neg P, \mathrm{Ax}_T \vdash H_{n-k+2}(\mathbf{T}_k)\end{matrix} \quad \begin{matrix}(\varphi_{k+1})\\ H_{n-k+2}(\mathbf{T}_k), H_{n-k+1}(\mathbf{T}) \vdash H_{n-k+1}(\mathbf{T}_{k+1})\end{matrix}}{P \wedge \neg P, \mathrm{Ax}_T, H_{n-k+1}(\mathbf{T}) \vdash H_{n-k+1}(\mathbf{T}_{k+1})}\ cut$$

Now the (only) non-atomic cut in $\hat{\rho}_k$ is with the cut formula $H_{n-k+2}(\mathbf{T}_k)$ which is introduced by $w\colon r$ in $\hat{\delta}_{k,m}$. By using iterated reduction of left-rank via the symmetric versions of 3.121.21 and 3.121.22 in Definition 5.1.6, the cut is eliminated and the proof χ_{n-k+1} "disappears" and the result is again of the form

$$\frac{\begin{array}{c}(\omega)\\ P \wedge \neg P \vdash\end{array}}{P \wedge \neg P, \mathrm{Ax}_T, H_{n-k+1}(\mathbf{T}) \vdash H_{n-k+1}(\mathbf{T}_{k+1})}\ w^* + p$$

The proof above is the result of a cut-elimination sequence $\hat{\rho}_{k,1}, \ldots, \hat{\rho}_{k,p}$ on $\hat{\rho}_k$ w.r.t. $>_G$. But then also δ_{k+1} is further reduced to a proof where $\hat{\rho}_k$ is replaced by $\hat{\rho}_{k,p}$; in this proof there is only one cut left (with the formula $H_{n-k+1}(\mathbf{T})$) and we may play the "weakening game" once more. Finally we obtain a proof $\hat{\delta}_{k,r}$ of the form

$$\frac{\begin{array}{c}(\omega)\\ P \wedge \neg P \vdash\end{array}}{P \wedge \neg P, \mathrm{Ax}_T \vdash H_{n-k+1}(\mathbf{T}_{k+1})}\ w^* + p$$

The conditions (1) and (2) are obviously fulfilled. This eventually gives $(*)$. After the reduction of φ_n to $\varphi_n[\hat{\delta}_{n,s}]_\lambda$, where λ is the position of $\hat{\delta}_n$ and $\hat{\delta}_{n,s}$ is the result of a Gentzen cut-elimination sequence on $\hat{\delta}_n$, there are only two cuts left. Again these cuts are swallowed by the proofs beginning with ω and followed by a sequence of weakenings plus a final permutation. Putting things together we obtain a cut-elimination sequence

$$\eta_n : \varphi_{n,1}, \ldots, \varphi_{n,q}$$

on φ_n w.r.t. $>_G$ with the properties:

(1) $\|\varphi_{n,i}\| \leq \|\varphi_n\|$ and

(2) $q \leq l(\varphi_n)$.

But then, by $(+)$,

$$\|\eta_n\| \leq q\|\varphi_n\| \leq \|\varphi_n\|^2 \leq 2^{2(dn+r)}.$$

Therefore η_n is a cut-elimination sequence on φ_n w.r.t. $>_G$ of elementary complexity.

For the other direction consider Tait's reduction method on the sequence φ_n. The cut formulas in φ_n fall into two categories;

- the new cut formula $\neg Q$ with $comp(\neg Q) = 1$ and

- the old cut formulas from γ_n.

Now let η be an arbitrary cut-elimination sequence on φ_n w.r.t. $>_T$. By definition of $>_T$ only cuts with maximal cut formulas can be selected in a reduction step w.r.t. $>_T$. Therefore η contains a proof ψ with $\varphi_n >_T^* \psi$ and $cutcomp(\psi) = 2$. As the new cut in φ_n with cut formula $\neg Q$ is of complexity 1, it is still present in ψ.
A straightforward proof transformation gives a proof χ s.t. $\gamma_n >_T^* \chi$, $cutcomp(\chi) = 2$, and $l(\chi) < l(\psi)$ (in some sense the Tait procedure does not "notice" the new cut). But every cut-free proof of γ_n has a length $> \frac{s(n)}{2}$ and cut-elimination of cuts with (fixed) complexity k is elementary [71]. More precisely there exists a k and a cut-elimination sequence θ on χ w.r.t. $>_T$ s.t.

$$\|\theta\| \leq e(k, l(\chi)).$$

This is only possible if there is no elementary bound on $l(\chi)$ in terms of $\|\varphi_n\|$ (otherwise we would get ACNFs of γ_n of length elementarily in $\|\gamma_n\|$). But then there is no elementary bound on $l(\psi)$ in terms of $\|\varphi_n\|$. Putting things together we obtain that for every k and for *every* cut-elimination sequence η on φ_n

$$\|\eta\| > e(k, \|\varphi_n\|) \text{ almost everywhere .}$$

□

Theorem 5.4.2 shows that there exist cut elimination sequences η_n on φ_n w.r.t. $>_G$ s.t. $\|\eta_n\|$ is elementarily bounded in n; however this does not mean that *every* cut-elimination sequence on φ_n w.r.t. $>_G$ is elementary. In fact $>_G$ is highly "unstable" in its different deterministic versions. Consider the subproof $\hat{\delta}_1$ in the proof of Theorem 5.4.2:

$$\cfrac{\cfrac{\begin{matrix}(\omega)\\ P \land \neg P \vdash\end{matrix}}{P \land \neg P \vdash \neg Q}\ w\colon r \qquad \cfrac{\begin{matrix}(\psi_{n+1})\\ \mathrm{Ax}_T \vdash H_{n+1}(\mathbf{T})\end{matrix}}{\neg Q, \mathrm{Ax}_T \vdash H_{n+1}(\mathbf{T})}\ w\colon l}{P \land \neg P, \mathrm{Ax}_T \vdash H_{n+1}(\mathbf{T})}\ cut$$

If, in $>_G$, we focus on the weakening $(w : l)$ in the right part of the cut and apply rule 3.113.2 we obtain $\hat{\delta}_1 >_G \mu$, where μ is the proof

$$\cfrac{\begin{matrix}(\psi_{n+1})\\ \mathrm{Ax}_T \vdash H_{n+1}(\mathbf{T})\end{matrix}}{P \land \neg P, \mathrm{Ax}_T \vdash H_{n+1}(\mathbf{T})}\ w\colon l$$

But μ contains the whole proof ψ_{n+1}. In the course of cut-elimination ψ_{n+1} is built into the produced proofs exactly as in the cut-elimination procedure on γ_n itself. The resulting proof in Φ_0 is in fact longer than γ_n^* (the corresponding cut-free proof of the n-th element of Statman's sequence) and thus is of nonelementary length! This tells us that there are different deterministic versions α_1 and α_2 of $>_G$ s.t. α_1 gives a nonelementary speed-up of α_2 on the input set $(\varphi_n)_{n \in \mathbb{N}}$.
In the introduction of additional cuts into Statman's proof sequence we use the weakening rule. Similar constructions can be carried out in versions of the Gentzen calculus without weakening. What we need is just a sequence of short **LK**-proofs of valid sequents containing "simple" redundant (in our case atomic) formulas on both sides serving as cut formulas. Note that **LK** without any redundancy (working with minimally valid sequents only) is not complete.

Remark: Though $>_G$ and $>_T$ are incomparable it is easy to define a new cut-elimination relation which NE-improves both $>_G$ and $>_T$. Just define $>_{GT} = >_G \cup >_T$. In fact, for any elimination order, we can find a method which NE-improves the former ones. ◇

Chapter 6

Cut-Elimination by Resolution

6.1 General Remarks

In Chapter 5 we analyzed methods which eliminate cuts by stepwise reduction of cut-complexity. These methods always identify the uppermost logical operator in the cut-formula and either eliminate it directly (grade reduction) or indirectly (rank reduction). Here it is typical that, during grade reduction, the cut formulas are "peeled" from outside. These methods are local in the sense that only a small part of the whole proof is analyzed, namely the derivation corresponding to the introduction of the uppermost logical operator. As a consequence many types of redundancy in proofs are left undetected in these reductive methods, leading to bad computational behavior.

In [18] we defined a method of cut-elimination which is based on an analysis of *all* cut-derivations in **LK**-proofs. The interplay of binary rules which produce ancestors of cut formulas and those which do not, defines a structure which can be represented as a set of clauses or as a clause term. This set of clauses is always unsatisfiable and admits a resolution refutation. The refutation thus obtained may serve as a skeleton of an **LK**-proof of the original sequent with only atomic cuts. The cut-free proof itself is obtained by replacing clauses in the resolution tree by so-called proof projections of the original proof. Though this method, cut-elimination by resolution, radically differs from Gentzen's reduction method defined in Chapter 5, it simulates all methods based on the Gentzen rules (see [20]) as will be shown in Section 6.8.

M. Baaz, A. Leitsch, *Methods of Cut-Elimination*, Trends in Logic 34,
DOI 10.1007/978-94-007-0320-9_6,

In contrast to Gentzen's method, cut-elimination by resolution requires the proof to be Skolemized. After cut-elimination the derivation can be transformed into another of the original (un-Skolemized) sequent. If the end-sequent is prenex then there exists a polynomial transformation into the original sequent.
Below we define structural Skolemization and the Skolemization of proofs and give some technical definitions.

6.2 Skolemization of Proofs

Originally Skolemization was considered a model theoretic method to replace quantifiers by function symbols. The soundness thereby depends on the application of the axiom of choice. The Skolem functions can be interpreted as functions over standard models. Historically such functions are connected to the necessity to calculate with objects after their existence has been established. In the proof theoretic context Skolemization is a transformation on first-order formulas which removes all strong quantifiers. This means that the correct choice of functions as Skolem functions depends on the ability to derive the obtained results without them. Thus special emphasis has to be laid on the elimination of Skolem functions within reasonable complexity bounds as already addressed in Hilbert and Bernays [51].
There are different types of Skolemizations which may strongly differ in the proof complexity of the transformed formula (see [16]). Below we define the structural Skolemization operator *sk*, which represents the immediate version of Skolemization.

Definition 6.2.1 (Skolemization) *sk* is a function which maps closed formulas into closed formulas; it is defined in the following way:

$$sk(F) \quad = \quad F \quad \text{if } F \text{ does not contain strong quantifiers.}$$

Otherwise assume that (Qy) is the first strong quantifier in F (in a tree ordering) which is in the scope of the weak quantifiers $(Q_1 x_1), \ldots, (Q_n x_n)$ (appearing in this order). Let f be an n-ary function symbol not occurring in F (f is a constant symbol for $n = 0$). Then $sk(F)$ is defined inductively as

$$sk(F) \quad = \quad sk(F_{(Qy)}\{y \leftarrow f(x_1, \ldots, x_n)\}).$$

where $F_{(Qy)}$ is F after omission of (Qy). $sk(F)$ is called the (structural) *Skolemization* of F. ◇

In model theory and automated deduction the definition of Skolemization mostly is dual to Definition 6.2.1, i.e. in case of prenex forms the existential quantifiers are eliminated instead of the universal ones. We call this kind of Skolemization *refutational Skolemization*. The dual kind of Skolemization (elimination of universal quantifiers) is frequently called "Herbrandization" [54]. The Skolemization of sequents, defined below, yields a more general framework covering both concepts.

Definition 6.2.2 (Skolemization of sequents) Let S be the sequent $A_1, \ldots, A_n \vdash B_1, \ldots, B_m$ consisting of closed formulas only and

$$sk((A_1 \wedge \ldots \wedge A_n) \to (B_1 \vee \ldots \vee B_m)) = (A'_1 \wedge \ldots \wedge A'_n) \to (B'_1 \vee \ldots \vee B'_m).$$

Then the sequent

$$S'\colon\ A'_1, \ldots, A'_n \vdash B'_1, \ldots, B'_m$$

is called the *Skolemization* of S. ◇

Example 6.2.1 Let S be the sequent $(\forall x)(\exists y)P(x,y) \vdash (\forall x)(\exists y)P(x,y)$. Then the Skolemization of S is S' : $(\forall x)P(x, f(x)) \vdash (\exists y)P(c,y)$ for a one-place function symbol f and a constant symbol c. Note that the Skolemization of the left-hand-side of the sequent corresponds to the refutational Skolemization concept for formulas. ◇

By a *Skolemized proof* we mean a proof of the Skolemized end sequent. Also proofs with cuts can be Skolemized, but the cut formulas themselves cannot. Only the strong quantifiers which are ancestors of the end sequent are eliminated. Skolemization does not increase the length of proofs. To measure proof length in a way which is reasonably independent of different structural versions of **LK** we choose the number of logical inferences and cuts.

Definition 6.2.3 Let φ be an arbitrary **LK**-proof. By $\|\varphi\|_l$ we denote the number of logical inferences and cuts in φ. Unary structural rules are not counted. ◇

Proposition 6.2.1 *Let φ be an* **LK**-*proof of S from an atomic axiom set $\mathcal{A}$. Then there exists a proof $sk(\varphi)$ of $sk(S)$ (the structural Skolemization of S) from $\mathcal{A}$ s.t. $\|sk(\varphi)\|_l \leq \|\varphi\|_l$.*

Proof: The transformation of φ to $sk(\varphi)$ is based on a technique described in [16], Lemma 4. There, however, it was applied to specific forms of cut-free proofs only. The extension to proofs containing cuts is not very difficult; for this purpose the method has to be restricted to formulas having successors in the end sequent S. In particular Skolemization has to be avoided on all occurrences of cut formulas. Below we present a transformation defined in [17].

Let us locate an innermost occurrence of a strong quantifier in a formula of the end-sequent S. Assume, e.g., that μ is such an occurrence which is a positive occurrence of $(\forall x)A(x)$. Then there are the following possibilities for the introduction of $(\forall x)A(x)$ in φ. In all cases (b)–(e) $(\forall x)A(x)$ occurs as a subformula in C.

(a)

$$\frac{\Gamma \vdash \Delta, A(\alpha)}{\Gamma \vdash \Delta, (\forall x)A(x)}\ \forall : r$$

(b)

$$\frac{\Gamma \vdash \Delta, B}{\Gamma \vdash \Delta, B \vee C}\ \vee : r$$

(c)

$$\frac{B, \Gamma \vdash \Delta}{B \wedge C, \Gamma \vdash \Delta}\ \wedge : l$$

(d)

$$\frac{\Gamma \vdash \Delta}{\Gamma \vdash \Delta, C}\ w : r$$

(e)

$$\frac{\Gamma \vdash \Delta}{C, \Gamma \vdash \Delta}\ w : l$$

The most interesting case is (a). The cases (b)–(e) can be handled in a very similar way. We first describe (a):

We locate the proof segment $\psi(\alpha)$ ending in $\Gamma \vdash \Delta, A(\alpha)$. Let $\rho[(\forall x)A(x)]$

be the path connecting $\Gamma \vdash \Delta, (\forall x)A(x)$ with S. We locate all introductions of weak quantifiers $(Q_i y_i)$ on $\rho[(\forall x)A(x)]$ which dominate the occurrence of $(\forall x)A(x)$; each of these introductions eliminates a term t_i. Let $t_1, \ldots, t_n$ be all of these terms. Then we introduce a new function symbol f_μ of arity n and replace $\psi(\alpha)$ by $\psi(f_\mu(t_1, \ldots, t_n))$, which is a proof of $\Gamma \vdash \Delta, A(f_\mu(t_1, \ldots, t_n))$.
Note that α is an eigenvariable and $\Gamma \vdash \Delta$ is left unchanged by this substitution. By observing appropriate regularity conditions no eigenvariable conditions are violated in $\psi(f_\mu(t_1, \ldots, t_n))$.
Next we skip the $\forall\colon r$-introduction of a) and replace

$$\rho[(\forall x)A(x)] \text{ by } \rho[A(f_\mu(t_1, \ldots, t_n))].$$

Note that, on $\rho[A(f_\mu(t_1, \ldots, t_n))]$, the terms $t_1, \ldots, t_n$ are eliminated successively. The intermediate occurrences of $A(\ldots)$ are thus of the form

$$A(f_\mu(y_1, \ldots, y_k, t_{k+1}, \ldots, t_n)).$$

The occurrence of $(\forall x)A(x)$ in the end sequent is thus transformed to $A(f(y_1, \ldots, y_n))$, which is precisely the Skolemized occurrence of $(\forall x)A(x)$ in S.
Note that, by this transformation, contractions are not disturbed as the function symbol is labeled by μ.
The case of a formula $(\exists x)A(x)$ occurring negatively in S is completely analogous.

We now turn to (b):
Let $\rho[B \vee C]$ be the path connecting $\Gamma \vdash \Delta, B \vee C$ with the end sequent. Let $y_1, \ldots, y_m$ be the weak quantifiers in C dominating $(\forall x)A(x)$ and $t_1, \ldots, t_n$ be the terms eliminated by weak quantifiers on the path down from $B \vee C$ to the end-sequent. Then we define

$$C' = C[A(f_\mu(t_1, \ldots, t_n, y_1, \ldots, y_m))]_\lambda$$

where λ is the occurrence of $(\forall x)A(x)$ in C and we replace $\rho[B \vee C]$ by $\rho[B \vee C']$.
We iterate this transformation until there is no strong quantifier left in the end-sequent.

Note that the cut formulas can merely be instantiated by these transformations; if the cuts are closed they simply remain unchanged. Moreover we can avoid any conflict with eigenvariable conditions by performing an appropriate renaming in advance.

The transformation drops quantifier-introductions and performs term substitution; none of these can lead to an increase in the number of logical inferences (instead we may obtain a strict decrease). Therefore if $sk(\varphi)$ is the final proof of the sequent $sk(S)$ then $sk(\varphi)$ is a proof from $\mathcal{A}$ (note that, by definition, axiom sets are closed under substitution and the axioms are quantifier-free), and

$$\|sk(\varphi)\|_l \leq \|\varphi\|_l. \ \square$$

Example 6.2.2 Let $\varphi =$

$$\dfrac{\dfrac{\dfrac{\dfrac{\dfrac{P(c,\alpha) \vdash P(c,\alpha) \quad Q(\alpha) \vdash Q(\alpha)}{P(c,\alpha), P(c,\alpha) \to Q(\alpha) \vdash Q(\alpha)} \to: l + p: l}{P(c,\alpha) \to Q(\alpha), (\forall x)P(c,x) \vdash Q(\alpha)} \forall: l + p: l}{P(c,\alpha) \to Q(\alpha), (\forall x)P(c,x) \vdash (\exists y)Q(y)} \exists: r}{(\exists y)(P(c,y) \to Q(y)), (\forall x)P(c,x) \vdash (\exists y)Q(y)} \exists: l}{(\forall x)(\exists y)(P(x,y) \to Q(y)), (\forall x)P(c,x) \vdash (\exists y)Q(y)} \forall: l$$

Then $sk(\varphi) =$

$$\dfrac{\dfrac{\dfrac{\dfrac{P(c,f(c)) \vdash P(c,f(c)) \quad Q(f(c)) \vdash Q(f(c))}{P(c,f(c)), P(c,f(c)) \to Q(f(c)) \vdash Q(f(c))} \to: l + p: l}{P(c,f(c)) \to Q(f(c)), (\forall x)P(c,x) \vdash Q(f(c))} \forall: l + p: l}{P(c,f(c)) \to Q(f(c)), (\forall x)P(c,x) \vdash (\exists y)Q(y)} \exists: r}{(\forall x)(P(x,f(x)) \to Q(f(x))), (\forall x)P(c,x) \vdash (\exists y)Q(y)} \forall: l$$

Note that $\|\varphi\|_l = 5$ and $\|sk(\varphi)\|_l = 4$. $\diamond$

We have seen in Proposition 6.2.1 that Skolemization of proofs require axioms to be atomic (or at least quantifier-free). Indeed the standard axiom set with axioms $A \vdash A$ for arbitrary A is not closed under Skolemization. On the other hand the axioms need not be of the form $A \vdash A$. Therefore the Skolemization method works for all proofs with atomic axioms. By the transformation $\pi(A)$ for formulas A defined in Chapter 4 we may transform proofs from axioms of the form $A \vdash A$ into proofs from the standard axiom set, and this transformation is linear (Lemma 4.1.1).

Definition 6.2.4 Let Φ^s be the set of all **LK**-derivations with Skolemized end sequents. $\Phi^s_\emptyset$ is the set of all cut-free proofs in Φ^s and, for all $i \geq 0$, Φ^s_i is the set of all proofs in Φ^s with cut-complexity $\leq i$. $\diamond$

On Skolemized proofs cut-elimination means (for us) to transform a derivation in Φ^s into a derivation in Φ^s_0. If it is possible to eliminate the remaining atomic cuts we do it as a postprocessing step.

6.3 Clause Terms

The information present in the axioms refuted by the cuts will be represented by a set of clauses (note that clauses are just atomic sequents). Every proof φ with cuts can be transformed into a proof φ' of the empty sequent by skipping inferences going into the end-sequent. The axioms of this refutation φ' can be compactly represented by clause terms.

Definition 6.3.1 (clause term) *Clause terms* are $\{\oplus, \otimes\}$-terms over clause sets. More formally:

- (Finite) sets of clauses are clause terms.
- If X, Y are clause terms then $X \oplus Y$ is a clause term.
- If X, Y are clause terms then $X \otimes Y$ is a clause term.

$\diamond$

Definition 6.3.2 (semantics of clause terms) We define a mapping $|\ |$ from clause terms to sets of clauses in the following way:

$$\begin{aligned} |\mathcal{C}| &= \mathcal{C} \text{ for a set of clauses } \mathcal{C}, \\ |X \oplus Y| &= |X| \cup |Y|, \\ |X \otimes Y| &= |X| \times |Y|, \end{aligned}$$

where $\mathcal{C} \times \mathcal{D} = \{C \circ D \mid C \in \mathcal{C}, D \in \mathcal{D}\}$. $\diamond$

We define clause terms to be equivalent if the corresponding sets of clauses are equal, i.e. $X \sim Y$ iff $|X| = |Y|$.
Clause terms are binary trees whose nodes are finite sets of clauses (instead of constants or variables). Therefore term occurrences are defined in the same way as for ordinary terms. When speaking about occurrences in clause terms we only consider nodes in *this* term tree, but not occurrences inside the leaves, i.e. within the sets of clauses on the leaves. In contrast we consider the internal structure of leaves in the concept of substitution:

Definition 6.3.3 Let θ be a substitution. We define the application of θ to clause terms as follows:

$$\begin{aligned} X\theta &= \mathcal{C}\theta \text{ if } X = \mathcal{C} \text{ for a set of clauses } \mathcal{C}, \\ (X \oplus Y)\theta &= X\theta \oplus Y\theta, \\ (X \otimes Y)\theta &= X\theta \otimes Y\theta. \end{aligned}$$

There are four binary relations on clause terms which will play a important role in the proof of our main result on cut-reduction.

Definition 6.3.4 Let X, Y be clause terms. We define

$X \subseteq Y$ iff $|X| \subseteq |Y|$ (i.e. iff $|X|$ is a subclause of $|Y|$),

$X \sqsubseteq Y$ iff for all $C \in |Y|$ there exists a $D \in |X|$ s.t. $D \sqsubseteq C$,

$X \leq_s Y$ iff there exists a substitution θ with $X\theta = Y$. [1]

◇

Remark: If $Y \subseteq X$ then $X \sqsubseteq Y$. Indeed, assume that $|Y| \subseteq |X|$ (every clause in $|Y|$ is also a clause in $|X|$); then, for every $C \in |Y|$, there exists a $D \in |X|$ (namely C itself) s.t. $D \sqsubseteq C$. ◇

The operators $\oplus$ and $\otimes$ are compatible with the relations $\subseteq$ and $\sqsubseteq$. This is formally proved in the following lemmas.

Lemma 6.3.1 *Let X, Y, Z be clause terms and $X \subseteq Y$. Then*

(1) $X \oplus Z \subseteq Y \oplus Z$,

(2) $Z \oplus X \subseteq Z \oplus Y$,

(3) $X \otimes Z \subseteq Y \otimes Z$,

(4) $Z \otimes X \subseteq Z \otimes Y$.

Proof: (2) follows from (1) because $\oplus$ is commutative, i.e. $X \oplus Z \sim Z \oplus X$. The cases (3) and (4) are analogous. Thus we only prove (1) and (3).

(1) $|X \oplus Z| = |X| \cup |Z| \subseteq |Y| \cup |Z| = |Y \oplus Z|$.

(3) Let $C \in |X \otimes Z|$. Then there exist clauses D, E with $D \in |X|$, $E \in |Z|$ and $C = D \circ E$. Clearly D is also in $|Y|$ and thus $C \in |Y \otimes Z|$. □

Lemma 6.3.2 *Let X, Y, Z be clause terms and $X \sqsubseteq Y$. Then*

(1) $X \oplus Z \sqsubseteq Y \oplus Z$,

(2) $Z \oplus X \sqsubseteq Z \oplus Y$,

[1] Note that $\leq_s$ is defined directly on the syntax of clause terms, and not via the semantics.

(3) $X \otimes Z \sqsubseteq Y \otimes Z$,

(4) $Z \otimes X \sqsubseteq Z \otimes Y$.

Proof: (1) and (2) are trivial, (3) and (4) are analogous. Thus we only prove (4):
Let $C \in |Z \otimes Y|$. Then $C \in |Z| \times |Y|$ and there exist $D \in |Z|$ and $E \in |Y|$ s.t. $C = D \circ E$. By definition of $\sqsubseteq$ there exists an $E' \in |X|$ with $E' \sqsubseteq E$. This implies $D \circ E' \in |Z \otimes X|$ and $D \circ E' \sqsubseteq D \circ E$. So $Z \otimes X \sqsubseteq Z \otimes Y$. □

We are now able to show that replacing subterms in a clause term preserves the relations $\subseteq$ and $\sqsubseteq$.

Lemma 6.3.3 *Let λ be an occurrence in a clause term X and $Y \preceq X.\lambda$ for $\preceq \in \{\subseteq, \sqsubseteq\}$. Then $X[Y]_\lambda \preceq X$.*

Proof: We proceed by induction on the term-complexity (i.e. number of nodes) of X.
If X is a set of clauses then λ is the top position and $X.\lambda = X$. Consequently $X[Y]_\lambda = Y$ and thus $X[Y]_\lambda \preceq X$.
Let X be $X_1 \odot X_2$ for $\odot \in \{\oplus, \otimes\}$. If λ is the top position in X then the lemma trivially holds. Thus we may assume that λ is a position in X_1 or in X_2. We consider the case that λ is in X_1 (the other one is completely symmetric): then there exists a position μ in X_1 s.t. $X.\lambda = X_1.\mu$. By induction hypothesis we get $X_1[Y]_\mu \preceq X_1$. By the Lemmas 6.3.1 and 6.3.2 we obtain

$$X_1[Y]_\mu \odot X_2 \quad \preceq \quad X_1 \odot X_2.$$

But

$$X_1[Y]_\mu \odot X_2 = (X_1 \odot X_2)[Y]_\lambda = X[Y]_\lambda$$

and therefore $X[Y]_\lambda \preceq X$. □

We will see in Section 6.8 that the relations $\subseteq, \sqsubseteq$ and $\leq_s$ are preserved under cut-reduction steps. Together they define a relation $\rhd$:

Definition 6.3.5 Let X and Y two clause terms. We define $X \rhd Y$ if (at least) one of the following properties is fulfilled:

(a) $Y \subseteq X$ or

(b) $X \sqsubseteq Y$ or

(c) $X \leq_s Y$.

Now we are in possession of the machinery to define the characteristic clause term of an **LK**-derivation.

6.4 The Method CERES

Below we introduce a method of cut-elimination which essentially uses only the semantic information of the refutability of the cuts after the rest of the proof has been deleted.

Definition 6.4.1 (characteristic term) Let φ be an **LK**-derivation of S and let Ω be the set of all occurrences of cut formulas in φ. We define the *characteristic (clause) term* $\Theta(\varphi)$ inductively via $\Theta(\varphi)/\nu$ for occurrences of sequents ν in φ:

Let ν be the occurrence of an initial sequent in φ. Then $\Theta(\varphi)/\nu = \{S(\nu, \Omega)\}$ (see Definition 3.2.14).

Let us assume that the clause terms $\Theta(\varphi)/\nu$ are already constructed for all sequent occurrences ν in φ with depth$(\nu) \leq k$. Now let ν be an occurrence with depth$(\nu) = k + 1$. We distinguish the following cases:

(a) ν is the conclusion of μ, i.e. a unary rule applied to μ gives ν. Here we simply define $\Theta(\varphi)/\nu = \Theta(\varphi)/\mu$.

(b) ν is the consequent of μ_1 and μ_2, i.e. a binary rule ξ applied to μ_1 and μ_2 gives ν.

(b1) The occurrences of the auxiliary formulas of ξ are ancestors of Ω, thus the formulas occur in $S(\mu_1, \Omega), S(\mu_2, \Omega)$. Then $\Theta(\varphi)/\nu = \Theta(\varphi)/\mu_1 \oplus \Theta(\varphi)/\mu_2$.

(b2) The occurrences of the auxiliary formulas of ξ are not ancestors of Ω. In this case we define $\Theta(\varphi)/\nu = \Theta(\varphi)/\mu_1 \otimes \Theta(\varphi)/\mu_2$.

Note that, in a binary inference, either the occurrences of both auxiliary formulas are ancestors of Ω or none of them.
Finally the characteristic term $\Theta(\varphi)$ is defined as $\Theta(\varphi)/\nu$ where ν is the occurrence of the end-sequent. ◇

Remark: If φ is a cut-free proof then there are no occurrences of cut formulas in φ and $\Theta(\varphi)$ is a term defined by $\otimes$ and $\{\vdash\}$ only. The property itself of a proof φ being cut-free could be defined in an extended sense by $|\Theta(\varphi)| = \{\vdash\}$. ◇

Definition 6.4.2 (characteristic clause set) Let φ be an **LK**-derivation and $\Theta(\varphi)$ be the characteristic term of φ. Then $\mathrm{CL}(\varphi)$, for $\mathrm{CL}(\varphi) = |\Theta(\varphi)|$, is called the *characteristic clause set* of φ. ◇

Example 6.4.1 Let φ be the derivation (for α, β free variables, a a constant symbol)

$$\frac{\varphi_1 \qquad \varphi_2}{(\forall x)(\neg P(x) \lor Q(x)) \vdash (\exists y)Q(y)}\ cut$$

where φ_1 is the **LK**-derivation:

$$\begin{array}{c}
\dfrac{\dfrac{P(\alpha)^\star \vdash Q(\alpha)^\star, P(\alpha)}{\neg P(\alpha), P(\alpha)^\star \vdash Q(\alpha)^\star}\ \neg : r \qquad Q(\alpha), P(\alpha)^\star \vdash Q(\alpha)^\star}{P(\alpha)^\star, \neg P(\alpha) \lor Q(\alpha) \vdash Q(\alpha)^\star}\ \lor : l + p : l \\
\hline
\neg P(\alpha) \lor Q(\alpha) \vdash Q(\alpha)^\star, \neg P(\alpha)^\star \quad \neg : r \\
\hline
\neg P(\alpha) \lor Q(\alpha) \vdash Q(\alpha)^\star, (\neg P(\alpha) \lor Q(\alpha))^\star \quad \lor : r \\
\hline
\neg P(\alpha) \lor Q(\alpha) \vdash (\neg P(\alpha) \lor Q(\alpha))^\star, (\neg P(\alpha) \lor Q(\alpha))^\star \quad p : r + \lor : r \\
\hline
\neg P(\alpha) \lor Q(\alpha) \vdash (\neg P(\alpha) \lor Q(\alpha))^\star \quad c : r \\
\hline
\neg P(\alpha) \lor Q(\alpha) \vdash (\exists y)(\neg P(\alpha) \lor Q(y))^\star \quad \exists : r \\
\hline
(\forall x)(\neg P(x) \lor Q(x)) \vdash (\exists y)(\neg P(\alpha) \lor Q(y))^\star \quad \forall : l \\
\hline
(\forall x)(\neg P(x) \lor Q(x)) \vdash (\forall x)(\exists y)(\neg P(x) \lor Q(y))^\star \quad \forall : r
\end{array}$$

and φ_2 is:

$$\begin{array}{c}
\dfrac{\dfrac{\vdash Q(\beta), P(a)^\star}{\neg P(a)^\star \vdash Q(\beta)}\ \neg : l \qquad Q(\beta)^\star \vdash Q(\beta)}{(\neg P(a) \lor Q(\beta))^\star \vdash Q(\beta)}\ \lor : l \\
\hline
(\neg P(a) \lor Q(\beta))^\star \vdash (\exists y)Q(y) \quad \exists : r \\
\hline
(\exists y)(\neg P(a) \lor Q(y))^\star \vdash (\exists y)Q(y) \quad \exists : l \\
\hline
(\forall x)(\exists y)(\neg P(x) \lor Q(y))^\star \vdash (\exists y)Q(y) \quad \forall : l
\end{array}$$

Let Ω be the set of the two occurrences of the cut formula in φ. The ancestors of Ω are marked by $\star$. We compute the characteristic clause term $\Theta(\varphi)$:

From the $\star$-marks in φ we first get the clause terms corresponding to the initial sequents:

$X_1 = \{P(\alpha) \vdash Q(\alpha)\}$, $X_2 = \{P(\alpha) \vdash Q(\alpha)\}$, $X_3 = \{\vdash P(a)\}$, $X_4 = \{Q(\beta) \vdash\}$.

The leftmost-uppermost inference in φ_1 is unary and thus the clause term X_1 corresponding to the conclusion does not change. The first binary inference in φ_1 (it is $\vee : l$) takes place on non-ancestors of Ω – the auxiliary formulas of the inference are not marked by $\star$. Consequently we obtain the term

$$X_1 \otimes X_2 = Y_1 = \{P(\alpha) \vdash Q(\alpha)\} \otimes \{P(\alpha) \vdash Q(\alpha)\}.$$

The following inferences in φ_1 are all unary and so we obtain

$$\Theta(\varphi)/\nu_1 = Y_1$$

for ν_1 being the position of the end sequent of φ_1 in φ.
Again the uppermost-leftmost inference in φ_2 is unary and thus X_3 does not change. The first binary inference in φ_2 takes place on ancestors of Ω (the auxiliary formulas are $\star$-ed) and we have to apply the $\oplus$ to X_3, X_4. So we get

$$Y_2 = \{\vdash P(a)\} \oplus \{Q(\beta) \vdash\}.$$

Like in φ_1 all remaining inferences in φ_2 are unary leaving the clause term unchanged. Let ν_2 be the occurrence of the end-sequent of φ_2 in φ. Then the corresponding clause term is

$$\Theta(\varphi)/\nu_2 = Y_2.$$

The last inference (cut) in φ takes place on ancestors of Ω and we have to apply $\oplus$ again. This eventually yields the characteristic term

$$\begin{aligned}\Theta(\varphi) &= Y_1 \oplus Y_2 = \\ &\quad (\{P(\alpha) \vdash Q(\alpha)\} \otimes \{P(\alpha) \vdash Q(\alpha)\}) \oplus (\{\vdash P(a)\} \oplus \{Q(\beta) \vdash\}).\end{aligned}$$

For the characteristic clause set we obtain

$$\mathrm{CL}(\varphi) = |\Theta(\varphi)| = \{P(\alpha), P(\alpha) \vdash Q(\alpha), Q(\alpha);\ \vdash P(a);\ Q(\beta) \vdash\}.$$

◇

It is easy to verify that the set of characteristic clauses $\mathrm{CL}(\varphi)$ constructed in the example above is unsatisfiable. This is not merely a coincidence, but a general principle expressed in the next proposition. For the proof we need the technical notion of context product which allows to extend a proof on all nodes by a clause.

Definition 6.4.3 (context product) Let C be a sequent and φ be an **LK**-derivation s.t. no free variable in C occurs as eigenvariable in φ. We define the left context product $C \star \varphi$ of C and φ (which gives a proof of $C \circ S$) inductively:

- If φ consists only of the root node ν and $Seq(\nu) = S$ then $C \star \varphi$ is a proof consisting only of a node μ s.t. $Seq(\mu) = C \circ S$.

- assume that φ is of the form

$$\cfrac{\begin{array}{c}(\varphi')\\ S'\end{array}}{S}\ \xi$$

where ξ is a unary rule. Assume also that $C \star \varphi'$ is already defined and is an **LK**-derivation of $C \circ S'$. Then we define $C \star \varphi$ as:

$$\cfrac{\cfrac{\cfrac{\begin{array}{c}C \star \varphi'\\ C \circ S'\end{array}}{S''}\ p^*}{S'''}\ \xi}{C \circ S}\ p^*$$

Note that for performing the rule ξ on $C \circ S'$ instead on S' we have to permute the sequent $C \circ S'$ first in order to make the rule applicable. If ξ is a right rule no additional permutation is required. $C \star \varphi$ is well defined also for the rules $\forall\colon r$ and $\forall\colon l$ as C does not contain free variables which are eigenvariables in φ.

- Assume that φ is of the form

$$\cfrac{\begin{array}{c}(\varphi_1)\\ S_1\end{array} \quad \begin{array}{c}(\varphi_2)\\ S_2\end{array}}{S}\ \xi$$

and $C \star \varphi_1$ is a proof of $C \circ S_1$, $C \star \varphi_2$ is a proof of $C \circ S_2$. Then we define the proof $C \star \varphi$ as

$$\cfrac{\cfrac{\cfrac{\begin{array}{c}(C \star \varphi_1)\\ C \circ S_1\end{array}}{S'_1}\ p^* \quad \cfrac{\begin{array}{c}(C \star \varphi_2)\\ C \circ S_2\end{array}}{S'_2}\ p^*}{S'}\ \xi}{C \circ S}\ s^*$$

Note that, if ξ is the cut rule, restoring the context after application of ξ might require weakening; otherwise s^* stands for structural derivations consisting of permutations and contractions.

The right context product $\varphi \star C$ is defined in the same way. $\diamond$

Example 6.4.2 Let φ be the proof

$$\dfrac{\dfrac{R(a) \vdash R(a) \quad Q(a) \vdash Q(a)}{R(a) \to Q(a), R(a) \vdash Q(a)} \to: l}{(\forall x)(R(x) \to Q(x)), R(a) \vdash Q(a)} \forall: l$$

and $C = P(y) \vdash Q(y)$. Then the context product $C \star \varphi$ is

$$\dfrac{\dfrac{\dfrac{\dfrac{\dfrac{\dfrac{P(y), R(a) \vdash Q(y), R(a)}{R(a), P(y) \vdash Q(y), R(a)} p: l \quad \dfrac{P(y), Q(a) \vdash Q(y), Q(a)}{Q(a), P(y) \vdash Q(y), Q(a)} p: l}{R(a) \to Q(a), R(a), P(y), P(y) \vdash Q(y), Q(y), Q(a)} \to: l}{P(y), R(a) \to Q(a), R(a) \vdash Q(y), Q(a)} s^*}{R(a) \to Q(a), P(y), R(a) \vdash Q(y), Q(a)} p: l}{(\forall x)(R(x) \to Q(x)), P(y), R(a) \vdash Q(y), Q(a)} \forall: l}{P(y), (\forall x)(R(x) \to Q(x)), R(a) \vdash Q(y), Q(a)} p: l$$

$\diamond$

Proposition 6.4.1 *Let φ be a regular* **LK***-proof of a closed sequent and $\varphi \in \Phi^s$. Then* $\mathrm{CL}(\varphi)$ *is unsatisfiable.*

Proof: Let $\Theta(\varphi)$ be the characteristic term (see Definition 6.4.1); for every node ν in φ we set $\mathcal{C}_\nu = \Theta(\varphi)/\nu$. Let Ω be the set of all occurrences of cut formulas in φ.

We prove by induction on the derivation that, for all nodes ν in φ,

(*) $S(\nu, \Omega)$ (see Definition 3.2.14) is **LK**-derivable from $\mathcal{C}_\nu$.

If ν_0 is the root node of φ then, clearly, $S(\nu_0, \Omega) = \{\vdash\}$, as in the end sequent there are no ancestors of cuts. So $\vdash$ is **LK**-derivable from $\mathrm{CL}(\varphi)$. As **LK** is sound, the set of clauses $\mathrm{CL}(\varphi)$ is unsatisfiable. So it remains to prove (*).

If ν is a leaf in φ then, by definition of $\Theta(\varphi)/\nu$, $\mathcal{C}_\nu = S(\nu, \Omega)$. Therefore $S(\nu, \Omega)$ itself is the **LK**-derivation of $S(\nu, \Omega)$ from $\mathcal{C}_\nu$.

(1) The node ν is a conclusion of a unary inference, i.e. $\varphi.\nu =$

$$\dfrac{\begin{array}{c}(\varphi.\mu)\\ Seq(\mu)\end{array}}{Seq(\nu)} \xi$$

By induction hypothesis there exists an **LK**-derivation $\psi(\mu)$ of $S(\mu, \Omega)$ from $\mathcal{C}_\mu$. By definition of the characteristic term we have $\mathcal{C}_\nu = \mathcal{C}_\mu$. Our aim is to extend $\psi(\mu)$ to a derivation $\psi(\nu)$ of $S(\nu, \Omega)$ from $\mathcal{C}_\nu$.

(1a) The principal formula of ξ is not an Ω-ancestor. Then $S(\mu,\Omega) = S(\nu,\Omega)$ and we define $\psi(\nu) = \psi(\mu)$.

(1b) The principal formula of ξ is an Ω-ancestor. Then we define $\psi(\nu) =$

$$\cfrac{\begin{matrix}(\psi(\mu))\\ S(\mu,\Omega)\end{matrix}}{S(\nu,\Omega)}\ \xi + p^*$$

where p^* symbolizes a sequence of permutations (at most 4 are needed). Clearly $\psi(\nu)$ is an **LK**-derivation of $S(\nu,\Omega)$ from $\mathcal{C}_\nu$.

(2) ν is a consequence of a binary inference. Then $\varphi.\nu =$

$$\cfrac{\begin{matrix}(\varphi.\mu_1)\\ Seq(\mu_1)\end{matrix} \quad \begin{matrix}(\varphi.\mu_2)\\ Seq(\mu_2)\end{matrix}}{Seq(\nu)}\ \xi$$

By induction hypothesis there exist **LK**-derivations

$\psi(\mu_1)$ of $S(\mu_1,\Omega)$ from $\mathcal{C}_{\mu_1}$ and

$\psi(\mu_2)$ of $S(\mu_2,\Omega)$ from $\mathcal{C}_{\mu_2}$.

(2a) The principal formula of ξ is not an Ω-ancestor. Then, by definition of the characteristic clause set

$$\mathcal{C}_\nu = \mathcal{C}_{\mu_1} \times \mathcal{C}_{\mu_2}.$$

We construct an **LK**-derivation of $S(\nu,\Omega)$ from $\mathcal{C}_\nu$:
Define, for every $C \in \mathcal{C}_{\mu_1}$, the derivation $C \star \psi(\mu_2)$. Note that this left context product is defined as C does not contain free variables which are eigenvariables in $\psi(\mu_2)$; indeed, φ is regular and, as the end-sequent is closed, every free variable in C is also an eigenvariable in the proof. $C \star \psi(\mu_2)$ is a proof of $C \circ S(\mu_2,\Omega)$ from $\{C\} \times \mathcal{C}_{\mu_2}$ which is a subset of $\mathcal{C}_{\mu_1} \times \mathcal{C}_{\mu_2}$; so $C \star \psi(\mu_2)$ is a proof of $C \circ S(\mu_2,\Omega)$ from $\mathcal{C}_{\mu_1} \times \mathcal{C}_{\mu_2}$.
Now consider the derivation $\psi_1 \colon \psi(\mu_1) \star S(\mu_2,\Omega)$ (ψ_1 is well defined by the regularity conditions); the initial sequents of ψ_1 are of the form $C \circ S(\mu_2,\Omega)$ for $C \in \mathcal{C}_{\mu_1}$. Replace every initial sequent in ψ_1 by the derivation $C \star \psi(\mu_2)$. The result is a proof χ of $S(\mu_1,\Omega) \circ S(\mu_2,\Omega)$ from $\mathcal{C}_{\mu_1} \times \mathcal{C}_{\mu_2}$.
We define $\psi(\nu) = \chi$. But $\mathcal{C}_\nu = \mathcal{C}_{\mu_1} \times \mathcal{C}_{\mu_2}$ and $S(\nu,\Omega) = S(\mu_1,\Omega) \circ S(\mu_2,\Omega)$; therefore $\psi(\nu)$ is a derivation of $S(\nu,\Omega)$ from $\mathcal{C}_\nu$.

(2b) The principal formula of ξ is an Ω-ancestor. Then, by definition of the characteristic clause term,

$$\mathcal{C}_\nu = \mathcal{C}_{\mu_1} \cup \mathcal{C}_{\mu_2}.$$

We just define $\psi(\nu) =$

$$\frac{\begin{array}{cc}(\psi(\mu_1)) & (\psi(\mu_2)) \\ S(\mu_1, \Omega) & S(\mu_2, \Omega)\end{array}}{S(\nu, \Omega)}\ \xi + s^*$$

Clearly $\psi(\nu)$ is an **LK**-derivation of $S(\nu, \Omega)$ from $\mathcal{C}_\nu$. □

Let $\varphi \in \Phi^s$ be a deduction of $S{:}\,\Gamma \vdash \Delta$ and $\mathrm{CL}(\varphi)$ be the characteristic clause set of φ. Then $\mathrm{CL}(\varphi)$ is unsatisfiable and, by the completeness of resolution (see [61, 69]), there exists a resolution refutation γ of $\mathrm{CL}(\varphi)$. By applying a ground projection to γ we obtain a ground resolution refutation γ' of $\mathrm{CL}(\varphi)$; by our definition of resolution, γ' is also an **LK**-deduction of $\vdash$ from (ground instances of) $\mathrm{CL}(\varphi)$. This deduction γ' may serve as a skeleton of an Φ^s_0-proof ψ of $\Gamma \vdash \Delta$ itself. The construction of ψ from γ' is based on *projections* replacing φ by cut-free deductions $\varphi(C)$ of $\bar{P}, \Gamma \vdash \Delta, \bar{Q}$ for clauses $C : \bar{P} \vdash \bar{Q}$ in $\mathrm{CL}(\varphi)$. Roughly speaking, the projections of the proof φ are obtained by skipping all the inferences leading to a cut. As a “residue” we obtain a characteristic clause in the end sequent. Thus a projection is a cut-free derivation of the end sequent S + some atomic formulas. For the application of projections it is vital to have a Skolemized end sequent, otherwise eigenvariable conditions could be violated.
For the definition of projection we define a technical notion:

Definition 6.4.4 Let φ be an **LK**-proof, ν a node in φ and Ω a set of formula occurrences in φ. Then we define $\bar{S}(\nu, \Omega)$ by

$$Seq(\nu) = S(\nu, \Omega) \circ \bar{S}(\nu, \Omega).$$

$\bar{S}(\nu, \Omega)$ is just the subsequent of $Seq(\nu)$ consisting of the *non-ancestors* of Ω, i.e. of the ancestors of the end-sequent. ◇

Lemma 6.4.1 *Let φ be a deduction in Φ^s of a sequent S from an axiom set $\mathcal{A}$ and let C be a clause in* $\mathrm{CL}(\varphi)$. *Then there exists a deduction $\varphi[C] \in \Phi^s_\emptyset$ of $C \circ S$ from $\mathcal{A}$ and*

$$\|\varphi[C]\|_l \le \|\varphi\|_l.$$

Proof: We construct, for every node ν and for every $C \in \mathcal{C}_\nu$, a cut-free proof $\varphi_\nu[C]$ of $C \circ \bar{S}(\nu, \Omega)$ from $\mathcal{A}$ where Ω is the set of all occurrences of cut formulas in φ. Moreover we show that $\|\varphi_\nu[C]\|_l \leq \|\varphi.\nu\|_l$. Then, as there are no ancestors of Ω in the end sequent, $\varphi_{\nu_0}[C]$ (for the root node ν_0 is a cut-free proof of $C \circ S$ from $\mathcal{A}$ and $\|\varphi[C]\|_l \leq \|\varphi\|_l$ for $\varphi[C] = \varphi_{\nu_0}[C]$.

We proceed by induction on the derivation.

(a) ν is a leaf. Then by definition of the characteristic clause set $\mathcal{C}_\nu = \{S(\nu, \Omega)\}$. By Definition 6.4.4 we have $Seq(\nu) = S(\nu, \Omega) \circ \bar{S}(\nu, \Omega)$. So we simply define

$$\varphi_\nu[C] \quad = \quad \nu.$$

Clearly $\|\varphi_\nu[C]\|_l = \|\varphi.\nu\|_l = 0$. Moreover $Seq(\nu)$ is an axiom in $\mathcal{A}$.

(b) Let φ_ν be of the form

$$\frac{\begin{array}{c}(\varphi.\mu)\\ Seq(\mu)\end{array}}{Seq(\nu)}\ \xi$$

By induction hypothesis we have for all $C \in \mathcal{C}_\mu$ a cut-free proof $\varphi_\mu[C]$ of $C \circ \bar{S}(\mu, \Omega)$ from $\mathcal{A}$ and $\|\varphi_\mu[C]\|_l \leq \|\varphi.\mu\|_l$. As ξ is a unary rule we have $\mathcal{C}_\nu = \mathcal{C}_\mu$.

(b1) The principal formula of ξ occurs in $S(\nu, \Omega)$. Then clearly $\bar{S}(\nu, \Omega) = \bar{S}(\mu, \Omega)$ and we simply define

$$\varphi_\nu[C] = \varphi_\mu[C].$$

clearly $\varphi_\nu[C]$ is a cut-free derivation of $C \circ \bar{S}(\mu, \Omega)$ from $\mathcal{A}$ for which we obtain

$$\|\varphi_\nu[C]\|_l \leq \|\varphi.\mu\|_l \leq \|\varphi.\nu\|_l.$$

(b2) The principal formula of ξ does not occur in $S(\nu, \Omega)$ (i.e. it occurs in $\bar{S}(\nu, \Omega)$). Then we define $\varphi_\nu[C] =$

$$\frac{\begin{array}{c}(\varphi_\mu[C])\\ C \circ \bar{S}(\mu, \Omega)\end{array}}{C \circ \bar{S}(\nu, \Omega)}\ \xi + p^*$$

Clearly $\varphi_\nu[C]$ is a cut-free derivation of $C \circ \bar{S}(\nu, \Omega)$ from $\mathcal{A}$.

If ξ is a nonlogical inference then

$$\begin{aligned} \|\varphi.\nu\|_l &= \|\varphi.\mu\|_l \text{ and} \\ \|\varphi_\nu[C]\|_l &= \|\varphi_\mu[C]\|_l. \end{aligned}$$

so $\|\varphi_\nu[C]\|_l \leq \|\varphi.\nu\|_l$.
If ξ is a logical inference then

$$\begin{aligned} \|\varphi.\nu\|_l &= \|\varphi.\mu\|_l + 1 \text{ and} \\ \|\varphi_\nu[C]\|_l &= \|\varphi_\mu[C]\|_l + 1. \end{aligned}$$

Again we get $\|\varphi_\nu[C]\|_l \leq \|\varphi.\nu\|_l$.
Note that $\varphi_\nu[C]$ is well-defined as no eigenvariable conditions are violated. Indeed, there are only *weak* quantifiers in the end sequent! Thus ξ can neither be $\forall : r$ nor $\exists : l$.

(c) $\varphi.\nu$ is of the form

$$\frac{\begin{array}{cc}(\varphi.\mu_1) & (\varphi.\mu_2) \\ Seq(\mu_1) & Seq(\mu_2)\end{array}}{Seq(\nu)}\ \xi$$

By induction hypothesis we have for every $C \in \mathcal{C}_{\mu_i}$ a cut-free proof $\varphi_{\mu_i}[C]$ of $C \circ \bar{S}(\mu_i, \Omega)$ from $\mathcal{A}$ and

$$\|\varphi_{\mu_i}[C]\|_l \leq \|\varphi.\mu_i\|_l.$$

(c1) The auxiliary formulas of ξ occur in $S(\mu_i, \Omega)$. Then, by definition of the characteristic clause set

$$\mathcal{C}_\nu = \mathcal{C}_{\mu_1} \cup \mathcal{C}_{\mu_2}.$$

Now let $C \in \mathcal{C}_{\mu_1}$. Then we define $\varphi_\nu[C] =$

$$\frac{\begin{array}{c}(\varphi_{\mu_1}[C]) \\ C \circ \bar{S}(\mu_1, \Omega)\end{array}}{C \circ \bar{S}(\nu, \Omega)}\ w^* + p^*$$

and for $C \in \mathcal{C}_{\mu_2}$ we define $\varphi_\nu[C] =$

$$\frac{\begin{array}{c}(\varphi_{\mu_2}[C]) \\ C \circ \bar{S}(\mu_2, \Omega)\end{array}}{C \circ \bar{S}(\nu, \Omega)}\ w^* + p^*$$

In both cases $\varphi_\nu[C]$ is a cut-free proof of $C \circ \bar{S}(\nu, \Omega)$ from $\mathcal{A}$ and

$$\|\varphi_\nu[C]\|_l \leq \|\varphi_{\mu_i}[C]\|_l \leq \|\varphi.\mu_i\|_l \leq \|\varphi.\nu\|_l.$$

(c2) The auxiliary formulas of ξ occur in $\bar{S}(\mu_i, \Omega)$. By definition of the characteristic clause set we have

$$\mathcal{C}_\nu = \mathcal{C}_{\mu_1} \times \mathcal{C}_{\mu_2}.$$

So let $C \in \mathcal{C}_\nu$. Then there exist $D_1 \in \mathcal{C}_{\mu_1}$ and $D_2 \in \mathcal{C}_{\mu_2}$ with $D_1 \circ D_2 = C$. So we define

$$\dfrac{\dfrac{\dfrac{(\varphi_{\mu_1}[D_1])}{D_1 \circ \bar{S}(\mu_1, \Omega)}}{S'_1}\, p^* \qquad \dfrac{\dfrac{(\varphi_{\mu_2}[D_2])}{D_2 \circ \bar{S}(\mu_2, \Omega)}}{S'_2}\, p^*}{D_1 \circ D_2 \circ \bar{S}(\nu, \Omega)}\ \xi + s^*$$

ξ must be a logical inference, as in the case of cut the auxiliary formulas must occur in $S(\mu_1, \Omega), S(\mu_2, \Omega)$. Therefore

$$\|\varphi_\nu[C]\|_l = \|\varphi_{\mu_1}[C]\|_l + \|\varphi_{\mu_1}[C]\|_l + 1$$

so, by induction hypothesis,

$$\|\varphi_\nu[C]\|_l \leq \|\varphi.\mu_1\|_l + \|\varphi.\mu_1\|_l + 1$$

But

$$\|\varphi.\nu\|_l = \|\varphi.\mu_1\|_l + \|\varphi.\mu_1\|_l + 1.$$

□

Definition 6.4.5 (projection) Let φ be a proof in Φ^s and $C \in \mathrm{CL}(\varphi)$. Then the **LK**-proof $\varphi[C]$, defined in Lemma 6.4.1, is called the *projection* of φ w.r.t. C. Let σ be an arbitrary substitution; then $\varphi[C\sigma]$ is defined as $\varphi[C]\sigma$ and is also called the projection of φ w.r.t. $C\sigma$ (i.e. instances of projections are also projections). ◇

Remark: Note that projections of the form $\varphi[C]\sigma$ are always well defined as (1) the end-sequent of φ is closed and (2) $\varphi[C]$ contains only weak quantifiers, and thus no eigenvariable condition can be violated. ◇

The construction of $\varphi[C]$ for a $C \in \mathrm{CL}(\varphi)$ is illustrated below.

Example 6.4.3 Let φ be the proof of the sequent

$$S : (\forall x)(\neg P(x) \vee Q(x)) \vdash (\exists y)Q(y)$$

as defined in Example 6.4.1. We have shown that

$$\mathrm{CL}(\varphi) = \{P(\alpha), P(\alpha) \vdash Q(\alpha), Q(\alpha);\ \ \vdash P(a);\ \ Q(\beta) \vdash\}.$$

We now define $\varphi[C_1]$, the projection of φ to C_1: $P(\alpha), P(\alpha) \vdash Q(\alpha), Q(\alpha)$. The problem can be reduced to a projection in φ_1 because the last inference in φ is a cut and

$$\Theta(\varphi)/\nu_1 \quad = \quad \{P(\alpha), P(\alpha) \vdash Q(\alpha), Q(\alpha)\}.$$

By skipping all inferences in φ_1 leading to the cut formulas we obtain the deduction

$$\dfrac{\dfrac{\dfrac{\dfrac{P(\alpha) \vdash P(\alpha), Q(\alpha)}{\neg P(\alpha), P(\alpha) \vdash Q(\alpha)}\ p{:}\,r + \neg{:}\,l \qquad Q(\alpha), P(\alpha) \vdash Q(\alpha)}{P(\alpha), \neg P(\alpha) \vee Q(\alpha) \vdash Q(\alpha)}\ \vee{:}\,l + p{:}\,l}{P(\alpha), (\forall x)(\neg P(x) \vee Q(x)) \vdash Q(\alpha)}\ \forall{:}\,l + p{:}\,l}{}$$

In order to obtain the end sequent we only need an additional weakening and $\varphi[C_1] =$

$$\dfrac{\dfrac{\dfrac{\dfrac{P(\alpha) \vdash P(\alpha), Q(\alpha)}{\neg P(\alpha), P(\alpha) \vdash Q(\alpha)}\ \neg{:}\,l + p{:}\,r \qquad Q(\alpha), P(\alpha) \vdash Q(\alpha)}{P(\alpha), \neg P(\alpha) \vee Q(\alpha) \vdash Q(\alpha)}\ \vee{:}\,l + p{:}\,l}{P(\alpha), (\forall x)(\neg P(x) \vee Q(x)) \vdash Q(\alpha)}\ \forall{:}\,l + p{:}\,l}{P(\alpha), (\forall x)(\neg P(x) \vee Q(x)) \vdash Q(\alpha), (\exists y)Q(y)}\ w{:}\,r$$

For $C_2 = \vdash P(a)$ we obtain the projection $\varphi[C_2]$:

$$\dfrac{\dfrac{\vdash P(a), Q(\beta)}{\vdash P(a), (\exists y)Q(y)}\ \exists{:}\,r}{(\forall x)(\neg P(x) \vee Q(x)) \vdash P(a), (\exists y)Q(y)}\ w{:}\,l$$

Similarly we obtain $\varphi[C_3]$:

$$\dfrac{\dfrac{Q(\beta) \vdash Q(\beta)}{Q(\beta) \vdash (\exists y)Q(y)}\ \exists{:}\,r}{Q(\beta), (\forall x)(\neg P(x) \vee Q(x)) \vdash (\exists y)Q(y)}\ w{:}\,l + p{:}\,l$$

We have seen that, in the projections, only inferences on non-ancestors of cuts are performed. If the auxiliary formulas of a binary rule are ancestors of cuts we have to apply weakening in order to obtain the required formulas from the second premise.

Definition 6.4.6 Let φ be a proof of a closed sequent S and $\varphi \in \Phi^s$. We define a set

$$\mathrm{PES}(\varphi) = \{S \circ C\sigma \mid C \in \mathrm{CL}(\varphi), \sigma \text{ a substitution}\}$$

which contains all end sequents of projections w.r.t. φ. ◇

Example 6.4.4 Let φ be the proof of

$$S \colon (\forall x)(P(x) \to Q(x)) \vdash (\exists y)Q(y)$$

as defined in Examples 6.4.1 and 6.4.3. Then

$$\mathrm{CL}(\varphi) = \{C_1 : P(\alpha), P(\alpha) \vdash Q(\alpha), Q(\alpha);\ C_2 : \vdash P(a);\ C_3 : Q(v) \vdash\}.$$

First we define a resolution refutation δ of $\mathrm{CL}(\varphi)$:

$$\dfrac{\dfrac{\vdash P(a) \quad P(\alpha), P(\alpha) \vdash Q(\alpha), Q(\alpha)}{\vdash Q(a), Q(a)}\,R \qquad Q(\beta) \vdash}{\vdash}\,R$$

From δ we define a ground resolution refutation γ:

$$\dfrac{\dfrac{\vdash P(a) \quad P(a), P(a) \vdash Q(a), Q(a)}{\vdash Q(a), Q(a)}\,cut \qquad Q(a) \vdash}{\vdash}\,cut$$

The ground substitution defining the ground projection is

$$\sigma : \{\alpha \leftarrow a, \beta \leftarrow a\}.$$

Let

$$\chi_1 = \varphi[C_1\sigma],\ \chi_2 = \varphi[C_2\sigma] \text{ and } \chi_3 = \varphi[C_3\sigma]$$

for the projections $\varphi[C_1], \varphi[C_2]$ and $\varphi[C_3]$ defined in Example 6.4.3. Moreover let us write B for $(\forall x)(P(x) \to Q(x))$ and C for $(\exists y)(P(a) \to Q(y))$. So the end-sequent is $B \vdash C$.

Then $\varphi(\gamma)$, the CERES normal form of φ w.r.t. γ is

$$\dfrac{\dfrac{\dfrac{\overset{(\chi_2)}{B \vdash C, P(a)} \quad \overset{(\chi_1)}{B, P(a) \vdash C, Q(a)}}{B, B \vdash C, C, Q(a)}\,cut}{B \vdash C, Q(a)}\,c^* \qquad \overset{(\chi_3)}{B, Q(a) \vdash C}}{\dfrac{B, B \vdash C, C}{B \vdash C}\,c^*}\,cut$$

◇

Finally we give the general definition of CERES (cut-elimination by resolution) as a whole. As an input we take proofs φ in Φ^s.

- construct $\mathrm{CL}(\varphi)$,
- compute the projections,
- construct a resolution refutation γ of $\mathrm{CL}(\varphi)$,
- compute a ground resolution refutation γ' from γ,
- construct $\varphi(\gamma')$.

Theorem 6.4.1 CERES *is a cut-elimination method, i.e., for every proof φ of a sequent S in Φ^s* CERES *produces a proof ψ of S s.t. $\psi \in \Phi^s_0$.*

Proof: By Proposition 6.4.1 $\mathrm{CL}(\varphi)$ is unsatisfiable. By the completeness of the resolution principle there exists a resolution refutation γ of $\mathrm{CL}(\varphi)$. By Lemma 6.4.1 there exists a projection of $\varphi[C]$ to every clause in $C \in \mathrm{CL}(\varphi)$. We compute the corresponding instances of $\varphi[C']$ of $\varphi[C]$ corresponding to the instances C' appearing in the ground projection γ' of γ. Finally replace every occurrence of a C' in a leaf of γ' by $\varphi[C']$. The resulting proof $\varphi(\gamma')$ is a proof with only atomic cuts of S. □

The proof constructed by the CERES-method is a specific type of proof in Φ^s which contains parts of the original proofs in form of projections; we call this type of proof a CERES normal form.

Definition 6.4.7 (CERES normal form) Let φ be a proof of S s.t. $\varphi \in \Phi^s$ and let γ be a ground resolution refutation of the (unsatisfiable) set of clauses $\mathrm{CL}(\varphi)$; note that γ is also a proof in Φ^s (from $\mathrm{CL}(\varphi)$). We first construct $\gamma' : S \star \gamma$; γ' is an **LK**-derivation of S from $\mathrm{PES}(\varphi)$. Now define $\varphi(\gamma')$ by replacing all initial clauses $C \circ S$ in γ' by the projections $\varphi[C]$. By definition, $\varphi(\gamma')$ is an **LK**-proof of S in Φ^s_0 and is called the CERES-*normal form of φ w.r.t. γ.* ◇

6.5 The Complexity of CERES

We have shown in Section 4.3 that cut-elimination is intrinsically nonelementary. Therefore also CERES, applied to Statman's sequence, produces a nonelementary blowup w.r.t. to the size of the input proof. The question remains what is the main source of complexity in CERES and which of the CERES parts behave nonelementarily on a worst-case sequence. We show first that the size of the characteristic clause set is not the main source of complexity. It is "merely" exponential in the size of the input proof.

Lemma 6.5.1 *Let t be a clause term then $|||t||| \leq 2^{\|t\|}$ (the symbolic size of set of clauses defined by a clause term is at most exponential in that of the term).*

Proof: We proceed by induction on $o(t)$, the number of occurrences of $\oplus$ and $\otimes$ in t.

(IB): $o(t) = 0$: then $|t| = t = \{\, C\}$ for a clause C and

$$|||t||| = \|C\| = \|t\| < 2^{\|t\|}.$$

(IH) Let us assume that for all clause terms t with $o(t) \leq n$ we have $|||t||| \leq 2^{\|t\|}$.

Now let t be a term s.t. $o(t) = n + 1$. We distinguish two cases:

(a) $t = t_1 \oplus t_2$. Then $o(t_1), o(t_2) \leq n$ and, by (IH),

$$|||t_1||| \leq 2^{\|t_1\|},\ |||t_2||| \leq 2^{\|t_2\|}.$$

Therefore we get, by $|t_1 \oplus t_2| = |t_1| \cup |t_2|$,

$$|||t||| \leq |||t_1||| + |||t_2||| \leq 2^{\|t_1\|} + 2^{\|t_2\|} < 2^{\|t\|}.$$

(b) $t = t_1 \otimes t_2$. Again we get $o(t_1), o(t_2) \leq n$ and, by (IH),

$$|||t_1||| \leq 2^{\|t_1\|},\ |||t_2||| \leq 2^{\|t_2\|}.$$

But here we have

$$|t_1 \otimes t_2| = \{C_1 \circ C_2 \mid C_1 \in |t_1|, C_2 \in |t_2|\}.$$

Hence

$$|||t||| \leq |||t_1||| * |||t_2||| \leq_{(IH)} 2^{\|t_1\|} * 2^{\|t_2\|} < 2^{\|t\|}.$$

This concludes the induction proof. □

Proposition 6.5.1 *For every* $\varphi \in \Phi^s$ $\|\mathrm{CL}(\varphi)\| \leq 2^{\|\varphi\|}$.

Proof: $\mathrm{CL}(\varphi) = |\Theta(\varphi)$. So, by Lemma 6.5.1,

$$\|\mathrm{CL}(\varphi)\| \leq 2^{\|\Theta(\varphi)\|}.$$

Obviously $\|\Theta(\varphi)\| \leq \|\varphi\|$ and therefore

$$\|\mathrm{CL}(\varphi)\| \leq 2^{\|\varphi\|}. \ \square$$

The essential source of complexity in the CERES-method is the length of the resolution refutation γ of $\mathrm{CL}(\varphi)$. Computing the global m.g.u. σ and a p-resolution refutation $\gamma'\colon \gamma\sigma$ from γ is at most exponential in $\|\gamma\|$.
For measuring the complexity of p-resolution refutations (ground resolution refutations) w.r.t. resolution refutations we need some auxiliary concepts.

Definition 6.5.1 (pre-resolvent) Let C and D be clauses of the form

$$\begin{array}{rcl} C & = & \Gamma \vdash \Delta_1, A_1, \ldots, \Delta_n, A_n, \Delta_{n+1}, \\ D & = & \Pi_1, B_1, \ldots, \Pi_m, B_m, \Pi_{m+1} \vdash \Lambda \end{array}$$

s.t. C and D do not share variables and the atoms $A_1, \ldots, A_n, B_1, \ldots, B_m$ share the same predicate symbol. Then the clause

$$R\colon\ \Gamma, \Pi_1, \ldots \Pi_{m+1} \vdash \Delta_1, \ldots, \Delta_{n+1}, \Lambda$$

is called a *pre-resolvent* of C and D. If there exists a m.g.u. σ of

$$W\colon \{A_1, \ldots, A_n, B_1, \ldots, B_m\}$$

then (by Definition 3.3.10) $R\sigma$ is a resolvent of C and D. In this case we call $R\sigma$ a resolvent corresponding to R. The set W is called the unification problem of (C, D, R). ◇

Remark: Pre-resolution is clearly unsound. It makes sense only if there exists a corresponding resolvent. The idea of pre-resolution is close to lazy unification (see, e.g. [41]). ◇

Definition 6.5.2 (pre-resolution derivation) A *pre-resolution deduction* γ is a labelled tree γ like a resolution tree where resolutions are replaced by pre-resolutions. A pre-resolution refutation is a pre-resolution deduction of $\vdash$. Let $W_1, \ldots, W_k$ be the unification problems of the pre-resolvents in γ. Then the simultaneous unification problem $\mathcal{W}\colon (W_1, \ldots, W_n)$ (see Definition 3.3.2) is called the *unification problem* of γ. If σ is a simultaneous unifier of $\mathcal{W}$ then σ is called the *total m.g.u.* of γ. ◇

Example 6.5.1 Let $\gamma =$

$$\cfrac{\vdash P(x) \qquad \cfrac{P(f(y)), P(f(a)) \vdash Q(y) \quad Q(z) \vdash}{P(f(y)), P(f(a)) \vdash}}{\vdash}$$

γ is a pre-resolution refutation. The unification problem of γ is

$$W = (\{P(x), P(f(y), P(f(a))\}, \{Q(y), Q(z)\}).$$

The simultaneous unifier

$$\sigma = \{x \leftarrow f(a), y \leftarrow a, z \leftarrow a\}$$

is the total m.g.u. of γ. ◇

Proposition 6.5.2 *Let γ be a pre-resolution refutation of a clause set $\mathcal{C}$ and σ be the total m.g.u. of γ. Them $\gamma\sigma$ is a p-resolution refutation of $\mathcal{C}$.*

Proof: σ solves all unification problems of γ simultaneously. After application of σ to the simultaneous unification problems all atoms in this problems are equal and so the pre-resolutions become p-resolutions; the pre-resolution tree becomes a p-resolution tree. □

Corollary 6.5.1 *Let γ be a pre-resolution refutation of a clause set $\mathcal{C}$, σ be the total m.g.u. of γ and $X = V(rg(\sigma))$. Let $X = \{x_1, \ldots, x_k\}$, c be a constant symbol and $\theta = \{x_1 \leftarrow c, \ldots x_k \leftarrow c\}$. Then $\gamma\sigma\theta$ is a ground resolution refutation of $\mathcal{C}$ and $\|\gamma\sigma\| = \|\gamma\sigma\theta\|$.*

Proof: Obvious. □

Lemma 6.5.2 *Let γ be a resolution refutation of a set of clauses $\mathcal{C}$. Then there exists a ground resolution refutation γ' of $\mathcal{C}$ with the following properties:*

(a) $l(\gamma) = l(\gamma')$,

(b) $\|\gamma\| \leq \|\gamma'\|$,

(c) $\|\gamma'\| \leq 5 * l(\gamma)^2 * \|\mathcal{C}\| * 2^{5*l(\gamma)^2*\|\mathcal{C}\|}$.

Proof: By omitting all applications of unifications in γ we obtain a pre-resolution tree γ_0. We solve the unification problem in γ_0 (it is solvable as γ exists!) and obtain a total m.g.u. σ of γ_0. Then, by Corollary 6.5.1 and for an appropriate substitution θ γ': $\gamma_0\sigma\theta$ is a ground refutation of $\mathcal{C}$. (a) and (b) are obvious; it remains to prove (c):
By a method defined in Section 3.3 the solution of a simultaneous unification problem $\mathcal{W}$ can be reduced to a unification problem of two terms $\{t_1, t_2\}$ s.t.

$$(+)\ \|t_1\| + \|t_2\| \leq 5 * \|\mathcal{W}\|.$$

By Theorem 3.3.2 we get for the maximal term t in the unifier σ of $\{t_1, t_2\}$:

$$(\star)\ \|t\| \leq 2^{\|t_1\|+\|t_2\|},$$

Now consider the unification problem $\mathcal{W}$ of γ_0. Clearly

$$\|\mathcal{W}\| \leq \|\gamma_0\| \leq l(\gamma_0)^2 * \|\mathcal{C}\|.$$

Note that, in γ_0, no substitutions are applied! It remains to apply (+) and ($\star$). □

Definition 6.5.3 Let γ be a resolution refutation of a set of clauses $\mathcal{C}$. Then γ' as constructed in Lemma 6.5.2 is called a *minimal ground projection* of γ. ◇

Remark: In general there are infinitely many ground projections of a resolution refutation. But those obtained by m.g.u.s and grounding by constants as defined in Corollary 6.5.1 are the shortest ones. ◇

Definition 6.5.4 We define two functions h and H:

$$\begin{aligned} h(n,m) &= 5n^2m * 2^{5n^2m}, \\ H(n,m) &= h(n, 2^m). \end{aligned}$$

for $n, m \in \mathbb{N}$. ◇

It is easy to see that both h and H are elementary functions.

Lemma 6.5.3 *Let $\varphi \in \Phi^s$ and let γ be a resolution refutation of* CL(φ). *Then there exists a ground resolution refutation γ' of $\mathcal{C}$ s.t.*

$$\|\gamma'\| \leq H(\|\varphi\|, l(\gamma)).$$

Proof: By Lemma 6.5.2 there exists a minimal ground projection γ' of γ s.t.

$$(+)\ \|\gamma'\| \leq h(l(\gamma), \|\mathrm{CL}(\varphi)\|).$$

By Proposition 6.5.1 we have

$$(++)\ \|\mathrm{CL}(\varphi)\| \leq 2^{\|\varphi\|}.$$

Putting (+) and (++) together we obtain

$$\|\gamma'\| \leq H(\|\varphi\|, l(\gamma)).\ \square$$

Let

$$r(\gamma') = \max\{\|t\| \mid t \text{ is a term occurring in } \gamma'\}.$$

Lemma 6.5.4 *Let $\varphi \in \Phi^s$ and γ be a resolution refutation of* CL(φ) *and γ' be a minimal ground projection of γ. Then for the* CERES*-normal form $\varphi(\gamma')$ we get*

$$\|\varphi(\gamma')\| \leq H(\|\varphi\|, l(\gamma)) * \|\varphi\| * r(\gamma').$$

Proof: Let $\gamma' = \gamma\sigma$ as defined in Lemma 6.5.2. Then, for any clause $C \in$ CL(φ) we have

$$\|C\sigma\| \leq \|C\| * r(\gamma').$$

Therefore we obtain

$$\|\varphi(C\sigma)\| \leq \|\varphi(C)\| * r(\gamma') \leq \|\varphi\| * r(\gamma')$$

for the instantiated proof projection to the clause C. As the CERES-normal form is obtained by inserting the projections into the p-resolution refutation γ', we obtain

$$\|\varphi(\gamma')\| \leq \|\gamma'\| * \|\varphi\| * r(\gamma').$$

Finally we apply Lemma 6.5.3. $\square$

Now we show that any sequence of resolution refutations of the characteristic clause sets of the Statman sequence $(\gamma_n)_{n\in\mathbb{N}}$ is of nonelementary size w.r.t. the proof complexity of the end sequents of γ_n. More formally

Proposition 6.5.3 *Let $(\varphi_n)_{n\in\mathbb{N}}$ be sequence of proofs in Φ^s. Assume that there exists an elementary function f and a sequence of resolution refutations $(\gamma_n)_{n\in\mathbb{N}}$ of $(\mathrm{CL}(\varphi_n))_{n\in\mathbb{N}}$ s.t.*

$$l(\gamma_n) \leq f(\|\varphi_n\|)$$

for $n \in \mathbb{N}$. Then there exists an elementary function g and a sequence of CERES *normal forms φ_n^* of φ_n s.t.*

$$\|\varphi_n^*\| \leq g(\|\varphi_n\|).$$

Proof: We have seen in Lemma 6.5.3 that, for any γ_n, there exists a minimal ground projection γ'_n of γ_n s.t.

$$\|\gamma'_n\| \leq H(\|\varphi_n\|, l(\gamma_n)).$$

By $r(\gamma') \leq \|\gamma'_n\|$ and by using Lemma 6.5.4 we obtain

$$\|\varphi_n(\gamma'_n)\| \leq \|\varphi_n\| * H(\|\varphi_n\|, l(\gamma_n))^2.$$

By assumption $l(\gamma_n) \leq f(\|\varphi_n\|)$ and so

$$\|\varphi_n(\gamma'_n)\| \leq \|\varphi_n\| * H(\|\varphi_n\|, f(\|\varphi_n\|))^2.$$

As H and f are elementary, so is the function g, defined as

$$g(n) = n * H(n, f(n))^2. \ \Box$$

We have seen in Section 4.3 that cut-elimination is inherently nonelementary. As a consequence also the complexity of CERES is nonelementary. The following proposition shows that, in the Statman proof sequence, the lengths of resolution refutations of the characteristic clause sets are the main source of complexity in the CERES method.

Proposition 6.5.4 *Let $(\gamma_n)_{n\in\mathbb{N}}$ be the sequence of proofs of $(S_n)_{n\in\mathbb{N}}$ defined in Section 4.3 and let $(\rho_n)_{n\in\mathbb{N}}$ be a sequence of resolution refutations of the sequence of clause sets $(\mathrm{CL}(\gamma_n))_{n\in\mathbb{N}}$. Then $(l(\rho_n))_{n\in\mathbb{N}}$ is nonelementary in $(\mathrm{PC}^{\mathcal{A}_e}(S_n))_{n\in\mathbb{N}}$.*

Proof: By Corollary 4.3.1 we know that $(\mathrm{PC}_0^{\mathcal{A}_e}(S_n))_{n\in\mathbb{N}}$ is nonelementary in $(\mathrm{PC}^{\mathcal{A}_e}(S_n))_{n\in\mathbb{N}}$. For all n let γ^*_n be a minimal CERES normal form based on ρ_n.
As $(\gamma^*_n)_{n\in\mathbb{N}}$ is a sequence of proofs in P^s_0 we have

$$\|\gamma^*_n\| \geq \mathrm{PC}_0^{\mathcal{A}_e}(S_n).$$

Therefore $(\|\gamma^*_n\|)_{n\in\mathbb{N}}$ is nonelementary in $(\mathrm{PC}^{\mathcal{A}_e}(S_n))_{n\in\mathbb{N}}$.
As $(\|\gamma_n\|)_{n\in\mathbb{N}}$ is at most exponential in $(\mathrm{PC}^{\mathcal{A}_e}(S_n))_{n\in\mathbb{N}}$ the sequence $(\|\gamma^*_n\|)_{n\in\mathbb{N}}$ is also nonelementary in $(\|\gamma_n\|)_{n\in\mathbb{N}}$.
By Proposition 6.5.3 we know that $(\|\gamma^*_n\|)_{n\in\mathbb{N}}$ is also elementary in $(\|\gamma_n\|)_{n\in\mathbb{N}}$ if $(l(\rho_n))_{n\in\mathbb{N}}$ is elementary in $(\|\gamma_n\|)_{n\in\mathbb{N}}$.
Therefore $(l(\rho_n))_{n\in\mathbb{N}}$ is nonelementary in $(\|\gamma_n\|)_{n\in\mathbb{N}}$ and thus in $(\mathrm{PC}^{\mathcal{A}_e}(S_n))_{n\in\mathbb{N}}$. $\Box$

6.6 Subsumption and p-Resolution

Definition 6.6.1 Let C and D be clauses. Then $C \leq_{cp} D$ if there exists an **LK**-derivation of a clause E from the only axiom C using only contractions and permutations s.t. $E \sqsubseteq D$. $\diamond$

Example 6.6.1 Consider the clauses $\vdash P(x), Q(y), P(x)$ and $R(y) \vdash Q(y), P(x), R(x)$. Then

$$\vdash P(x), Q(y), P(x) \leq_{cp} R(y) \vdash Q(y), P(x), R(x).$$

The corresponding **LK**-derivation is

$$\cfrac{\cfrac{\vdash P(x), Q(y), P(x)}{\vdash Q(y), P(x), P(x)}\; p:r}{\vdash Q(y), P(x)}\; c:r$$

and $\vdash Q(y), P(x) \sqsubseteq R(y) \vdash Q(y), P(x), R(x)$.

$\diamond$

Remark: The relation $\leq_{cp}$ is a sub-relation of subsumption to be defined below. $\diamond$

Definition 6.6.2 (subsumption) A clause C *subsumes* a clause D ($C \leq_{ss} D$) if there exists a substitution σ s.t. $C\sigma \leq_{cp} D$.
We extend the relation $\leq_{ss}$ to sets of clauses $\mathcal{C}$, $\mathcal{D}$ in the following way: $\mathcal{C} \leq_{ss} \mathcal{D}$ if for all $D \in \mathcal{D}$ there exists a $C \in \mathcal{C}$ s.t. $C \leq_{ss} D$. $\diamond$

Example 6.6.2 Let

$$C = \vdash P(x), Q(y), P(f(y)) \text{ and } D = R(z) \vdash Q(f(z)), P(f(f(z)))), R(f(z)).$$

Then, for $\sigma = \{x \leftarrow f(f(z)), y \leftarrow f(z)\}$,

$$C\sigma = \vdash P(f(f(z))), Q(f(z)), P(f(f(z)))$$

and $C\sigma \leq_{cp} D$; so $C \leq_{ss} D$. $\diamond$

Subsumption can also be extended to clause terms:

Definition 6.6.3 A clause term X subsumes a clause term Y if $|X| \leq_{ss} |Y|$. $\diamond$

Lemma 6.6.1 *The relation $\leq_{cp}$ is reflexive and transitive. Moreover $C \leq_{cp} C'$ and $D \leq_{cp} D'$ implies $C \circ D \leq_{cp} C' \circ D'$.*

Proof: reflexivity and transitivity are trivial.
Now assume that $C \leq_{cp} C'$ by a derivation ψ of a clause C_0 from C s.t. $C_0 \sqsubseteq C'$; similarly let χ be a derivation of D_0 from D s.t. $D_0 \sqsubseteq D'$. Then ψ can be modified to a derivation ψ_1 of $C_0 \circ D$ from $C \circ D$. Similarly χ is modified to a derivation χ_1 of $C_0 \circ D_0$ from $C_0 \circ D$. All inferences in ψ_1 and χ_1 are contractions and permutations. Then the derivation

$$\frac{\psi_1}{\chi_1}$$

is a derivation of $C_0 \circ D_0$ from $C \circ D$ using contractions and permutations only. But $C_0 \circ D_0 \sqsubseteq C' \circ D'$. □

Proposition 6.6.1 *Subsumption is reflexive and transitive.*

Proof:

- reflexivity: $C\theta \leq_{cp} C$ for $\theta = \epsilon$.
- transitivity: assume $C \leq_{ss} D$ and $D \leq_{ss} E$. Then there are substitutions θ, λ s.t.

  $$C\theta \leq_{cp} D \text{ and } D\lambda \leq_{cp} E.$$

  But then also

  $$C\theta\lambda \leq_{cp} D\lambda \text{ and } D\lambda \leq_{cp} E.$$

  So $C\theta\lambda \leq_{cp} E$ by the transitivity of $\leq_{cp}$ and $C \leq_{ss} E$. □

Remark: In contrast to $\leq_{cp}$, $C \leq_{ss} C'$ and $D \leq_{ss} D'$ does not imply $C \circ D \leq_{ss} C' \circ D'$ in general. Just take

$$C = \vdash P(x),\ C' = \vdash P(f(x)),\ D = \vdash Q(x),\ D' = \vdash Q(f(f(x))).$$

Then $C\theta_1 \leq_{cp} C'$ and $D\theta_2 \leq_{cp} D'$ for $\theta_1 = \{x \leftarrow f(x)\}, \theta_2 = \{x \leftarrow f(f(x))\}$; so $C \leq_{ss} C'$ and $D \leq_{ss} D'$ (by different substitutions). Clearly there is no substitution θ s.t. $(C \circ C')\theta \leq_{cp} D \circ D'$, and so $C \circ C' \not\leq_{ss} D \circ D'$. ◇

It is essential for the CERES method that the transitive closure $\rhd^*$ of the relation $\rhd$ (see Definition 6.3.5) can be considered as a weak form of subsumption.

Proposition 6.6.2 *Let X and Y be clause terms s.t. $X \rhd^* Y$. Then $X \leq_{ss} Y$.*

Proof: As the relation $\leq_{ss}$ is reflexive and transitive it suffices to show that $\rhd$ is a subrelation of $\leq_{ss}$.

a. $Y \subseteq X$: $X \leq_{ss} Y$ is trivial.

b. $X \sqsubseteq Y$: For all $C \in |Y|$ there exists a $D \in |X|$ with $D \sqsubseteq C$. But then also $D \leq_{ss} C$. The definition of the subsumption relation for sets yields $X \leq_{ss} Y$.

c. $X \leq_s Y$: $X \leq_{ss} Y$ is trivial. □

Subsumption plays a key role in automated deduction where it is used as a deletion method (see [61]). In fact, during proof search, a huge number of clauses is usually generated; most of them, however, are redundant in the sense that they are subsumed by previously generated clauses. As elimination of subsumed clauses does not affect the completeness of resolution theorem provers subsumed clauses can be safely eliminated.

We have seen that $\leq_{cp}$ is a sub-relation of $\leq_{ss}$. Like for $\leq_{ss}$, the relation $C \leq_{cp} D$ implies that D is "redundant". This redundancy is inherited by resolvents.

Lemma 6.6.2 *Let C, C', D, D' be clauses s.t. $C \leq_{cp} C'$, $D \leq_{cp} D'$ and let E' be a p-resolvent of C' and D'. Then either*

(1) *$C \leq_{cp} E'$, or*

(2) *$D \leq_{cp} E'$, or*

(3) *there exists a p-resolvent E of C and D s.t. $E \leq_{cp} E'$.*

Proof: As $\leq_{cp}$ is invariant under permutations we may assume that

$$\begin{aligned} C' &= \Gamma_1' \vdash \Gamma_2', A^n, \\ D' &= A^m, \Delta_1' \vdash \Delta_2', \text{ and} \\ E' &= \Gamma_1', \Delta_1' \vdash \Gamma_2', \Delta_2'. \end{aligned}$$

If $C \leq_{cp} C'$, and $C \leq_{cp} \Gamma_1' \vdash \Gamma_2'$, then clearly $C \leq_{cp} E'$; similarly for $D \leq_{cp} D'$, where $D \leq_{cp} \Delta_1' \vdash \Delta_2'$.
So let us assume that

$$C \leq_{cp} C', \; D \leq_{cp} D'$$

hold, but

$$\begin{array}{lll} C & \not\leq_{cp} & \Gamma'_1 \vdash \Gamma'_2 \text{ and} \\ D & \not\leq_{cp} & \Delta'_1 \vdash \Delta'_2. \end{array}$$

Then C and D must be of the form

$$\begin{array}{l} C = \Gamma_1 \vdash \Gamma_2, A^k, \text{ for } k > 0, \\ D = A^l, \Delta_1 \vdash \Delta_2, \text{ for } l > 0, \text{ and} \\ \Gamma_1 \vdash \Gamma_2 \leq_{cp} \Gamma'_1 \vdash \Gamma'_2, \quad \Delta_1 \vdash \Delta_2 \leq_{cp} \Delta'_1 \vdash \Delta'_2. \end{array}$$

But then the clause

$$E\colon \Gamma_1, \Delta_1 \vdash \Gamma_2, \Delta_2$$

is a p-resolvent of C and D and $E \leq_{cp} E'$.

Note that $S_1 \leq_{cp} S'_1$ and $S_2 \leq_{cp} S'_2$ implies

$$S_1 \circ S_2 \leq_{cp} S'_1 \circ S'_2$$

By Lemma 6.6.1. □

An analogous result holds for subsumption. As an auxiliary result we need the lifting theorem, one of the key results in automated deduction.

Theorem 6.6.1 (lifting theorem) *Let C, D be clauses with $C \leq_s C'$ and $D \leq_s D'$. Assume that C' and D' have a resolvent E'. Then there exists a resolvent E of C and D s.t. $E \leq_s E'$.*

Proof: In [61] page 79.

Lemma 6.6.3 *Let C, C', D, D' be clauses s.t. C and D are variable disjoint, $C \leq_{ss} C'$, $D \leq_{ss} D'$ and let E' be a resolvent of C' and D'. Then either*

(1) *$C \leq_{ss} E'$, or*

(2) *$D \leq_{ss} E'$, or*

(3) *there exists a resolvent E of C and D s.t. $E \leq_{ss} E'$.*

Proof: By definition of subsumption there are substitutions θ, λ with

$$(\star)\ C\theta \leq_{cp} C',\ \ D\lambda \leq_{cp} D'.$$

As E' is a resolvent of C' and D' there exists a substitution σ (which is a most general unifier of atoms) s.t. E' is a p-resolvent of $C'\sigma$ and $D'\sigma$. Note that the property $(\star)$ is closed under substitution and we obtain

$$(+)\ C\theta\sigma \leq_{cp} C'\sigma,\ \ D\lambda\sigma \leq_{cp} D'\sigma.$$

By Lemma 6.6.2 either

(1) $C\theta\sigma \leq_{ss} E'$, or

(2) $D\lambda\sigma \leq_{ss} E'$, or

(3) there exists a p-resolvent E of $C\theta\sigma$ and $D\lambda\sigma$ s.t. $E \leq_{cp} E'$.

In case (1) we also have $C \leq_{ss} E'$ as $C \leq_{ss} C\theta\sigma$ and $\leq_{ss}$ is transitive.
In case (2) we obtain $D \leq_{ss} E'$ in the same way.

In case of (3) we apply the lifting theorem and find a resolvent E_0 of C and D with $E_0\mu = E$ for some substitution μ. But $E_0\mu = E$ implies $E_0 \leq_{ss} E$; we also have $E \leq_{ss} E'$ and transitivity of subsumption yields $E_0 \leq_{ss} E'$. $\square$

Example 6.6.3 Let

$$\begin{array}{rcl} C & = & P(x), P(f(y)) \vdash Q(y), \\ C' & = & P(f(z)) \vdash Q(z), \\ D & = & Q(f(u)), Q(z) \vdash R(z), \\ D' & = & Q(f(a)) \vdash R(f(a)). \end{array}$$

Then $C \leq_{ss} C'$ and $D \leq_{ss} D'$. Moreover

$$E'\colon\ P(f(f(a))) \vdash R(f(a))$$

is a resolvent of C' and D'. Neither $C \leq_{ss} E'$ nor $D \leq_{ss} E'$ hold. But there exists the resolvent

$$E\colon\ P(x), P(f(f(u))) \vdash R(f(u))$$

of C and D (with m.g.u. $= \{y \leftarrow f(u), z \leftarrow f(u)\}$) and $E \leq_{ss} E'$. $\diamond$

The subsumption relation can also be extended to resolution derivations.

Definition 6.6.4 Let γ and δ be resolution derivations. We define $\gamma \leq_{ss} \delta$ by induction on the number of nodes in δ:

If δ consists of a single node labelled with a clause D then $\gamma \leq_{ss} \delta$ if γ consists of a single node labelled with C and $C \leq_{ss} D$.

Let δ be

$$\frac{\begin{array}{cc}(\delta_1) & (\delta_2)\\ D_1 & D_2\end{array}}{D}\ R$$

and γ_1 be a derivation of C_1 with $\gamma_1 \leq_{ss} \delta_1$, γ_2 be a derivation of C_2 with $\gamma_2 \leq_{ss} \delta_2$. Then we distinguish the following cases:

$C_1 \leq_{ss} D$: then $\gamma_1 \leq_{ss} \delta$.

$C_2 \leq_{ss} D$: then $\gamma_2 \leq_{ss} \delta$.

Otherwise let C be a resolvent of C_1 and C_2 which subsumes D. Such a resolvent exists by Lemma 6.6.3. For every γ of the form

$$\frac{\begin{array}{cc}(\gamma_1) & (\gamma_2)\\ C_1 & C_2\end{array}}{C}\ R$$

we get $\gamma \leq_{ss} \delta$.

◇

Proposition 6.6.3 *Let $\mathcal{C}, \mathcal{D}$ be sets of clauses with $\mathcal{C} \leq_{ss} \mathcal{D}$ and let δ be a resolution derivation from $\mathcal{D}$. Then there exists a resolution derivation γ from $\mathcal{C}$ s.t. $\gamma \leq_{ss} \delta$.*

Proof: By Lemma 6.6.3 and by Definition 6.6.4. □

Corollary 6.6.1 *Let $\mathcal{C}, \mathcal{D}$ be sets of clauses with $\mathcal{C} \leq_{ss} \mathcal{D}$ and let δ be a resolution refutation of $\mathcal{D}$. Then there exists a resolution refutation γ of $\mathcal{C}$ s.t. $\gamma \leq_{ss} \delta$.*

Proof: Obvious.

Example 6.6.4 Let $\mathcal{C}$ and $\mathcal{D}$ be sets of clauses for

$$\mathcal{C} = \{\vdash P(x)\ ;\ P(f(y)) \vdash Q(y)\ ;\ \vdash R(z)\},$$
$$\mathcal{D} = \{\vdash P(f(a))\ ;\ P(f(a)) \vdash Q(a), R(f(a))\ ;\ \vdash R(z), Q(z)\ ;\ Q(u) \vdash R(u)\}.$$

Then $\mathcal{C} \leq_{ss} \mathcal{D}$. We consider two resolution derivations δ_1 and δ_2 from $\mathcal{D}$. $\delta_1 =$

$$\frac{\vdash P(f(a)) \quad P(f(a)) \vdash Q(a), R(f(a))}{\vdash Q(a), R(f(a))}$$

$\delta_2 =$

$$\frac{\dfrac{\vdash P(f(a)) \quad P(f(a)) \vdash Q(a), R(f(a))}{\vdash Q(a), R(f(a))} \qquad Q(u) \vdash R(u)}{\vdash R(f(a)), R(a)}$$

Let $\gamma_1 =$

$$\frac{\vdash P(x) \quad P(f(y)) \vdash Q(y)}{\vdash Q(y)}$$

and $\gamma_2 = \vdash R(z)$. Then $\gamma_1 \leq_{ss} \delta_1$ and $\gamma_2 \leq_{ss} \delta_2$. $\diamond$

6.7 Canonic Resolution Refutations

If $\psi \in \Phi_0^s$ then there exists a *canonic* resolution refutation $\mathrm{RES}(\psi)$ of the set of clauses $\mathrm{CL}(\psi)$. $\mathrm{RES}(\psi)$ is "the" resolution proof corresponding to ψ. Indeed, as ψ is a deduction with atomic cuts only, the part of ψ ending in the cut formulas is roughly a p-resolution refutation. For the construction of $\mathrm{RES}(\psi)$ we need some technical definitions:

Definition 6.7.1 Let γ be a p-resolution derivation of a clause C from a set of clauses $\mathcal{C}$ and let D be a clause. We define a p-resolution deduction $\gamma \cdot D$ from $\mathcal{C} \times \{D\}$ in the following way:

(1) replace all initial clauses S in γ by $S \circ D$.

(2) Apply the cuts as in γ and leave the inference nodes unchanged (note that this is possible by our definition of the cut rule).

$\diamond$

Remark: Note that $\gamma \cdot D$ is not identical to the right context product $\gamma \star D$. Indeed, $\gamma \cdot D$ contains exactly as many rules as γ, while $\gamma \star D$ may contain additional structural rules. $\diamond$

Example 6.7.1 Let $\gamma =$

$$\frac{P(a) \vdash R(x) \quad R(x), R(x) \vdash Q(x)}{P(a) \vdash Q(x)}\ cut$$

and $D = R(x) \vdash S(x)$. Then $\gamma \cdot D =$

$$\frac{P(a), R(x) \vdash R(x), S(x) \quad R(x), R(x), R(x) \vdash Q(x), S(x)}{P(a), R(x), R(x) \vdash Q(x), S(x), S(x)}\ cut$$

Lemma 6.7.1 *Let γ be a p-resolution deduction of C from $\mathcal{C}$ and let δ be a p-resolution deduction of D from $\mathcal{D}$. Then there exists a p-resolution deduction ρ of a clause E from $\mathcal{C} \times \mathcal{D}$ s.t.*

(1) $C \circ D \sqsubseteq E$,

(2) $E \leq_{cp} C \circ D$ *(see Definition 6.6.1), and*

(3) $l(\rho) \leq l(\gamma) * l(\delta)$.

Proof: By induction on the number of inference nodes in δ. If $\delta = D$ for a clause D then we define

$$\gamma \odot \delta = \gamma \cdot D.$$

Then, by definition of $\gamma \cdot D$ in Definition 6.7.1, $\gamma \odot \delta$ is a derivation of $C \circ D^n$ from $\mathcal{C} \times \mathcal{D}$ (as $D \in \mathcal{D}$). Clearly $C \circ D \sqsubseteq C \circ D^n$ ((1) holds), but also $C \circ D^n \leq_{cp} C \circ D$ ((2) holds). Moreover (3) holds by

$$l(\gamma \odot \delta) = l(\gamma \cdot D) = l(\gamma) = l(\gamma) * l(\delta) \text{ by } l(\delta) = 1.$$

Now let δ be a derivation of D from $\mathcal{D}$ with $l(\delta) > 1$. Then δ is of the form

$$\frac{\begin{matrix}(\delta_1) \\ \Gamma_1 \vdash \Delta_1, A^n\end{matrix} \quad \begin{matrix}(\delta_2) \\ A^m, \Gamma_2 \vdash \Delta_2\end{matrix}}{\Gamma_1, \Gamma_2 \vdash \Delta_1, \Delta_2}\ cut$$

Note, that for simplicity we assumed the cut-atoms to be right- and leftmost. By induction hypothesis we have derivations $\gamma \odot \delta_1$ and $\gamma \odot \delta_2$ of clauses E_1 and E_2 fulfilling (1),(2),(3) above.
By (1)

$$\begin{array}{rcl} \Gamma_1 \vdash \Delta_1, A^n & \sqsubseteq & E_1, \\ A^m, \Gamma_2 \vdash \Delta_2 & \sqsubseteq & E_2 \text{ and also} \\ E_1 & \leq_{cp} & \Gamma_1 \vdash \Delta_1, A^n, \\ E_2 & \leq_{cp} & A^m, \Gamma_2 \vdash \Delta_2. \end{array}$$

Therefore the end sequents of δ_i are subclauses of the E_i, but all atoms in the E_i occur also in the end sequents of δ_i (on the appropriate side of the sequent). Therefore we can simulate the cut above by cutting E_1 and E_2. If A does not occur in Γ_1, Γ_2 or in Δ_1, Δ_2 possibly more occurrences of A have to be cut out in E_1 and E_2. So we define $\gamma \odot \delta =$

$$\frac{\gamma \odot \delta_1 \quad \gamma \odot \delta_2}{E}\ cut$$

Then (1) and (2) obviously hold. For (3) we observe that

$$\begin{array}{rcl} l(\gamma \odot \delta_i) & \leq & l(\gamma) * l(\delta_i) \text{ by (IH) and} \\ l(\gamma \odot \delta) & = & l(\gamma \odot \delta_1) + l(\gamma \odot \delta_2) + 1, \text{ so} \\ l(\gamma \odot \delta) & \leq & l(\gamma) * (l(\delta_1) + l(\delta_2)) + 1 \\ & \leq & l(\gamma) * (l(\delta_1) + l(\delta_2) + 1) = l(\gamma) * l(\delta). \ \Box \end{array}$$

Example 6.7.2 Let γ be

$$\frac{P(a) \vdash R(x) \quad R(x), R(x) \vdash Q(x)}{P(a) \vdash Q(x)}\ cut$$

as in Example 6.7.1 and $\delta =$

$$\frac{R(x) \vdash S(x) \quad S(x) \vdash}{R(x) \vdash}\ cut$$

Then γ_1: $\gamma \cdot (R(x) \vdash S(x)) =$

$$\frac{P(a), R(x) \vdash R(x), S(x) \quad R(x), R(x), R(x) \vdash Q(x), S(x)}{P(a), R(x), R(x) \vdash Q(x), S(x), S(x)}\ cut$$

and γ_2: $\gamma \cdot (S(x) \vdash) =$

$$\frac{P(a), S(x) \vdash R(x) \quad R(x), R(x), S(x) \vdash Q(x)}{P(a), S(x), S(x) \vdash Q(x)}\ cut$$

Consequently $\gamma \odot \delta =$

$$\frac{\overset{(\gamma_1)}{P(a), R(x), R(x) \vdash Q(x), S(x), S(x)} \quad \overset{(\gamma_2)}{P(a), S(x), S(x) \vdash Q(x)}}{P(a), P(a), R(x), R(x) \vdash Q(x), Q(x)}\ cut$$

If ψ is in Φ_0^s then there exists something like a *canonic* resolution refutation of $\mathrm{CL}(\psi)$. The definition of this refutation follows the steps of the definition of the characteristic clause term.

Definition 6.7.2 Let ψ be an **LK**-derivation in Φ_0^s , Ω be the set of occurrences of the (atomic) cut formulas in ψ and $\mathcal{C} = \mathrm{CL}(\psi)$. For simplicity we write $\mathcal{C}_\nu$ for the set of clauses $|\Theta(\psi)/\nu|$ defined by the characteristic terms as in Definition 6.4.1. Clearly $\mathcal{C} = \mathcal{C}_{\nu_0}$ for the root node ν_0 in ψ.
We proceed inductively and define a p-resolution derivation γ_ν for every node ν in ψ s.t. γ_ν is a derivation of a clause C_ν from $\mathcal{C}_\nu$ s.t.

(I) $C_\nu \leq_{cp} S(\nu, \Omega)$.

Assume that we have already constructed all derivations s.t. (I) holds for all nodes. Then, for ν_0, we have $S(\nu_0, \Omega) = \vdash$ and therefore $C(\nu_0) = \vdash$; so γ_{ν_0} is a refutation of $\mathrm{CL}(\psi)$.

If ν is a leaf in ψ then we define γ_ν as $S(\nu, \Omega)$ and $C(\nu) = S(\nu, \Omega)$. By definition of $\mathcal{C}$ we have $\mathcal{C}_\nu = \{S(\nu, \Omega)\}$. Clearly γ_ν is p-resolution derivation of $C(\nu)$ from $\mathcal{C}_\nu$ and $C(\nu) \leq_{cp} S(\nu, \Omega)$.

(1) Let γ_μ be already defined for a node μ in ψ s.t. γ_μ is a p-resolution derivation of C_μ from $\mathcal{C}_\mu$ s.t. $C_\mu \leq_{cp} S(\mu, \Omega)$. Moreover let ξ be a unary inference in ψ with premise μ and conclusion ν. Then we define

$$\gamma_\nu = \gamma_\mu, \quad \text{and } C_\nu = C_\mu.$$

If ξ goes into the end-sequent then, by definition, $S(\mu, \Omega) = S(\nu, \Omega)$. So let us assume that ξ goes into a cut. As ψ is in Φ_0 ξ is either a contraction, a weakening, or a permutation. Therefore, either $S(\nu, \Omega) = S(\mu, \Omega)$ or $S(\nu, \Omega)$ is obtained from $S(\mu, \Omega)$ by one of the rules $c\colon l, c\colon r, w\colon l, w\colon r, p\colon l, p\colon r$. In all cases we have

$$S(\mu, \Omega) \leq_{cp} S(\nu, \Omega).$$

By assumption we have $C_\mu \leq_{cp} S(\mu, \Omega)$, and by definition $C_\nu = C_\mu$. As $\leq_{cp}$ is transitive we get

$$C_\nu \leq_{cp} S(\nu, \Omega).$$

Now γ_ν is a derivation of C_ν from $\mathcal{C}_\mu$; but, according to the definition of the characteristic clause term we have $\mathcal{C}_\nu = \mathcal{C}_\mu$. This concludes the construction for the unary case.

(2) Assume that γ_{μ_i} are p-resolution derivations of C_{μ_i} from $\mathcal{C}_{\mu_i}$ for $i = 1, 2$ s.t.

$$C_{\mu_1} \leq_{cp} S(\mu_1, \Omega),$$
$$C_{\mu_2} \leq_{cp} S(\mu_2, \Omega).$$

Let ν be an inference node in ψ with premises μ_1, μ_2 and the corresponding binary rule ξ. We distinguish two cases:

(2a) The auxiliary formulas of ξ are in $S(\mu_1, \Omega)$ and $S(\mu_2, \Omega)$.
Then ξ must be a cut (there are no other binary inferences leading to Ω).
We distinguish three cases:
(2a.1) $C_{\mu_1} \leq_{cp} S(\nu, \Omega)$,
(2a.2) $C_{\mu_2} \leq_{cp} S(\nu, \Omega)$,
(2a.3) neither (2a.1) nor (2a.2) holds.
In case (2a.1) we define $\gamma_\nu = \gamma_{\mu_1}$ and $C_\nu = C_{\mu_1}$.
In case (2a.2) we define $\gamma_\nu = \gamma_{\mu_2}$ and $C_\nu = C_{\mu_2}$.
In case (2a.3) we know from Lemma 6.6.2 that there exists a p-resolvent E of $C(\mu_1)$ and $C(\mu_2)$ s.t. $E \leq_{cp} S(\nu, \Omega)$ (note that $S(\nu, \Omega)$ is a p-resolvent of $S(\mu_1, \Omega)$ and $S(\mu_2, \Omega)$!). So we define $\gamma_\nu =$

$$\frac{\gamma_{\mu_1} \quad \gamma_{\mu_2}}{E}\ cut$$

and set $C_\nu = E$.

(2b) The auxiliary formulas of ξ are not in $S(\mu_1, \Omega)$ and $S(\mu_2, \Omega)$.
In this case we define

$$\gamma_\nu = \gamma_{\mu_1} \odot \gamma_{\mu_2} \text{ and}$$

$$C_\nu = C_{\mu_1} \circ C_{\mu_2}.$$

By definition of $\odot$ the derivation γ_ν is a p-resolution derivation of $C_{\mu_1} \circ C_{\mu_2}$ from $\mathcal{C}_{\mu_1} \times \mathcal{C}_{\mu_2}$. By induction we have

$$C_{\mu_1} \leq_{cp} S(\mu_1, \Omega),$$
$$C_{\mu_2} \leq_{cp} S(\mu_2, \Omega).$$

By Lemma 6.6.1

$$C_{\mu_1} \circ C_{\mu_2} \leq_{cp} S(\mu_1, \Omega) \circ S(\mu_1, \Omega).$$

But

$$S(\nu,\Omega) = S(\mu_1,\Omega) \circ S(\mu_2,\Omega)$$

and, by definition of the characteristic term, $\mathcal{C}_\nu = \mathcal{C}_{\mu_1} \times \mathcal{C}_{\mu_2}$.

Finally we define $\mathrm{RES}(\psi) = \gamma_{\nu_0}$ where ν_0 is the root node in ψ. $\diamond$

Below we show that, for an AC-derivation ψ the number of nodes in $\mathrm{RES}(\psi)$ may be exponential in the number of nodes of ψ. But note that, in general, resolution refutations of $\mathrm{CL}(\psi)$ are of nonelementary length (see Section 6.5). Thus the proofs $\mathrm{RES}(\psi)$ for AC-derivations ψ can be considered "small".

Proposition 6.7.1 *Let ψ be an* **LK**-*derivation in Φ_0^s. Then*

$$l(\mathrm{RES}(\psi)) \ \leq \ 2^{2*l(\psi)}.$$

Proof: Let $\Theta(\psi)$ be the characteristic term of ψ. We write Θ_ν for $\Theta(\psi)/\nu$ (see Definition 6.4.1) and $|\Theta_\nu|$ for the number of subterms occurring in Θ_ν. We proceed by induction on the definition of γ_ν in Definition 6.7.2, in particular we prove that for all nodes ν in ψ

$$(*)\ l(\gamma_\nu) \ \leq \ 2^{l(\psi.\nu)+|\Theta_\nu|}.$$

For leaves ν we have $l(\gamma_\nu) = 1$ and $(*)$ is trivial.
So let us assume that $(*)$ holds for the node μ and ν is the conclusion of a unary inference with premise μ. Then by definition of γ_ν:

$$\begin{array}{rcl} l(\gamma_\nu) & = & l(\gamma_\mu), \\ \Theta_\nu & = & \Theta_\mu, \\ l(\psi.\nu) & = & l(\psi.\mu) + 1 \text{ and by assumption on } \mu \end{array}$$

$$l(\gamma_\nu) = l(\gamma_\mu) \leq 2^{l(\psi.\mu)+|\Theta_\mu|} < 2^{l(\psi.\nu)+|\Theta_\nu|}.$$

Assume that $(*)$ holds for nodes μ_1, μ_2 and ν is the conclusion of a binary inference with premises μ_1, μ_2.
Then, by definition of Θ_ν,

$$|\Theta_\nu| = |\Theta_{\mu_1}| + |\Theta_{\mu_2}| + 1,$$

no matter whether $\Theta_\nu = \Theta_{\mu_1} \oplus \Theta_{\mu_2}$ or $\Theta_\nu = \Theta_{\mu_1} \otimes \Theta_{\mu_2}$.

If the inference takes place on ancestors of Ω then

$$\begin{aligned} l(\gamma_\nu) &= l(\gamma_{\mu_1}) + l(\gamma_{\mu_2}) + 1, \text{ and} \\ l(\psi.\nu) &= l(\psi.\mu_1) + l(\psi.\mu_2) + 1. \end{aligned}$$

By the assumptions on μ_1, μ_2 we have

$$l(\gamma_{\mu_1}) \leq 2^{l(\psi.\mu_1)+|\Theta_{\mu_1}|},$$
$$l(\gamma_{\mu_2}) \leq 2^{l(\psi.\mu_2)+|\Theta_{\mu_1}|},$$

and therefore

$$\begin{aligned} l(\gamma_\nu) &= l(\gamma_{\mu_1}) + l(\gamma_{\mu_2}) + 1 \\ &\leq 2^{l(\psi.\mu_1)+|\Theta_{\mu_1}|} + 2^{l(\psi.\mu_2)+|\Theta_{\mu_2}|} + 1 \\ &\leq 2^{l(\psi.\mu_1)+|\Theta_{\mu_1}|+l(\psi.\mu_2)+|\Theta_{\mu_2}|+1} \\ &\leq 2^{l(\psi.\nu)+|\Theta_\nu|}. \end{aligned}$$

If the inference takes place on non-ancestors of Ω then

$$\begin{aligned} l(\gamma_\nu) &\leq l(\gamma_{\mu_1}) * l(\gamma_{\mu_2}), \\ l(\psi.\nu) &= l(\psi.\mu_1) + l(\psi.\mu_2) + 1. \end{aligned}$$

and, by the assumptions on μ_1, μ_2,

$$\begin{aligned} l(\gamma_\nu) &\leq l(\gamma_{\mu_1}) * l(\gamma_{\mu_2}) \\ &\leq 2^{l(\psi.\mu_1)+|\Theta_{\mu_1}|} * 2^{l(\psi.\mu_2)+|\Theta_{\mu_2}|} \\ &= 2^{l(\psi_{\mu_1})+l(\psi_{\mu_2})+|\Theta_{\mu_1}|+|\Theta_{\mu_2}|} \\ &< 2^{l(\psi.\nu)+|\Theta_\nu|}. \end{aligned}$$

Thus by induction and choosing the root node for ν we obtain

$$(I)\ l(\mathrm{RES}(\psi)) \leq 2^{l(\psi)+|\Theta(\psi)|}.$$

Obviously $|\Theta(\psi)| \leq l(\psi)$ (indeed the term tree of $\Theta(\psi)$ has $1 + n$ nodes, where n is the number of binary inferences in ψ), and so we obtain

$$(I)\ l(\mathrm{RES}(\psi)) \leq 2^{2*l(\psi)}.$$ □

The results of this chapters show that proofs in ACNF are well-behaving under CERES, similarly as for cut-free proofs.

6.8 Characteristic Terms and Cut-Reduction

In this section we show that methods of cut-elimination based on the set of rules $\mathcal{R}$ are redundant w.r.t. the results of the CERES method. It will turn out that the characteristic clause set $\mathrm{CL}(\varphi')$ of a Gentzen normal form φ' of a proof φ is subsumed by the original characteristic clause set $\mathrm{CL}(\varphi)$. In this sense every $\mathcal{R}$-reduction step on a proof is redundant in the sense of clause logic.

Lemma 6.8.1 *Let φ, φ' be* **LK***-derivations with $\varphi >_{\mathcal{R}} \varphi'$ for a cut reduction relation $>_{\mathcal{R}}$ based on $\mathcal{R}$. Then $\Theta(\varphi) \rhd \Theta(\varphi')$.*

Proof: We construct a proof by cases on the definition of $>_{\mathcal{R}}$. To this aim we consider sub-derivations ψ of φ of the form

$$\frac{\begin{array}{cc}(\rho, X) & (\sigma, Y)\\ \Gamma \vdash \Delta & \Pi \vdash \Lambda\end{array}}{\Gamma, \Pi^* \vdash \Delta^*, \Lambda}\ cut(A)$$

where $X = \Theta(\varphi)/\lambda$ for the occurrence λ corresponding to the derivation ρ and $Y = \Theta(\varphi)/\mu$ for the occurrence μ corresponding to σ. By ν we denote the occurrence of ψ in φ. That means we do not only indicate the sub-derivations ending in the cut, but also the corresponding clause terms. Note that by definition of the characteristic term we have $\Theta(\varphi)/\nu = X \oplus Y$.
If $\psi >_{\mathcal{R}} \chi$ then, by definition of the reduction relation $>_{\mathcal{R}}$, we get $\varphi = \varphi[\psi]_\nu >_{\mathcal{R}} \varphi[\chi]_\nu$. For the remaining part of the proof we denote $\varphi[\chi]_\nu$ by φ'. Our aim is to prove that $\Theta(\varphi) \rhd \Theta(\varphi')$.

(I) $\mathrm{rank}(\psi) = 2$:

(Ia) ψ is of the form

$$\frac{\dfrac{\begin{array}{c}(\rho', X)\\ \Gamma \vdash \Delta\end{array}}{\Gamma \vdash \Delta, A}\ w\colon r \qquad \begin{array}{c}(\sigma, Y)\\ \Pi \vdash \Lambda\end{array}}{\Gamma, \Pi^* \vdash \Delta, \Lambda}\ cut(A)$$

By definition of $\mathcal{R}$ we have $\psi >_{\mathcal{R}} \chi$ for $\chi =$

$$\frac{\begin{array}{c}(\rho', X)\\ \Gamma \vdash \Delta\end{array}}{\Gamma, \Pi^* \vdash \Delta, \Lambda}\ s^*$$

Therefore also $\varphi[\psi]_\nu >_{\mathcal{R}} \varphi[\chi]_\nu$, i.e. $\varphi >_{\mathcal{R}} \varphi'$. But $\Theta(\varphi')/\nu = X$ and $\Theta(\varphi)/\nu = X \oplus Y$. Clearly $X \subseteq X \oplus Y$ and so $X \oplus Y \rhd X$; by Lemma 6.3.3 we conclude that $\Theta(\varphi) \rhd \Theta(\varphi')$.

(Ib) $A = \neg B$ and ψ is of the form

$$\dfrac{\dfrac{\begin{matrix}(\rho', X)\\ B, \Gamma \vdash \Delta\end{matrix}}{\Gamma \vdash \Delta, \neg B}\ \neg{:}\,r \qquad \dfrac{\begin{matrix}(\sigma', Y)\\ \Pi \vdash \Lambda, B\end{matrix}}{\neg B, \Pi \vdash \Lambda}\ \neg{:}\,l}{\Gamma, \Pi \vdash \Delta, \Lambda}\ cut(A)$$

Then $\psi >_{\mathcal{R}} \chi$ for $\chi =$

$$\dfrac{\dfrac{\begin{matrix}(\sigma', Y)\\ \Pi \vdash \Lambda, B\end{matrix} \quad \begin{matrix}(\rho', X)\\ B, \Gamma \vdash \Delta\end{matrix}}{\Gamma^*, \Pi \vdash \Delta, \Lambda^*}\ cut(B)}{\Gamma, \Pi \vdash \Delta, \Lambda}\ s^*$$

Here we have

$$\begin{aligned}\Theta(\varphi)/\nu &= X \oplus Y,\\ \Theta(\varphi')/\nu &= Y \oplus X.\end{aligned}$$

Clearly, by $Y \oplus X \subseteq X \oplus Y$, $X \oplus Y \rhd Y \oplus X$ (we even have $X \oplus Y \sim Y \oplus X$) and by Lemma 6.3.3 we obtain $\Theta(\varphi) \rhd \Theta(\varphi')$.

(Ic) $A = B \wedge C$ and ψ is of the form

$$\dfrac{\dfrac{\begin{matrix}(\rho_1, X_1)\\ \Gamma \vdash \Delta, B\end{matrix} \quad \begin{matrix}(\rho_2, X_2)\\ \Gamma \vdash \Delta, C\end{matrix}}{\Gamma \vdash \Delta, B \wedge C}\ \wedge{:}\,r \qquad \dfrac{\begin{matrix}(\sigma', Y)\\ B, \Pi \vdash \Lambda\end{matrix}}{B \wedge C, \Pi \vdash \Lambda}\ \wedge{:}\,l}{\Gamma, \Pi \vdash \Delta, \Lambda}\ cut(A)$$

Then $\psi >_{\mathcal{R}} \chi$ for $\chi =$

$$\dfrac{\dfrac{\begin{matrix}(\rho_1, X_1)\\ \Gamma \vdash \Delta, B\end{matrix} \quad \begin{matrix}(\sigma', Y)\\ B, \Pi \vdash \Lambda\end{matrix}}{\Gamma, \Pi^* \vdash \Delta^*, \Lambda}\ cut(B)}{\Gamma, \Pi \vdash \Delta, \Lambda}\ s^*$$

In this case we have

$$\begin{aligned}\Theta(\varphi)/\nu &= (X_1 \oplus X_2) \oplus Y,\\ \Theta(\varphi')/\nu &= X_1 \oplus Y.\end{aligned}$$

Clearly, $X_1 \oplus Y \subseteq (X_1 \oplus X_2) \oplus Y$ and thus $(X_1 \oplus X_2) \oplus Y \rhd X_1 \oplus Y$. By application of Lemma 6.3.3 we obtain $\Theta(\varphi) \rhd \Theta(\varphi')$.

The case where $B \wedge C$ is inferred from C is completely symmetric.

(Id) $A = B \vee C$: symmetric to (Ib).

(Ie) $A = B \to C$. Then ψ is of the form

$$\dfrac{\dfrac{\begin{array}{c}(\rho', X)\\ B, \Gamma \vdash \Delta, C\end{array}}{\Gamma \vdash \Delta, B \to C}\ {\to}{:}\,r \qquad \dfrac{\begin{array}{cc}(\sigma_1, Y_1) & (\sigma_2, Y_1)\\ \Pi_1 \vdash \Lambda_1, B & C, \Pi_2 \vdash \Lambda_2\end{array}}{B \to C, \Pi_1, \Pi_2 \vdash \Lambda_1, \Lambda_2}\ {\to}{:}\,l}{\Gamma, \Pi_1, \Pi_2 \vdash \Delta, \Lambda_1, \Lambda_2}\ cut(A)$$

Here we have $\psi >_{\mathcal{R}} \chi$ for $\chi =$

$$\dfrac{\dfrac{\dfrac{\begin{array}{cc}(\sigma_1, Y_1) & (\rho', X)\\ \Pi_1 \vdash \Lambda_1, B & B, \Gamma \vdash \Delta, C\end{array}}{\Pi_1, \Gamma \vdash \Lambda_1, \Delta, C}\ cut \qquad \begin{array}{c}(\sigma_2, Y_1)\\ C, \Pi_2 \vdash \Lambda_2\end{array}}{\Pi_1, \Gamma, \Pi_2 \vdash \Lambda_1, \Delta, \Lambda_2}\ cut}{\Gamma, \Pi_1, \Pi_2 \vdash \Delta, \Lambda_1, \Lambda_2}\ s^*$$

Here we obtain

$$\begin{array}{rcl}\Theta(\varphi)/\nu &=& X \oplus (Y_1 \oplus Y_2),\\ \Theta(\varphi')/\nu &=& (Y_1 \oplus X) \oplus Y_2.\end{array}$$

By $(Y_1 \oplus X) \oplus Y_2 \subseteq X \oplus (Y_1 \oplus Y_2)$ we get $X \oplus (Y_1 \oplus Y_2) \rhd (Y_1 \oplus X) \oplus Y_2$. Again, by application of Lemma 6.3.3, we obtain $\Theta(\varphi) \rhd \Theta(\varphi')$.

(If) $A = (\forall x)B$. Then ψ is of the form

$$\dfrac{\dfrac{\begin{array}{c}(\rho'(x/y), X(x/y))\\ \Gamma \vdash \Delta, B(x/y)\end{array}}{\Gamma \vdash \Delta, (\forall x)B(x)}\ \forall{:}\,r \qquad \dfrac{\begin{array}{c}(\sigma', Y)\\ B(x/t), \Pi \vdash \Lambda\end{array}}{(\forall x)B(x), \Pi \vdash \Lambda}\ \forall{:}\,l}{\Gamma, \Pi \vdash \Delta, \Lambda}\ cut(A)$$

$\psi >_{\mathcal{R}} \chi$ for

$$\dfrac{\dfrac{\begin{array}{cc}(\rho'(x/t), X(x/t)) & (\sigma', Y)\\ \Gamma \vdash \Delta, B(t) & B(x/t), \Pi \vdash \Lambda\end{array}}{\Gamma, \Pi^* \vdash \Delta^*, \Lambda}\ cut(B(x/t)}{\Gamma, \Pi \vdash \Delta, \Lambda}\ s^*$$

By definition of the characteristic terms we have

$$\begin{array}{rcl}\Theta(\varphi)/\nu &=& X(x/y) \oplus Y,\\ \Theta(\varphi')/\nu &=& X(x/t) \oplus Y.\end{array}$$

By assumption φ is regular and the variable y only occurs in the sub-derivation ρ. Therefore

$$\begin{array}{rcl} \Theta(\varphi')/\nu & = & (X(x/y) \oplus Y)\{y \leftarrow t\} \text{ and even} \\ \Theta(\varphi') & = & \Theta(\varphi)\{y \leftarrow t\}. \end{array}$$

But this means $\Theta(\varphi) \leq_s \Theta(\varphi')$ and therefore $\Theta(\varphi) \rhd \Theta(\varphi')$.

(Ig) $A = (\exists x)B$: symmetric to (Id).

(II) $\mathrm{rank}(\psi) > 2$.

We assume that $\mathrm{rank}_r(\psi) > 1$ (the case $\mathrm{rank}_l(\psi) > 1$ is symmetric).

(IIa) A occurs in Γ. Then $\psi >_{\mathcal{R}} \chi$ for $\chi =$

$$\frac{\begin{array}{c}(\sigma, Y)\\ \Pi \vdash \Lambda\end{array}}{\Gamma, \Pi^* \vdash \Delta^*, \Lambda}\ s^*$$

In this case

$$\begin{array}{rcl} \Theta(\varphi)/\nu & = & X \oplus Y, \\ \Theta(\varphi')/\nu & = & Y. \end{array}$$

Clearly $Y \subseteq X \oplus Y$ and thus $X \oplus Y \rhd Y$; by Lemma 6.3.3 $\Theta(\varphi) \rhd \Theta(\varphi')$.

(IIb) A does not occur in Γ.

(IIb.1) ξ is one of the inferences $w\colon l$ or $c\colon l$ and ψ is of the form:

$$\frac{\begin{array}{c}(\rho, X)\\ \Gamma \vdash \Delta\end{array} \quad \dfrac{\begin{array}{c}(\sigma', Y)\\ \Sigma \vdash \Lambda\end{array}}{\Pi \vdash \Lambda}\ \xi}{\Gamma, \Pi^* \vdash \Delta^*, \Lambda}\ cut(A)$$

Then $\psi >_{\mathcal{R}} \chi$ for $\chi =$

$$\frac{\dfrac{\begin{array}{c}(\rho, X)\\ \Gamma \vdash \Delta\end{array} \quad \begin{array}{c}(\sigma', Y)\\ \Sigma \vdash \Lambda\end{array}}{\Gamma, \Sigma^* \vdash \Delta^*, \Lambda}\ cut(A)}{\Gamma, \Pi^* \vdash \Delta^*, \Lambda}\ s^*$$

It is obvious that $\Theta(\varphi) = \Theta(\varphi')$ and so $\Theta(\varphi) \rhd \Theta(\varphi')$.

(IIb.2) ξ is a unary inference, $\xi \notin \{w\colon l, c\colon l\}$ and ψ is of the form

$$\dfrac{\begin{matrix}(\rho, X) \\ \Gamma \vdash \Delta\end{matrix} \quad \dfrac{\begin{matrix}(\sigma', Y) \\ B, \Pi \vdash \Lambda_1\end{matrix}}{C, \Pi \vdash \Lambda_2}\ \xi}{\Gamma, C^*, \Pi^* \vdash \Delta^*, \Lambda_2}\ cut(A)$$

where C^* is empty for $C = A$ and $C^* = C$ for $C \neq A$. We consider the derivation τ:

$$\dfrac{\dfrac{\dfrac{\begin{matrix}(\rho, X) \\ \Gamma \vdash \Delta\end{matrix} \quad \begin{matrix}(\sigma', Y) \\ B, \Pi \vdash \Lambda_1\end{matrix}}{\Gamma, B^*, \Pi^* \vdash \Delta^*, \Lambda_1}\ cut(A)}{\Gamma, B, \Pi^* \vdash \Delta^*, \Lambda_1}\ s^*}{\Gamma, C, \Pi^* \vdash \Delta^*, \Lambda_2}\ \xi + s^*$$

It is easy to see that

$$\begin{aligned}\Theta(\varphi[\tau]_\nu)/\nu &= X \oplus Y \text{ and} \\ \Theta(\varphi) &= \Theta(\varphi[\tau]_\nu).\end{aligned}$$

Indeed changing the order of unary inferences does not affect characteristic terms. If $A \neq C$ then, by definition of $>_{\mathcal{R}}$, we have $\chi = \tau$ and $\Theta(\varphi) = \Theta(\varphi')$.

If $A = C$ and $A \neq B$ we have $\chi =$

$$\dfrac{\dfrac{\begin{matrix}(\rho, X) \\ \Gamma \vdash \Delta\end{matrix} \quad \begin{matrix}(\tau, X \oplus Y) \\ \Gamma, A, \Pi^* \vdash \Delta^*, \Lambda_2\end{matrix}}{\Gamma, \Gamma^*, \Pi^* \vdash \Delta^*, \Delta^*, \Lambda_2}\ cut(A)}{\Gamma, \Pi^* \vdash \Delta^*, \Lambda_2}\ s^*$$

Now we have

$$\begin{aligned}\Theta(\varphi)/\nu &= X \oplus Y, \\ \Theta(\varphi')/\nu &= X \oplus (X \oplus Y).\end{aligned}$$

But $X \oplus Y \sim X \oplus (X \oplus Y)$ and thus also $X \oplus Y \rhd X \oplus (X \oplus Y)$. Therefore, using Lemma 6.3.3 again, we obtain $\Theta(\varphi) \rhd \Theta(\varphi')$.

If $A = B = C$ then $\Lambda_1 \neq \Lambda_2$ and χ is defined as

$$\dfrac{\dfrac{\begin{matrix}(\rho, X) \\ \Gamma \vdash \Delta\end{matrix} \quad \begin{matrix}(\sigma', Y) \\ A, \Pi \vdash \Lambda_1\end{matrix}}{\Gamma, \Pi^* \vdash \Delta^*, \Lambda_1}\ cut(A)}{\Gamma, \Pi^* \vdash \Delta^*, \Lambda_2}\ \xi$$

In this case, clearly, $\Theta(\varphi') = \Theta(\varphi)$ and thus $\Theta(\varphi) \rhd \Theta(\varphi')$.

(IIb.3) The last inference in σ is a binary one.

(IIb.3.1) The last inference in σ is $\wedge : r$. Then ψ is of the form

$$\dfrac{\begin{matrix}(\rho, X)\\ \Gamma \vdash \Delta\end{matrix} \quad \dfrac{\begin{matrix}(\sigma_1, Y_1)\\ \Pi \vdash \Lambda, B\end{matrix} \quad \begin{matrix}(\sigma_2, Y_2)\\ \Pi \vdash \Lambda, C\end{matrix}}{\Pi \vdash \Lambda, B \wedge C}\ \wedge{:}\,r}{\Gamma, \Pi^* \vdash \Delta^*, \Lambda, B \wedge C}\ cut(A)$$

Clearly A occurs in Π and ψ reduces to the following proof χ via cross-cut:

$$\dfrac{\dfrac{\begin{matrix}(\rho, X)\\ \Gamma \vdash \Delta\end{matrix} \quad \begin{matrix}(\sigma_1, Y_1)\\ \Pi \vdash \Lambda, B\end{matrix}}{\Gamma, \Pi^* \vdash \Delta^*, \Lambda, B}\ cut(A) \quad \dfrac{\begin{matrix}(\rho, X)\\ \Gamma \vdash \Delta\end{matrix} \quad \begin{matrix}(\sigma_2, Y_2)\\ \Pi \vdash \Lambda, C\end{matrix}}{\Gamma, \Pi^* \vdash \Delta^*, \Lambda, C}\ cut(A)}{\Gamma, \Pi^* \vdash \Delta^*, \Lambda, B \wedge C}\ \wedge{:}\,r$$

Now we have to distinguish two cases:

case a: $B \wedge C$ is ancestor of (another) cut in φ.

Then

$$\begin{aligned}\Theta(\varphi)/\nu &= X \oplus (Y_1 \oplus Y_2),\\ \Theta(\varphi')/\nu &= (X \oplus Y_1) \oplus (X \oplus Y_2).\end{aligned}$$

Clearly

$$X \oplus (Y_1 \oplus Y_2) \sim (X \oplus Y_1) \oplus (X \oplus Y_2)$$

and therefore $\Theta(\varphi') \sim \Theta(\varphi)$, thus $\Theta(\varphi) \rhd \Theta(\varphi')$.

case b: $B \wedge C$ is not an ancestor of a cut in φ.

Then

$$\begin{aligned}\Theta(\varphi)/\nu &= X \oplus (Y_1 \otimes Y_2),\\ \Theta(\varphi')/\nu &= (X \oplus Y_1) \otimes (X \oplus Y_2).\end{aligned}$$

But by using elementary properties of $\cup$ and $\times$ we obtain

$$X \oplus (Y_1 \otimes Y_2) \ \sqsubseteq \ (X \oplus Y_1) \otimes (X \oplus Y_2)$$

That means $\Theta(\varphi)/\nu \sqsubseteq \Theta(\varphi')/\nu$ and by application of Lemma 6.3.3 we again get $\Theta(\varphi) \sqsubseteq \Theta(\varphi')$, thus also $\Theta(\varphi) \rhd \Theta(\varphi')$.

(IIb.3.2) The last inference in σ is $\vee\colon l$. Then ψ is of the form

$$\cfrac{\begin{matrix}(\rho, X)\\ \Gamma \vdash \Delta\end{matrix} \quad \cfrac{\begin{matrix}(\sigma_1, Y_1)\\ B, \Pi \vdash \Lambda\end{matrix} \quad \begin{matrix}(\sigma_2, Y_2)\\ C, \Pi \vdash \Lambda\end{matrix}}{B \vee C, \Pi \vdash \Lambda}\ \vee\colon l}{(B \vee C)^*, \Gamma, \Pi^* \vdash \Delta^*, \Lambda}\ cut(A)$$

Note that A is in Π; for otherwise $A = B \vee C$ and $\text{rank}_r(\psi) = 1$, contradicting the assumption.

We first define the following derivation τ:

$$\cfrac{\cfrac{\cfrac{\begin{matrix}(\rho, X)\\ \Gamma \vdash \Delta\end{matrix} \quad \begin{matrix}(\sigma_1, Y_1)\\ B, \Pi \vdash \Lambda\end{matrix}}{B^*, \Gamma, \Pi^* \vdash \Delta^*, \Lambda}\ cut(A)}{B, \Gamma, \Pi^* \vdash \Delta^*, \Lambda}\ w^* \quad \cfrac{\cfrac{\begin{matrix}(\rho, X)\\ \Gamma \vdash \Delta\end{matrix} \quad \begin{matrix}(\sigma_2, Y_2)\\ C, \Pi \vdash \Lambda\end{matrix}}{C^*, \Gamma, \Pi^* \vdash \Delta^*, \Lambda}\ cut(A)}{C, \Gamma, \Pi^* \vdash \Delta^*, \Lambda}\ w^*}{(B \vee C), \Gamma, \Pi^* \vdash \Delta^*, \Lambda}\ \vee\colon l$$

As in IIb.3.1 we have to distinguish the case where $B \vee C$ is an ancestor of another cut in φ or not. So if we replace ψ by τ in φ we either get

$$\begin{aligned}\Theta(\varphi)/\nu &= X \oplus (Y_1 \oplus Y_2),\\ \Theta(\varphi[\tau]_\nu)/\nu &= (X \oplus Y_1) \oplus (X \oplus Y_2).\end{aligned}$$

or

$$\begin{aligned}\Theta(\varphi)/\nu &= X \oplus (Y_1 \otimes Y_2),\\ \Theta(\varphi[\tau]_\nu)/\nu &= (X \oplus Y_1) \otimes (X \oplus Y_2).\end{aligned}$$

Thus the situation is analogous to (IIb.3.1) and we get $\Theta(\varphi) \rhd \Theta(\varphi[\tau]_\nu)$.

If $A \neq B \vee C$ then $\chi = \tau$ and therefore $\Theta(\varphi) \rhd \Theta(\varphi')$.

If $A = B \vee C$ we define $\chi =$

$$\cfrac{\cfrac{\begin{matrix}(\rho, X)\\ \Gamma \vdash \Delta\end{matrix} \quad \begin{matrix}(\tau, (X \oplus Y_1) \oplus (X \oplus Y_2))\\ (B \vee C), \Gamma, \Pi^* \vdash \Delta^*, \Lambda\end{matrix}}{\Gamma, \Gamma^*, \Pi^* \vdash \Delta^*, \Delta^*, \Lambda}\ cut(A)}{\Gamma, \Pi^* \vdash \Delta^*, \Lambda}\ s^*$$

In this case

$$\begin{aligned}\Theta(\varphi)/\nu &= X \oplus (Y_1 \oplus Y_2),\\ \Theta(\varphi')/\nu &= X \oplus ((X \oplus Y_1) \oplus (X \oplus Y_2)).\end{aligned}$$

and we obtain

$$\Theta(\varphi)/\nu \sim \Theta(\varphi')/\nu$$

Once more Lemma 6.3.3 gives us $\Theta(\varphi) \rhd \Theta(\varphi')$.

(IIb.3.3) The last inference in σ is $\rightarrow: l$. Then ψ is of the form

$$\dfrac{\begin{array}{c}(\rho, X)\\ \Gamma \vdash \Delta\end{array} \quad \dfrac{\begin{array}{c}(\sigma_1, Y_1)\\ \Pi_1 \vdash \Lambda_1, B\end{array} \quad \begin{array}{c}(\sigma_2, Y_2)\\ C, \Pi_2 \vdash \Lambda_2\end{array}}{B \rightarrow C, \Pi_1, \Pi_2 \vdash \Lambda_1, \Lambda_2}\ \rightarrow: l}{\Gamma, (B \rightarrow C)^*, \Pi_1^*, \Pi_2^* \vdash \Delta^*, \Lambda_1, \Lambda_2}\ cut(A)$$

We have to consider various cases:

- A occurs in Π_1 and in Π_2.
  Like in IIb.3.2 we consider a proof τ:

$$\dfrac{\dfrac{\begin{array}{c}(\rho, X)\\ \Gamma \vdash \Delta\end{array} \quad \begin{array}{c}(\sigma_1, Y_1)\\ \Pi_1 \vdash \Lambda_1, B\end{array}}{\Gamma, \Pi_1^* \vdash \Delta^*, \Lambda_1, B}\ cut(A) \quad \dfrac{\dfrac{\begin{array}{c}(\rho, X)\\ \Gamma \vdash \Delta\end{array} \quad \begin{array}{c}(\sigma_2, Y_2)\\ C, \Pi_2 \vdash \Lambda_2\end{array}}{C^*, \Gamma, \Pi_2^* \vdash \Delta^*, \Lambda_2}\ cut(A)}{C, \Gamma, \Pi_2^* \vdash \Delta^*, \Lambda_2}\ \xi}{B \rightarrow C, \Gamma, \Pi_1^*, \Gamma, \Pi_2^* \vdash \Delta^*, \Lambda_1, \Delta^*, \Lambda_2}\ \rightarrow: l$$

If $(B \rightarrow C)^* = B \rightarrow C$ then ψ is transformed to τ + some unary structural rule applications. As in case IIb.3.2 we have to distinguish whether $B \rightarrow C$ is an ancestor of a cut or not. So by replacing ψ by τ in φ we either get

$$\begin{array}{rcl}\Theta(\varphi)/\nu & = & X \oplus (Y_1 \oplus Y_2),\\ \Theta(\varphi[\tau]_\nu)/\nu & = & (X \oplus Y_1) \oplus (X \oplus Y_2).\end{array}$$

or

$$\begin{array}{rcl}\Theta(\varphi)/\nu & = & X \oplus (Y_1 \otimes Y_2),\\ \Theta(\varphi[\tau]_\nu)/\nu & = & (X \oplus Y_1) \otimes (X \oplus Y_2).\end{array}$$

Obviously the terms are the same as in case IIb.3.2.

If $(B \rightarrow C)^*$ is empty then ψ is transformed to $\chi =$

$$\dfrac{\dfrac{\begin{array}{c}(\rho, X)\\ \Gamma \vdash \Delta\end{array} \quad (\tau, (X \oplus Y_1) \oplus (X \oplus Y_2))}{\Gamma, \Gamma, \Pi_1^*, \Gamma, \Pi_2^* \vdash \Delta, \Delta^*, \Lambda_1, \Delta^*, \Lambda_2}\ cut(A)}{\Gamma, \Pi_1^*, \Pi_2^* \vdash \Delta^*, \Lambda_1, \Lambda_2}\ s^*$$

Note that, for $(B \rightarrow C)^*$ empty, B, C and $B \rightarrow C$ are ancestors of a cut (namely the last cut in ψ), and so the term for τ is $(X \oplus Y_1) \oplus (X \oplus Y_2)$. Therefore we obtain

$$\begin{array}{rcl} \Theta(\varphi)/\nu & = & X \oplus (Y_1 \oplus Y_2), \\ \Theta(\varphi')/\nu & = & X \oplus (X \oplus Y_1) \oplus (X \oplus Y_2). \end{array}$$

Again, the terms are the same as case II.b.2.

- A occurs in Π_2, but not in Π_1. As in the previous case we obtain $\tau =$

$$\dfrac{\begin{matrix} (\sigma_1, Y_1) \\ \Pi_1 \vdash \Lambda_1, B \end{matrix} \quad \dfrac{\dfrac{\begin{matrix} (\rho, X) \\ \Gamma \vdash \Delta \end{matrix} \quad \begin{matrix} (\sigma_2, Y_2) \\ C, \Pi_2 \vdash \Lambda_2 \end{matrix}}{C^*, \Gamma, \Pi_2^* \vdash \Delta^*, \Lambda_2}\, cut(A)}{C, \Gamma, \Pi_2^* \vdash \Delta^*, \Lambda_2}\, \xi}{B \rightarrow C, \Pi_1, \Gamma, \Pi_2^* \vdash \Lambda_1, \Delta^*, \Lambda_2}\, \rightarrow: l$$

Again we distinguish the cases $B \rightarrow C = A$ and $B \rightarrow C \neq A$ and define the transformation χ exactly like above.

For $B \rightarrow C \neq A$ we obtain

$$\begin{array}{rcl} \Theta(\varphi)/\nu & = & X \oplus (Y_1 \oplus Y_2), \\ \Theta(\varphi[\tau]_\nu)/\nu & = & Y_1 \oplus (X \oplus Y_2). \end{array}$$

or

$$\begin{array}{rcl} \Theta(\varphi)/\nu & = & X \oplus (Y_1 \otimes Y_2), \\ \Theta(\varphi[\tau]_\nu)/\nu & = & Y_1 \otimes (X \oplus Y_2). \end{array}$$

In the first case we have

$$X \oplus (Y_1 \oplus Y_2) \sim Y_1 \oplus (X \oplus Y_2),$$

in the second

$$X \oplus (Y_1 \otimes Y_2) \sqsubseteq Y_1 \otimes (X \oplus Y_2).$$

In both cases we obtain

$$\Theta(\varphi)/\nu \sim \Theta(\varphi')/\nu, \quad \text{and}$$

by Lemma 6.3.3

$$\Theta(\varphi) \rhd \Theta(\varphi').$$

If $A = B \to C$ the proof χ is defined like in II.b.2 and we obtain the terms

$$\begin{array}{rcl}\Theta(\varphi)/\nu & = & X \oplus (Y_1 \oplus Y_2),\\ \Theta(\varphi')/\nu & = & X \oplus (Y_1 \oplus (X \oplus Y_2)).\end{array}$$

Clearly $\Theta(\varphi)/\nu \sim \Theta(\varphi')/\nu$, therefore $\Theta(\varphi)/\nu \rhd \Theta(\varphi')/\nu$ and, by Lemma 6.3.3,

$$\Theta(\varphi) \rhd \Theta(\varphi').$$

- A occurs in Π_1, but not in Π_2: analogous to the last case.

(IIb.3.4) The last inference in σ is a cut. Then ψ is of the form

$$\dfrac{\begin{array}{c}(\rho, X)\\ \Gamma \vdash \Delta\end{array} \quad \dfrac{\begin{array}{cc}(\sigma_1, Y_1) & (\sigma_2, Y_2)\\ \Pi_1 \vdash \Lambda_1 & \Pi_2 \vdash \Lambda_2\end{array}}{\Pi_1, \Pi_2^+ \vdash \Lambda_1^+, \Lambda_2}\, cut(B)}{\Gamma, \Pi_1^*, \Pi_2^{+*} \vdash \Delta^*, \Lambda_1^+, \Lambda_2}\, cut(A)$$

If A occurs in Π_1 and in Π_2 then $\chi =$

$$\dfrac{\dfrac{\dfrac{\begin{array}{cc}(\rho, X) & (\sigma_1, Y_1)\\ \Gamma \vdash \Delta & \Pi_1 \vdash \Lambda_1\end{array}}{\Gamma, \Pi_1^* \vdash \Delta^*, \Lambda_1}\, cut(A) \quad \dfrac{\begin{array}{cc}(\rho, X) & (\sigma_2, Y_2)\\ \Gamma \vdash \Delta & \Pi_2 \vdash \Lambda_2\end{array}}{\Gamma, \Pi_2^* \vdash \Delta^*, \Lambda_2}\, cut(A)}{\Gamma, \Gamma^+, \Pi_1^*, \Pi_2^{+*} \vdash \Delta^{*+}, \Delta^*, \Lambda_1^+, \Lambda_2}\, cut(B)}{\Gamma, \Pi_1^*, \Pi_2^{+*} \vdash \Delta^*, \Lambda_1^+, \Lambda_2}\, s^*$$

In this case we have

$$\begin{array}{rcl}\Theta(\varphi)/\nu & = & X \oplus (Y_1 \oplus Y_2),\\ \Theta(\varphi')/\nu & = & (X \oplus Y_1) \oplus (X \oplus Y_2).\end{array}$$

Clearly $X \oplus (Y_1 \oplus Y_2) \sim (X \oplus Y_1) \oplus (X \oplus Y_2)$ and so

$$X \oplus (Y_1 \oplus Y_2) \rhd (X \oplus Y_1) \oplus (X \oplus Y_2).$$

By Lemma 6.3.3 we get $\Theta(\varphi) \rhd \Theta(\varphi')$.

If A occurs in Π_1 and not in Π_2 then $\chi =$

$$\dfrac{\dfrac{\begin{array}{cc}(\rho, X) & (\sigma_1, Y_1)\\ \Gamma \vdash \Delta & \Pi_1 \vdash \Lambda_1\end{array}}{\Gamma, \Pi_1^* \vdash \Delta^*, \Lambda_1}\, cut(A) \quad \begin{array}{c}(\sigma_2, Y_2)\\ \Pi_2 \vdash \Lambda_2\end{array}}{\Gamma, \Pi_1^*, \Pi_2^+ \vdash \Delta^*, \Lambda_1^+, \Lambda_2}\, cut(B)$$

Here we have

$$\begin{array}{rcl}\Theta(\varphi)/\nu & = & X \oplus (Y_1 \oplus Y_2),\\ \Theta(\varphi')/\nu & = & (X \oplus Y_1) \oplus Y_2.\end{array}$$

and $\Theta(\varphi) \rhd \Theta(\varphi')$ is trivial.

The case where A is in Π_2, but not in Π_1 is completely symmetric. □

Theorem 6.8.1 *Let φ be an* **LK**-*derivation and ψ be an ACNF of φ under a cut reduction relation $>_{\mathcal{R}}$ based on $\mathcal{R}$. Then $\Theta(\varphi) \leq_{ss} \Theta(\psi)$.*

Proof: $\varphi >^*_{\mathcal{R}} \psi$. By Lemma 6.8.1 we get $\Theta(\varphi) \rhd^* \Theta(\psi)$. By Proposition 6.6.2 we obtain $\Theta(\varphi) \leq_{ss} \Theta(\psi)$. □

Theorem 6.8.2 *Let φ be an* **LK**-*derivation and ψ be an ACNF of φ under a cut reduction relation $>_{\mathcal{R}}$ based on $\mathcal{R}$. Then there exists a resolution refutation γ of* $\mathrm{CL}(\varphi)$ *s.t.* $\gamma \leq_{ss} \mathrm{RES}(\psi)$.

Proof: By Theorem 6.8.1 $\Theta(\varphi) \leq_{ss} \Theta(\psi)$ and therefore $\mathrm{CL}(\varphi) \leq_{ss} \mathrm{CL}(\psi)$. By Definition 6.7.2, $\mathrm{RES}(\psi)$ is a resolution refutation of $\mathrm{CL}(\psi)$; by Proposition 6.6.3 there exists a resolution refutation γ of $\mathrm{CL}(\varphi)$ s.t. $\gamma \leq_{ss} \mathrm{RES}(\psi)$. □

Corollary 6.8.1 *Let φ be an* **LK**-*derivation and ψ be an ACNF of φ under a cut reduction relation $>_{\mathcal{R}}$ based on $\mathcal{R}$. Then there exists a resolution refutation γ of* $\mathrm{CL}(\varphi)$ *s.t.*

$$l(\gamma) \leq l(\mathrm{RES}(\psi)) \leq l(\psi) * 2^{2*l(\psi)}.$$

Proof: By Theorem 6.8.1 there exists a resolution refutation γ with $\gamma \leq_{ss} \mathrm{RES}(\psi)$. By definition of subsumption of proofs (see Definition 6.6.4) we have $l(\gamma) \leq l(\mathrm{RES}(\psi))$. Finally the result follows from Proposition 6.7.1. □

Corollary 6.8.2 *Let φ be an* **LK**-*derivation and ψ be an ACNF of φ under a cut reduction relation $>_{\mathcal{R}}$ based on $\mathcal{R}$. Then there exists an ACNF χ of φ under* CERES *s.t.*

$$\|\chi\|_l \leq l(\varphi) * l(\psi) * 2^{2*l(\psi)}.$$

Proof: If γ is a resolution refutation of CL(φ) then an ACNF χ of φ can be obtained by CERES. As the **LK**-derivations in the projections are not longer than φ itself we get

$$\|\chi\|_l \leq \|\varphi\|_l * l(\gamma) \leq l(\varphi) * l(\gamma).$$

Then the inequality follows from Corollary 6.8.1. □

Corollary 6.8.3 *Let φ be an* **LK**-*derivation and ψ be an ACNF of φ under Gentzen's or Tait's method. Then there exists an ACNF χ of φ under* CERES *s.t.*

$$\|\chi\|_l \leq l(\varphi) * l(\psi) * 2^{2*l(\psi)}.$$

Proof: Gentzen's and Tait's methods are reduction methods based on $\mathcal{R}$. □

The methods of this section allow to describe a large class of cut-elimination methods in a uniform way. Hereby the order of reduction steps (e.g. in the methods of Gentzen and Tait–Schütte) does not matter. This arguments make the CERES method an essential tool for proving negative results about cut-elimination, e.g. that a certain cut-free proof is not obtainable by a given one. Consider, for example, an ACNF ψ s.t. CL(ψ) is not subsumed by CL(φ), where φ is the original proof; then we can be sure that ψ cannot be obtained by any sequence of cut-reduction rules from $\mathcal{R}$.

Note that the replacement of the of Skolem functions by the original quantifiers in an ACNF may lead to an exponential increase in terms of the symbolic complexity of the original end-sequent and of the ACNF (in case of prenex end-sequents the increase of complexity is even linear, c.f. the algorithm in [51] which selects the maximal Skolem term and replaces it by an eigenvariable).

6.9 Beyond $\mathcal{R}$: Stronger Pruning Methods

At the first glimpse it might appear that all cut-reduction methods based on a set of rules yield characteristic terms which are subsumed by the characteristic term of the original proof. However, Theorems 6.8.1 and 6.8.2 are not valid in general. In particular they do not hold when we eliminate atomic cuts. But even if we allow atomic cuts there exists a set of cut-reduction rules $\mathcal{R}'$ for which the theorems above are not valid.

Definition 6.9.1 **($\mathcal{R}'$)** Let $\mathcal{R}$ be the set of cut-reduction rules defined in Definition 5.1.6. With the exception of the rule in case 3.121.232 (right-rank > 1, case $\vee : l$) the rules in $\mathcal{R}'$ are the same as those in $\mathcal{R}$. We only modify the case where the cut formula A is identical to B (which is one of the auxiliary formulas of the $\vee : l$-inference). In this case the derivation ψ in case 3.121.232 is of the form:

$$\dfrac{\begin{matrix}(\rho)\\ \Gamma\vdash\Delta\end{matrix} \qquad \dfrac{\begin{matrix}(\sigma_1)\\ B,\Pi\vdash\Lambda\end{matrix}\quad \begin{matrix}(\sigma_2)\\ C,\Pi\vdash\Lambda\end{matrix}}{B\vee C,\Pi\vdash\Lambda}\ \vee{:}\,l}{\Gamma, B\vee C,\Pi^*\vdash\Delta^*,\Lambda}\ cut(B)$$

We define $\psi >_{\mathcal{R}'} \chi$ for $\chi =$

$$\dfrac{\dfrac{\begin{matrix}(\rho)\\ \Gamma\vdash\Delta\end{matrix}\quad \begin{matrix}(\sigma_1)\\ B,\Pi\vdash\Lambda\end{matrix}}{\Gamma,\Pi^*\vdash\Delta^*,\Lambda}\ cut(B)}{\Gamma, B\vee C,\Pi^*\vdash\Delta^*,\Lambda}\ s^*$$

◇

Theorem 6.9.1 *There exists an* **LK**-*derivation* φ *s.t. for all ACNFs* ψ *under* $\mathcal{R}'$*:*

(1) $\Theta(\varphi) \not\leq_{ss} \Theta(\psi)$,

(2) $\gamma \not\leq_{ss} \mathrm{RES}(\psi)$ *for all resolution refutations* γ *of* $\mathrm{CL}(\varphi)$.

Proof: In the **LK**-derivations below we mark all ancestors of cuts by $*$. Let P, Q, R be arbitrary atomic formulas and φ be the derivation

$$\dfrac{\dfrac{\vdash P^* \quad \vdash P^*}{\vdash (P\wedge P)^*}\ \wedge{:}\,r \qquad \dfrac{\dfrac{\dfrac{P,P^*\vdash P}{P,(P\wedge P)^*\vdash P}\ \wedge{:}\,l+p^*}{P\wedge P,(P\wedge P)^*\vdash P}\ \wedge{:}\,l \quad \dfrac{\dfrac{\dfrac{P^*\vdash Q^* \quad Q^*\vdash P}{P^*\vdash P}\ cut(Q)}{R,P^*\vdash P}\ w{:}\,l}{R,(P\wedge P)^*\vdash P}\ \wedge{:}\,l+p^*}{(P\wedge P)\vee R,(P\wedge P)^*\vdash P}\ \vee{:}\,l}{(P\wedge P)\vee R\vdash P}\ cut(P\wedge P)$$

Then

$$\begin{aligned}\Theta(\varphi) &= (\{\vdash P\}\oplus\{\vdash P\})\oplus((\{P\vdash\}\otimes(\{P\vdash Q\}\oplus\{Q\vdash\}),\\ \mathrm{CL}(\varphi) &= \{\vdash P;\quad P,P\vdash Q;\quad P,Q\vdash\}.\end{aligned}$$

There exists only one non-atomic cut in φ. By definition of $\mathcal{R}'$ we get $\varphi >_{\mathcal{R}'} \chi$ (and this is the only one-step reduction) for $\chi =$

$$\dfrac{\dfrac{\vdash P^* \quad \vdash P^*}{\vdash (P \wedge P)^*}\wedge : r \qquad \dfrac{\dfrac{P^*, P^* \vdash P}{P^*, (P \wedge P)^* \vdash P}\wedge : l + p^*}{(P \wedge P)^*, (P \wedge P)^* \vdash P}\wedge : l}{\dfrac{\vdash P}{(P \wedge P) \vee R \vdash P}w : l}cut(P \wedge P)$$

It is easy to see that the only ACNF of χ (under $\mathcal{R}$ and $\mathcal{R}'$) is ψ for $\psi =$

$$\dfrac{\dfrac{\vdash P^* \quad P^*, P^* \vdash P}{\vdash P}cut(P)}{(P \wedge P) \vee R \vdash P}w : l$$

But

$$\begin{array}{rcl} \Theta(\psi) & = & \{\vdash P\} \oplus \{P, P \vdash\}, \\ \mathrm{CL}(\psi) & = & \{\vdash P;\ P, P \vdash\}. \end{array}$$

There exists no clause $C \in \mathrm{CL}(\varphi)$ with $C \leq_{ss} P, P \vdash$, therefore $\mathrm{CL}(\varphi) \not\leq_{ss} \mathrm{CL}(\psi)$ and $\Theta(\varphi) \not\leq_{ss} \Theta(\psi)$. This proves (1).
By definition of RES we obtain $\mathrm{RES}(\psi) =$

$$\dfrac{\vdash P \quad P, P \vdash}{\vdash}cut.$$

As $\mathrm{CL}(\varphi) \not\leq_{ss} \{P, P \vdash\}$ there exists no refutation γ of $\mathrm{CL}(\varphi)$ with $\gamma \leq_{ss} \mathrm{RES}(\psi)$. This proves (2). □

Remark: Our choice of $\mathcal{R}'$ was in fact a minimal one, aimed to falsify Theorem 6.8.1. It is obvious that the principle can be extended to the case where $A = C$, and to the symmetric situation of left-rank > 1 and $\wedge : r$. Indeed there are several simple ways for further improving cut-elimination methods based on $\mathcal{R}$. All these stronger methods of pruning the proof trees during cut-reduction do not fulfil the properties expressed in Theorem 6.8.1 and in Theorem 6.8.2. ◇

6.10 Speed-Up Results

In this section we prove that CERES NE-improves both Gentzen and Tait-Schütte reductions. On the other hand *no* reductive cut-elimination method based on $\mathcal{R}$ NE-improves CERES. In this sense CERES is uniformly better that $>_G$ and $>_T$. As CERES and the reductive methods are structurally different we have to adapt our definition of NE-improvement to the CERES method.

Definition 6.10.1 Let $>_x$ be a proof reduction relation based on $\mathcal{R}$. We say that CERES NE-improves $>_x$ if there exists a sequence of proofs $(\varphi_n)_{n\in\mathbb{N}}$ s.t.

- there exists a sequence of resolution refutations $(\gamma_n)_{n\in\mathbb{N}}$ of $\mathrm{CL}(\varphi_n)$ s.t. $(l(\gamma_n))_{n\in\mathbb{N}}$ is elementary in $(\|\varphi_n\|)_{n\in\mathbb{N}}$.
- For all $k \in \mathbb{N}$ there exists a number m s.t. for all $n \geq m$ and for every cut-elimination sequence θ on φ_n we have $\|\theta\| > e(k, \|\varphi_n\|)$.

Similarly we define that $>_x$ NE-improves CERES if there exists a sequence of proofs $(\varphi_n)_{n\in\mathbb{N}}$ s.t.

- there exists a sequence of cut-elimination sequences $(\theta_n)_{n\in\mathbb{N}}$ s.t. $(\|\theta_n\|)_{n\in\mathbb{N}}$ is elementary in $(\|\varphi_n\|)_{n\in\mathbb{N}}$.
- For all $k \in \mathbb{N}$ there exists a number m s.t. for all $n \geq m$ and for all resolution refutations γ of $\mathrm{CL}(\varphi_n)$ we get $\|\gamma\| > e(k, \|\varphi_n\|)$.

◇

Remark: The definition above looks asymmetric as for CERES we use the measure l of length and for the reductive methods the symbolic norm $\| \ \|$. But this definition is justified by Proposition 6.5.3 which proves that it does not matter whether we use the measure $l(\gamma_n)$ or $\|\varphi_n^*\|$ for CERES normal forms φ_n^* based on γ_n for measuring the asymptotic complexity. ◇

Theorem 6.10.1 CERES *NE-improves* $>_G$.

Proof: Let $\Psi\colon (\psi_n)_{n\in\mathbb{N}}$ be the sequence of proofs defined in the proof of Theorem 5.4.1. We have shown that $>_T$ NE-improves $>_G$ on Ψ. We prove now that CERES is fast (i.e. elementary) on Ψ – and thus NE-improves $>_G$. By Proposition 6.5.3 it suffices to construct an elementary function f and a sequence ρ_n of resolution refutations of $\mathrm{CL}(\psi_n)$ s.t.

$$l(\rho_n) \leq f(\|\psi_n\|).$$

Recall the sequence $\psi_n =$

$$
\cfrac{\cfrac{\begin{matrix}(\pi_{g(n)}) \\ A_{g(n)} \vdash A_{g(n)}\end{matrix} \quad \begin{matrix}(\gamma_n) \\ \Delta_n \vdash D_n\end{matrix}}{A_{g(n)}, \Delta_n \vdash A_{g(n)} \wedge D_n}\ \wedge : r \qquad \cfrac{\cfrac{\cfrac{\begin{matrix}(\pi_{g(n)}) \\ A_{g(n)} \vdash A_{g(n)}\end{matrix} \quad A \vdash A}{A_{g(n)} \rightarrow A, A_{g(n)} \vdash A}\ \rightarrow : l}{A_{g(n)}, A_{g(n)} \rightarrow A \vdash A}\ p : l}{A_{g(n)} \wedge D_n, A_{g(n)} \rightarrow A \vdash A}\ \wedge : l}{A_{g(n)}, \Delta_n, A_{g(n)} \rightarrow A \vdash A}\ cut
$$

where γ_n is Statman's sequence defined in Chapter 4 and the π_m are the proofs from Definition 5.4.3.
By definition of the formula sequence $A_{g(n)}$ we get

$$\mathrm{CL}(\psi_n) = \{\vdash A;\ A \vdash\} \cup \mathrm{CL}(\gamma_n).$$

Trivially every $\mathrm{CL}(\psi_n)$ has the resolution refutation $\rho =$

$$\frac{\vdash A \quad A \vdash}{\vdash}$$

which is of constant length and, by defining $\rho_n = \rho$ for all n we get $l(\rho_n) = 3$. So we may define f as $f(n) = 3$ for all n, which is (of course) elementary.

□

Theorem 6.10.2 CERES *NE-improves* $>_T$.

Proof: In Theorem 5.4.2 we defined a proof sequence ϕ_n s.t. $>_G$ NE-improves $>_T$ on ϕ_n. Recall the definition of the sequence ϕ_n: Consider again Statman's sequence γ_n. Locate the uppermost proof δ_1 in γ_n; note that δ_1 is identical to ψ_{n+1}. In γ_n we first replace the proof δ_1 (or ψ_{n+1}) of the sequent $\Gamma_{n+1} \vdash H_{n+1}(\mathbf{T})$ by the proof $\hat{\delta}_1$:

$$
\cfrac{\cfrac{\begin{matrix}(\omega) \\ P \wedge \neg P \vdash\end{matrix}}{P \wedge \neg P \vdash \neg Q}\ w : r \qquad \cfrac{\begin{matrix}(\psi_{n+1}) \\ \mathrm{Ax}_T \vdash H_{n+1}(\mathbf{T})\end{matrix}}{\neg Q, \mathrm{Ax}_T \vdash H_{n+1}(\mathbf{T})}\ w : l}{P \wedge \neg P, \mathrm{Ax}_T \vdash H_{n+1}(\mathbf{T})}\ cut
$$

where ω is a proof of $P \wedge \neg P \vdash$ of constant length. Furthermore we use the same inductive definition in defining $\hat{\delta}_k$ as that of δ_k in Chapter 4. Finally we obtain a proof ϕ_n in place of γ_n. Note that ϕ_n differs from γ_n only by an additional cut on the formula $\neg Q$ and the formula $P \wedge \neg P$ in the antecedents of the sequents. Note that the cut-formula $\neg Q$ is introduced by weakening.

We know that the characteristic terms $\Theta(\gamma_n)$ of the Statman sequence contain no product $\otimes$; in fact there exists no binary logical operator in the end-sequents of γ_n. Let ν be the node corresponding to $\mathrm{Ax}_T \vdash H_{n+1}(\mathbf{T})$ and

$$\Theta(\gamma_n) = \Theta(\gamma_n)[t]_\nu.$$

Then, by construction,

$$\Theta(\phi_n) = \Theta(\gamma_n)[\{\vdash\} \oplus t]_\nu.$$

Indeed, the cut-formula has no ancestors in the axioms and so contributes only $\vdash$ to the clause term. Therefore, as $\Theta(\gamma_n)$ (and thus also $\Theta(\phi_n)$) contains no products we obtain

$$\mathrm{CL}(\phi_n) = \mathrm{CL}(\gamma_n) \cup \{\vdash\}.$$

Obviously, for all n, $\rho_n{:}\vdash$ is a resolution refutation of $\mathrm{CL}(\phi_n)$ and $l(\rho_n) = 1$. Hence

$$\mathrm{CL}(\phi_n) \leq f(\|\phi_n\|) \text{ for } f(n) = 1 \text{ for all } n.$$

As $>_T$ is nonelementary on ϕ_n we see that CERES NE-improves $>_T$. □

Theorem 6.10.3 *No reductive method based on $\mathcal{R}$ NE-improves* CERES*; in particular $>_\mathcal{R}$ does not NE-improve* CERES.

Proof: Assume, for contradiction, that $>_x$ is a reduction relation based on $\mathcal{R}$ which NE-improves CERES. By Definition 6.10.1 there exists a sequence of proofs φ_n s.t. there exists a $k \in \mathbb{N}$ and a $>_x$-normal forms φ_n^* with

(a) $\|\varphi_n^*\| \leq e(k, \|\varphi_n\|)$ and

(b) for all k there exists an m s.t. for all $n \geq m$ and all resolution refutations γ of $\mathrm{CL}(\varphi_n)$ we have $l(\gamma) > e(k, \|\varphi_n\|)$.

By Corollary 6.8.1 we know that there exists a sequence ρ_n of resolution refutations ρ_n of $\mathrm{CL}(\varphi_n)$ s.t.

$$l(\rho_n) \leq g(l(\varphi_n^*)) \text{ for } g = \lambda n.n * 2^{2*n}.$$

But $l(\varphi_n^*) \leq \|\varphi_n^*\|$ and therefore, by (a),

$$l(\rho_n) \leq g(e(k, \|\varphi_n\|)).$$

But

$$n * 2^{2*n} \leq e(3, n) \text{ and } e(3, e(k, n)) \leq e(k+3, n).$$

Therefore

$$l(\rho_n) \leq e(k+3, \|\varphi_n\|) \text{ for all } n,$$

which contradicts (b). □

Chapter 7

Extensions of CERES

In Chapter 6 the CERES method was defined as a cut-elimination method for **LK**-proofs. But the method is potentially much more general and can be extended to a wide range of first-order cacluli. First of all CERES is a semantic method, in the sense that it works for all sound sequent calculi with a definable ancestor relation and a semantically complete clausal calculus. In this chapter we first show that a CERES method can be defined for virtually any sound sequent calculus. Second, we define extensions of **LK** by equality and definitions rules which are useful for formalizing mathematical theorems and show how to adapt CERES to these extensions of **LK**. The extension **LKDe** defined in Section 7.3 will then be used for the analysis of mathematical proofs in Section 8.5.

7.1 General Extensions of Calculi

We defined the method CERES as a cut-elimination method for a specific version of the calculus **LK** (just for the original version of Gentzen [38]). This calculus is a mixture of additive rules (the contexts are contracted) like

$$\frac{A, \Gamma \vdash \Delta \quad B, \Gamma \vdash \Delta}{A \lor B, \Gamma \vdash \Delta} \lor : l$$

and of multiplicative rules (the contexts are merged) like

$$\frac{\Gamma \vdash \Delta, A \quad B, \Pi \vdash \Lambda}{A \to B, \Gamma, \Pi \vdash \Delta, \Lambda} \to : l$$

As Gentzen defined two calculi **LK** (for classical logic) and **LJ** (for intuitionistic logic) simultaneously, this mixture is quite economic and facilitates the

M. Baaz, A. Leitsch, *Methods of Cut-Elimination*, Trends in Logic 34, DOI 10.1007/978-94-007-0320-9_7, © Springer Science+Business Media B.V. 2011

presentation. As most of this book is on classical logic only, it does not matter whether we define **LK** in a purely additive, purely multiplicative or mixed version. The question remains how much the CERES-method changes when we change the version of **LK**. Obviously unary structural rules (like contraction and weakening) have no influence on the characteristic clause term, which is a immediate consequence of Definition 6.4.1. Also the projections are defined exactly in the same way. We see that our definition of CERES is the same for all these structural variants of **LK**. Note that this is not the case for reductive methods: when we change the structural version of **LK** we need new cut reduction rules and the whole proof of cut-elimination has to be redone.

CERES is not only robust under changes of structural rules, its definition also hardly changes when we consider arbitrary sound logical rules, provided we can identify auxiliary and main formulas and classify whether a inference goes into the end sequent or not. Consider for example the rule

$$\frac{\Gamma \vdash \Delta_1, A_1, \ldots, \Delta_n, A_n, \Delta_{n+1} \quad \Pi_1, B_1, \ldots, \Pi_n, B_m, \Pi_{m+1} \vdash \Lambda}{\Gamma, \Pi_1, \ldots, \Pi_{m+1} \vdash \Delta_1, \ldots, \Delta_{n+1}, \Lambda}\ \mathit{pseudocut}$$

for $n, m \geq 1$ and for formulas A_i, B_j s.t.

$$(\star)\ (A_1 \vee \cdots \vee A_n) \to (B_1 \wedge \cdots \wedge B_m) \text{ is valid.}$$

The rule above becomes ordinary cut if there exists a formula A s.t. $A = A_i = B_j$ for all i, j, in which case the formula $(\star)$ is logically equivalent to $A \to A$. Obviously the pseudo-cut rule is sound, but there would be no way to eliminate these cuts via reductive methods as the syntactic forms of the formulas A_i and B_j can be strongly different. On the other hand, CERES handles pseudo-cut exactly as cut: as the rule is sound (i.e. $(\star)$ holds) the characteristic clause (defined exactly in the same way) set will be unsatisfiable and thus refutable by resolution; also the projections are defined exactly in the same way.

It is also easy to generalize CERES to arbitrary sound n-ary logical rules for $n > 2$. Consider, e.g. the n-ary rule

$$\frac{\Gamma \vdash \Delta, A_1 \quad \Gamma \vdash \Delta, A_2 \cdots \Gamma \vdash \Delta, A_n}{\Gamma \vdash \Delta, A_1 \wedge (A_2 \wedge \cdots (A_{n-1} \wedge A_n) \cdots)}\ \wedge_n\colon r$$

Let the cut rule be pseudo-cut as defined above. For constructing a characteristic clause term we have to know whether such a inference $\wedge_n\colon r$ goes into a cut or not – which can be easily checked in the n-ary deduction tree.

This fact determines whether we apply union or merge, which can both be generalized to higher arities. Below we generalize the concepts of clause term and characteristic clause term to calculi with n-ary logical rules. The n-logical rules have n premises, where in each premise formulas are marked as auxiliary formulas; one formula in the consequent is marked as main formula. Furthermore we only require that the rule is propositionally sound (e.g. it need not respect the subformula property). To avoid problems with eigenvariables we do not change the quantifier rules. This way we also preserve the common version of Skolemization which is needed to ensure the soundness of the proof projections. Also the method of proof skolemization as defined in Proposition 6.2.1 can be carried over to these new calculi.

Definition 7.1.1 (clause term) The signature of clause terms consists of sets of clauses and the operators $\oplus^n$ and $\otimes^n$ for $n \geq 2$.

- (Finite) sets of clauses are clause terms.
- If $X_1, \ldots, X_n$ are clause terms then $\oplus^n(X_1, \ldots, X_n)$ is a clause term.
- If $X_1, \ldots, X_n$ are clause terms then $\otimes^n(X_1, \ldots, X_n)$ is a clause term.

$\diamond$

Like in Section 6.3 clause terms denote sets of clauses; the following definition gives the precise semantics.

Definition 7.1.2 We define a mapping $|\ |$ from clause terms to sets of clauses in the following way:

$$\begin{aligned} |\mathcal{S}| &= \mathcal{C} \text{ for sets of clauses } \mathcal{S}, \\ |\oplus^n(X_1, \ldots, X_n)| &= \bigcup_{i=1}^{n} |X_i|, \\ |\otimes^n(X_1, \ldots, X_n)| &= \odot(|X_1|, \ldots, |X_n|), \end{aligned}$$

where

$$\odot(\mathcal{S}_1, \ldots, \mathcal{S}_n) = \{S_1 \circ \cdots \circ S_n \mid S_1 \in \mathcal{S}_1, \ldots S_n \in \mathcal{S}_n\}.$$

We define clause terms to be equivalent if the corresponding sets of clauses are equal, i.e. $X \sim Y$ iff $|X| = |Y|$. $\diamond$

Definition 7.1.3 (characteristic term) Let $\mathcal{L}$ be a first-order sequent calculus, ϕ be a skolemized proof of S in $\mathcal{L}$ and let Ω be the set of all

occurrences of pseudo-cut formulas in ϕ. Like in Definition 6.4.1 we denote by $S(\nu, \Omega)$ the subsequent of the sequent at node ν consisting of the ancestors of Ω.
We define the *characteristic (clause) term* $\Theta(\phi)$ inductively:

Let ν be the occurrence of an initial sequent S' in ϕ. Then $\Theta(\phi)/\nu = \{S(\nu, \Omega)\}$.

Let us assume that the clause terms $\Theta(\phi)/\nu$ are already constructed for all nodes ν in ϕ with $\text{depth}(\nu) \leq k$. Now let ν be a node with $\text{depth}(\nu) = k+1$. We distinguish the following cases:

(a) ν is the consequent of μ, i.e. a unary rule applied to μ gives ν. Here we simply define
$$\Theta(\varphi)/\nu = \Theta(\varphi)/\mu.$$

(b) ν is the consequent of $\mu_1, \ldots, \mu_n$, for $n \geq 2$, i.e. an n-ary rule x applied to $\mu_1, \ldots, \mu_n$ gives ν.

(b1) The auxiliary formulas of x are ancestors of occurrences in Ω, i.e. the formulas occur in $S(\mu_i, \Omega)$ for all $i = 1, \ldots, n$. Then
$$\Theta(\phi)/\nu = \oplus^n(\Theta(\varphi)/\mu_1, \ldots, \Theta(\varphi)/\mu_n).$$

(b2) The auxiliary formulas of x are not ancestors of occurrences in Ω. In this case we define
$$\Theta(\phi)/\nu = \otimes^n(\Theta(\varphi)/\mu_1, \ldots, \Theta(\varphi)/\mu_n).$$

Note that, in an n-ary inference, either all auxiliary formulas are ancestors of Ω or none of them.
Finally the characteristic term $\Theta(\phi)$ of ϕ is defined as $\Theta(\phi)/\nu_0$ where ν_0 is the root node of ϕ. ◇

Definition 7.1.4 (characteristic clause set) Let ϕ be a proof in a first-order calculus $\mathcal{L}$ and $\Theta(\phi)$ be the characteristic term of ϕ. Then $\text{CL}(\phi)$, defined as $\text{CL}(\phi) = |\Theta(\phi)|$, is called the characteristic clause set of ϕ. ◇

As the clause logic of $\mathcal{L}$ is the same as for **LK** we can refute $\text{CL}(\phi)$ by resolution as usual. The projections to the clauses of the characteristic clause set are defined in the same way as for **LK**: we just drop all inferences going into the cut (in case of binary rules apply weakening) and we perform all rules going into the end-sequent. Plugging the projections into the leaves of the resolution refutation works exactly as for **LK**.

Example 7.1.1 We define a calculus **LK**$'$ from **LK** in the following way:

- we replace cut by pseudo-cut,
- we add the following rules:
  - $\wedge_3\colon r$ and
  - the de Morgan rule $dm\colon r$ below:

$$\frac{\Gamma \vdash \Delta, \neg(A \wedge B)}{\Gamma \vdash \Delta, \neg A \vee \neg B}\ dm\colon r$$

Let

$$\begin{array}{rcl} A & \equiv & (\forall x)((\neg Q(x) \wedge P(x)) \wedge Q(x)), \\ C_1 & \equiv & (\forall x)(\neg P(x) \vee \neg Q(x)) \wedge (\forall x)(P(x) \wedge R(x)), \\ C_2 & \equiv & \neg(\exists x)(P(x) \wedge Q(x)) \wedge (\exists x)(P(x) \wedge R(x)), \\ B & \equiv & B_1 \wedge (B_2 \wedge B_3). \end{array}$$

for $B_1 \equiv P(c) \to \neg Q(c),\ B_2 \equiv (\exists x)P(x),\ B_3 \equiv (\exists x)R(x))$.

We consider the following proof φ in **LK**$'$ (the cut-ancestors are marked by $\star$):

$$\cfrac{\cfrac{\begin{array}{c}(\varphi_{1,1})\\ A \vdash (\forall x)(\neg P(x) \vee \neg Q(x))^\star\end{array} \quad \begin{array}{c}(\varphi_{1,2})\\ A \vdash (\forall x)(P(x) \wedge R(x))^\star\end{array}}{A \vdash C_1^\star}\ \wedge\colon r \qquad \begin{array}{c}(\varphi_2)\\ C_2^\star \vdash B\end{array}}{A \vdash B}\ pseudocut$$

where $\varphi_{1,1} =$

$$\cfrac{\cfrac{\cfrac{\cfrac{\cfrac{\cfrac{\cfrac{\cfrac{Q(\alpha_1)^\star \vdash Q(\alpha_1)}{P(\alpha_1) \wedge Q(\alpha_1)^\star \vdash Q(\alpha_1)}\ \wedge\colon l_2}{P(\alpha_1) \wedge Q(\alpha_1)^\star, \neg Q(\alpha_1) \vdash}\ \neg\colon l}{\neg Q(\alpha_1) \vdash \neg(P(\alpha_1) \wedge Q(\alpha_1))^\star}\ \neg\colon r}{\neg Q(\alpha_1) \wedge P(\alpha_1) \vdash \neg(P(\alpha_1) \wedge Q(\alpha_1))^\star}\ \wedge\colon l_1}{(\neg Q(\alpha_1) \wedge P(\alpha_1)) \wedge R(\alpha_1) \vdash \neg(P(\alpha_1) \wedge Q(\alpha_1))^\star}\ \wedge\colon l_1}{A \vdash \neg(P(\alpha_1) \wedge Q(\alpha_1))^\star}\ \forall\colon l}{A \vdash \neg P(\alpha_1) \vee \neg Q(\alpha_1)^\star}\ dm\colon r}{A \vdash (\forall x)(\neg P(x) \vee \neg Q(x))^\star}\ \forall\colon r$$

and $\varphi_{1,2} =$

$$\dfrac{\dfrac{\dfrac{\dfrac{P(\alpha_2) \vdash P(\alpha_2)^\star}{\neg Q(\alpha_2) \wedge P(\alpha_2) \vdash P(\alpha_2)^\star}\ \wedge\colon l_2}{(\neg Q(\alpha_2) \wedge P(\alpha_2)) \wedge R(\alpha_2) \vdash P(\alpha_2)^\star}\ \wedge\colon l_1}{A \vdash P(\alpha_2)^\star}\ \forall\colon l \qquad \dfrac{\dfrac{R(\alpha_2) \vdash R(\alpha_2)^\star}{(\neg Q(\alpha_2) \wedge P(\alpha_2)) \wedge R(\alpha_2) \vdash R(\alpha_2)^\star}\ \wedge\colon l_2}{A \vdash R(\alpha_2)^\star}\ \forall\colon l}{\dfrac{A \vdash P(\alpha_2) \wedge R(\alpha_2)^\star}{A \vdash (\forall x)(P(x) \wedge R(x))^\star}\ \forall\colon r}\ \wedge\colon r$$

$\varphi_2 =$

$$\dfrac{\begin{matrix}(\varphi_{2,1}) \\ C_2^\star \vdash B_1\end{matrix} \quad \begin{matrix}(\varphi_{2,2}) \\ C_2^\star \vdash B_2\end{matrix} \quad \begin{matrix}(\varphi_{2,3}) \\ C_2^\star \vdash B_3\end{matrix}}{C_2^\star \vdash B_1 \wedge (B_2 \wedge B_3)}\ \wedge_3\colon r$$

$\varphi_{2,1} =$

$$\dfrac{\dfrac{\dfrac{\dfrac{\dfrac{\dfrac{\dfrac{P(c) \vdash P(c)^\star}{Q(c), P(c) \vdash P(c)^\star}\ w\colon l \qquad \dfrac{Q(c) \vdash Q(c)^\star}{Q(c), P(c) \vdash Q(c)^\star}\ s^*}{Q(c), P(c) \vdash P(c) \wedge Q(c)^\star}\ \wedge\colon r}{P(c) \vdash \neg Q(c), P(c) \wedge Q(c)^\star}\ \neg\colon r}{\vdash P(c) \to \neg Q(c), P(c) \wedge Q(c)^\star}\ \to\colon r + s^*}{\vdash P(c) \to \neg Q(c), (\exists x)(P(x) \wedge Q(x))^\star}\ \exists\colon r}{\neg(\exists x)(P(x) \wedge Q(x))^\star \vdash P(c) \to \neg Q(c)}\ \neg\colon l}{C_2^\star \vdash P(c) \to \neg Q(c)}\ \wedge\colon l_1$$

$\varphi_{2,2} =$

$$\dfrac{\dfrac{\dfrac{\dfrac{P(\alpha_3)^\star \vdash P(\alpha_3)}{P(\alpha_3) \wedge R(\alpha_3)^\star \vdash P(\alpha_3)}\ \wedge\colon l_1}{P(\alpha_3) \wedge R(\alpha_3)^\star \vdash (\exists x)P(x)}\ \exists\colon r}{(\exists x)(P(x) \wedge R(x))^\star \vdash (\exists x)P(x)}\ \exists\colon l}{C_2^\star \vdash (\exists x)P(x)}\ \wedge\colon l_2$$

$\varphi_{2,3} =$

$$\dfrac{\dfrac{\dfrac{\dfrac{R(\alpha_4)^\star \vdash R(\alpha_4)}{P(\alpha_4) \wedge R(\alpha_4)^\star \vdash R(\alpha_3)}\ \wedge\colon l_2}{P(\alpha_4) \wedge R(\alpha_4)^\star \vdash (\exists x)R(x)}\ \exists\colon r}{(\exists x)(P(x) \wedge R(x))^\star \vdash (\exists x)R(x)}\ \exists\colon l}{C_2^\star \vdash (\exists x)R(x)}\ \wedge\colon l_2$$

Let $\nu_{i,j}$ be the root node of the tree $\varphi_{i,j}$. Then we get the following clause terms

$$\Theta(\varphi)/\nu_{1,1} \quad = \quad \{Q(\alpha_1) \vdash\},$$

$$\begin{array}{rcl}\Theta(\varphi)/\nu_{1,2} & = & \oplus^2(\{\vdash P(\alpha_2)\}, \{\vdash R(\alpha_2)\}) \\ \Theta(\varphi)/\nu_{2,1} & = & \oplus^2(\{\vdash P(c)\}, \{\vdash Q(c)\}) \\ \Theta(\varphi)/\nu_{2,2} & = & \{P(\alpha_3) \vdash\} \\ \Theta(\varphi)/\nu_{2,3} & = & \{R(\alpha_4) \vdash\}.\end{array}$$

and

$$\Theta(\varphi) = \oplus^2(\oplus^2(\Theta(\varphi)/\nu_{1,1}, \Theta(\varphi)/\nu_{1,2}), \otimes^3(\Theta(\varphi)/\nu_{2,1}, \Theta(\varphi)/\nu_{2,2}, \Theta(\varphi)/\nu_{2,3})).$$

For the characteristic clause set we obtain

$$\begin{array}{rcl}\mathrm{CL}(\varphi) & = & \{Q(\alpha_1) \vdash;\ \vdash P(\alpha_2);\ \vdash R(\alpha_2); \\ & & P(\alpha_3), R(\alpha_4) \vdash P(c);\ P(\alpha_3), R(\alpha_4) \vdash Q(c)\}.\end{array}$$

$\mathrm{CL}(\varphi)$ can be refuted by the following ground resolution refutation $\gamma =$

$$\cfrac{\cfrac{\vdash R(\alpha_2) \quad \cfrac{\vdash P(\alpha_2) \quad P(\alpha_3), R(\alpha_4) \vdash Q(c)}{R(\alpha_4) \vdash Q(c)}}{\vdash Q(c)} \qquad Q(\alpha_1) \vdash}{\vdash}$$

We define the projection to the clause $P(\alpha_3), R(\alpha_4) \vdash Q(c)$:

$$\cfrac{\cfrac{\cfrac{\cfrac{\cfrac{Q(c) \vdash Q(c)}{Q(c), P(c) \vdash Q(c)}\ s^*}{P(c) \vdash Q(c), \neg Q(c)}\ \neg\colon r}{\vdash Q(c), P(c) \to \neg Q(c)}\ \to\colon r \quad \cfrac{P(\alpha_3) \vdash P(\alpha_3)}{P(\alpha_3) \vdash (\exists x)P(x)}\ \exists\colon r \quad \cfrac{R(\alpha_4) \vdash R(\alpha_4)}{R(\alpha_4) \vdash (\exists x)R(x)}\ \exists\colon r}{P(\alpha_3), R(\alpha_4) \vdash Q(c), B_1 \wedge (B_2 \wedge B_3)}\ \wedge_3\colon r}{A, P(\alpha_3), R(\alpha_4) \vdash Q(c), B}\ w\colon l$$

◇

7.2 Equality Inference

Gentzen's **LK** is the original calculus for which cut-elimination was defined. The original version of CERES is based on **LK** and several variants of it (we just refer to [18, 20]). In formalizing mathematical proofs it turns out that **LK** (and also natural deduction) are not sufficiently close to real mathematical inference.

First of all, the calculus **LK** lacks a specific handling of equality (in fact equality axioms have to be added to the end-sequent). Due to the importance of equality this defect was apparent to proof theorists; e.g. Takeuti

[74] gave an extension of **LK** to a calculus $\mathbf{LK}_=$, adding atomic equality axioms to the standard axioms of the form $A \vdash A$. The advantage of $\mathbf{LK}_=$ over **LK** is that no new axioms have to be added to the end-sequent; on the other hand, in presence of the equality axioms, full cut-elimination is no longer possible, but merely reduction to *atomic cut*. But still $\mathbf{LK}_=$ uses the same rules as **LK**; in fact, in $\mathbf{LK}_=$, equality is *axiomatized*, i.e. additional atomic (non-tautological) sequents are admitted as axioms. On the other hand, in formalizing mathematical proofs, using equality as a *rule* is much more natural and concise. For this reason we choose the most natural equality rule, which is strongly related to paramodulation in automated theorem proving. Our approach differs from this in [33], where a unary equality rule is used (which does not directly correspond to paramodulation). In the *equality rules* below we mark the auxiliary formulas by $+$ and the principal formula by $*$.

$$\frac{\Gamma_1 \vdash \Delta_1, s = t^+ \quad A[s]^+_\Lambda, \Gamma_2 \vdash \Delta_2}{A[t]^*_\Lambda, \Gamma_1, \Gamma_2 \vdash \Delta_1, \Delta_2} =: l1 \qquad \frac{\Gamma_1 \vdash \Delta_1, t = s^+ \quad A[s]^+_\Lambda, \Gamma_2 \vdash \Delta_2}{A[t]^*_\Lambda, \Gamma_1, \Gamma_2 \vdash \Delta_1, \Delta_2} =: l2$$

for inference on the left and

$$\frac{\Gamma_1 \vdash \Delta_1, s = t^+ \quad \Gamma_2 \vdash \Delta_2, A[s]^+_\Lambda}{\Gamma_1, \Gamma_2 \vdash \Delta_1, \Delta_2, A[t]^*_\Lambda} =: r1 \qquad \frac{\Gamma_1 \vdash \Delta_1, t = s^+ \quad \Gamma_2 \vdash \Delta_2, A[s]^+_\Lambda}{\Gamma_1, \Gamma_2 \vdash \Delta_1, \Delta_2, A[t]^*_\Lambda} =: r2$$

on the right, where Λ denotes a set of positions of subterms where replacement of s by t has to be performed. We call $s = t$ the *active equation* of the rules.

Furthermore, as the only axiomatic extension, we need the set of reflexivity axioms

$$\mathrm{REF} = \vdash s = s$$

for all terms s.

Definition 7.2.1 The calculus **LKe** is **LK** extended by the axioms REF and by the rules $=: l1, =: l2, =: r1, =: r2$. ◇

In CERES it is crucial that all nonlogical rules (which also work on atomic sequents) correspond to clausal inference rules in automated deduction. While cut and contraction correspond to resolution (and factoring, dependent on the version of resolution), the equality rules $=: l1, =: l2, =: r1, =: r2$ correspond to paramodulation, which is the most efficient equality rule in automated deduction [66]. Indeed, when we compute the most general unifiers and apply them to the paramodulation rule, then it becomes one of the rules $=: l1, =: l2, =: r1, =: r2$.

The extensions defined above, can easily be built in without affecting the clarity and efficiency of the method. Similarly to the inference rules of **LK**, we still distinguish binary and unary nodes. The characteristic clause term and the characteristic clause sets are defined exactly as in the Definitions 6.4.1 and 6.4.2. Furthermore **LKe**-proofs can be skolemized like **LK**-proofs.

Theorem 7.2.1 *Let φ be a skolemized proof in* **LKe**. *Then the clause set* $\mathrm{CL}(\varphi)$ *is equationally unsatisfiable.*

Remark: A clause set $\mathcal{C}$ is equationally unsatisfiable if $\mathcal{C}$ does not have a model where $=$ is interpreted as equality over a domain. $\diamond$

Proof: The proof is essentially the same as in Proposition 6.4.1. Let ν be a node in φ and $S'(\nu)$ the subsequent of $S(\nu)$ which consists of the ancestors of Ω (where Ω is the set of occurrences of all cut formulas). It is shown by induction that $S'(\nu)$ is **LKe**-derivable from $\mathcal{C}_\nu$. If ν_0 is the root then, clearly, $S'(\nu_0) = \vdash$ and the empty sequent $\vdash$ is **LKe**-derivable from the axiom set $\mathcal{C}_{\nu_0}$, which is just $\mathrm{CL}(\varphi)$. As all inferences in **LKe** are sound over equational interpretations, $\mathrm{CL}(\varphi)$ is equationally unsatisfiable. Note that, in proofs without the rules $=: l$ and $=: r$, the set $\mathrm{CL}(\varphi)$ is just unsatisfiable. But,clearly, the rules $=: l$ and $=: r$ are sound only over equational interpretations. $\square$

Note that, for proving Theorem 7.2.1, we just need the soundness of **LKe**, not its completeness.

The next steps in CERES are again:

(1) the computation of the proof projections $\varphi[C]$ w.r.t. clauses $C \in \mathrm{CL}(\varphi)$,

(2) the refutation of the set $\mathrm{CL}(\varphi)$, resulting in an RP-tree γ, i.e. in a deduction tree defined by the inferences of resolution and paramodulation, and

(3) "inserting" the projections $\varphi[C]$ into the leaves of γ.

Step (1) is done like in CERES for **LK**, i.e. we skip in φ all inferences where the auxiliary resp. main formulas are ancestors of a cut. Instead of the end-sequent S we get $S \circ C$ for a $C \in \mathrm{CL}(\varphi)$. The construction does not differ from this in Section 6.4 as the form of the rules do not matter.

Step (2) consists in ordinary theorem proving by resolution and paramodulation (which is equationally complete). For refuting $\mathrm{CL}(\varphi)$ any first-order prover like Vampire,[1] SPASS[2] or Prover9[3] can be used. By the completeness of the methods we find a refutation tree γ as $\mathrm{CL}(\varphi)$ is unsatisfiable by Theorem 7.2.1.
Step (3) makes use of the fact that, after application of the simultaneous most general unifier of the inferences in γ, the resulting tree γ' is actually a *derivation in* **LKe**! Indeed, after computation of the simultaneous unifier, paramodulation becomes $=: l$ and $=: r$ and resolution becomes cut in **LKe**. Now for every leaf ν in γ', which is labeled by a clause C' (an instance of a clause $C \in \mathrm{CL}(\varphi)$) we insert the proof projection $\varphi[C']$. The result is a proof with only atomic cuts.

There are calculi (characterized by the term *deduction modulo*) which separate the equational reasoning from the rest of first-order inference(see [34, 35]). These calculi also admit cut-elimination and are good candidates for applying the CERES method.

7.3 Extension by Definition

The *definition rules* directly correspond to the *extension principle* (see [36]) in predicate logic. It simply consists in introducing new predicate and function symbols as abbreviations for formulas and terms. Nowadays there exist several calculi which make use of this powerful principle; we just mention [29]. Let A be a first-order formula with the free variables $x_1, \ldots, x_k$ (denoted by $A[x_1, \ldots, x_k]$ and P be a *new* k-ary predicate symbol (corresponding to the formula A). Then the rules are:

$$\frac{A[t_1, \ldots, t_k], \Gamma \vdash \Delta}{P(t_1, \ldots, t_k), \Gamma \vdash \Delta}\ def(P){:}\,l \qquad \frac{\Gamma \vdash \Delta, A[t_1, \ldots, t_k]}{\Gamma \vdash \Delta, P(t_1, \ldots, t_k)}\ def(P){:}\,r$$

for arbitrary sequences of terms $t_1, \ldots, t_k$. Definition introduction is a simple and very powerful tool in mathematical practice. Note that the introduction of important concepts and notations like groups, integrals etc. can be formally described by introduction of new symbols. There are also definition introduction rules for new function symbols which are of similar type. Note that the rules above are only sound if the interpretation of the new predicate

[1] http://www.vampire.fm/
[2] http://spass.mpi-sb.mpg.de/
[3] http://www-unix.mcs.anl.gov/AR/prover9/

symbols is subjected to the constraint

$$P(x_1, \ldots, x_k) \leftrightarrow A[x_1, \ldots, x_k].$$

Definition 7.3.1 **LKDe** is **LKe** extended by the rules $def(\;){:}\,l$ and $def(\;){:}\,r$. $\diamond$

Remark: The *axiom system* for **LKDe** may be an arbitrary set of atomic sequents containing the standard axiom set. The only axioms which have to be added for equality are $\vdash s = s$ where s is an arbitrary term. $\diamond$

Clearly the extensions of **LK** to **LKe** and **LKDe** do not increase the logical expressivity of the calculus, but they make it much more compact and natural. To illustrate the rules defined above we give a simple example. The aim is to prove the (obvious) theorem that a number divides the square of a number b if it divides b itself. In the formalization below a and b are constant symbols and the predicate symbol D stands for "divides" and is defined by

$$D(x, y) \leftrightarrow (\exists z) x * z = y.$$

The active equations are written in boldface.

$$\dfrac{\dfrac{\dfrac{\dfrac{\dfrac{\dfrac{\vdash (\mathbf{a} * \mathbf{z_0}) * \mathbf{b} = \mathbf{a} * (\mathbf{z_0} * \mathbf{b}) \qquad \dfrac{a * z_0 = b \vdash \mathbf{a} * \mathbf{z_0} = \mathbf{b} \qquad \vdash b * b = b * b}{a * z_0 = b \vdash (a * z_0) * b = b * b}\;{=}{:}\,r2}{a * z_0 = b \vdash a * (z_0 * b) = b * b}\;{=}{:}\,r1}{a * z_0 = b \vdash (\exists z) a * z = b * b}\;\exists r}{(\exists z) a * z = b \vdash (\exists z) a * z = b * b}\;\exists{:}\,l}{(\exists z) a * z = b \vdash D(a, b * b)}\;def(D){:}\,r}{D(a, b) \vdash D(a, b * b)}\;def(D){:}\,l$$

$$\dfrac{D(a, b) \vdash D(a, b * b)}{\vdash D(a, b) \rightarrow D(a, b * b)}\;\rightarrow{:}\,r$$

The axioms of the proof are: (1) an instance of the associativity law, (2) the equational axiom $\vdash b * b = b * b$ and the tautological standard axiom $a * z_0 = b \vdash a * z_0 = b$.

Chapter 8

Applications of CERES

CERES has applications to complexity theory, proof theory and to general mathematics. We first characterize classes of proofs which admit fast cut-elimination due to the resulting structure of the characteristic clause sets. Furthermore CERES can be applied to the efficient constructions of interpolants in classical logic and other logics for which CERES-methods can be defined. CERES is also suitable for calculating most general proofs from proof examples. Finally we demonstrate that CERES is also an efficient tool for the in-depth analysis of mathematical proofs.

8.1 Fast Cut-Elimination Classes

In this section we use CERES as a tool to prove fast cut-elimination for several subclasses of **LK**-proofs. By using the structure of the characteristic clause sets in the CERES-method we characterize fast classes of cut-elimination; these classes are either defined by restrictions on the use of inference rules or on the syntax of formulas occurring in the proofs. In the analysis of the classes, resolution refinements play a major role, in particular those refinements which can be shown terminating on the corresponding classes of characteristic clause sets. In fact, the satisfiability problem of all natural clause classes $\mathcal{X}$ decidable by a resolution refinement is of elementary complexity and so are the resolution refutations of the clause sets in $\mathcal{X}$.
We show now that, for a class of proofs $\mathcal{P}$, the complexity of resolution on $\mathrm{CL}(\varphi)$ for $\varphi \in \mathcal{P}$ characterizes the complexity of cut-elimination on $\mathcal{P}$.

Definition 8.1.1 Let $\mathcal{C}$ be an unsatisfiable set of clauses. Then the *resolu-*

M. Baaz, A. Leitsch, *Methods of Cut-Elimination*, Trends in Logic 34, 175
DOI 10.1007/978-94-007-0320-9_8, © Springer Science+Business Media B.V. 2011

tion complexity of $\mathcal{C}$ is defined as

$$rc(\mathcal{C}) = \min\{\|\gamma\| \mid \gamma \text{ is a resolution refutation of } \mathcal{C}\}.$$

◇

Clearly, by the undecidability of clause logic, there is no recursive bound on $rc(\mathcal{C})$ in terms of $\|\mathcal{C}\|$. However, the resolution complexity of characteristic clause sets is always bounded by a primitive recursive function; this because CERES cannot be outperformed by Tait's method (see [20, 73]) for which there exists a primitive recursive, though nonelementary, bound (see [40]).

Definition 8.1.2 Let $\mathcal{K}$ be a class of skolemized proofs. We say that CERES is *fast on* $\mathcal{K}$ if there exists an elementary function f s.t. for all φ in $\mathcal{K}$:

$$rc(\mathrm{CL}(\varphi)) \leq f(\|\varphi\|).$$

◇

By Proposition 6.5.3 and by $l(\gamma) \leq \|\gamma\|$ for all resolution deductions γ, the run time of the whole algorithm CERES which constructs $\mathrm{CL}(\varphi)$, computes the resolution refutation γ and the p-resolution refutation γ', the projections and eventually $\varphi(\gamma')$, is bounded by an elementary function – provided CERES is fast as defined above. Note that CERES is fast on $\mathcal{K}$ iff there exists a $k \in \mathbb{N}$ s.t. for all $\varphi \in \mathcal{K}$: $rc(\mathrm{CL}(\varphi)) \leq e(k, \|\varphi\|)$.
The main goal of this section is to identify classes $\mathcal{X}$ where CERES is fast, thus giving proofs of elementary cut-elimination on $\mathcal{K}$. In one case we even show that CERES is fast on a class of proofs where all Gentzen-type methods of cut-elimination are of nonelementary complexity. Moreover, the simulation of Gentzen type methods by CERES shown in Section 6.8 indicates that the use of Gentzen type methods in proving the property of fast cut-elimination cannot be successful if CERES fails.

It is well known that the complexity of cut-elimination does not only depend on the complexity of cut formulas but also on the syntactic form of the end sequent. The first class we are presenting here is known to have an elementary cut-elimination, but the proof of this property is trivial via the CERES-method, thus giving a flavor of our approach.

Definition 8.1.3 UIE is the class of all skolemized **LK**-proofs from the standard axiom set where all inferences going into the end-sequent are unary. ◇

Proposition 8.1.1 CERES *is fast on* **UIE**. *In particular cut-elimination is at most exponential on* **UIE**.

Proof: Let φ be a proof of $\Gamma \vdash \Delta$ in **UIE**. As there are no binary inferences going into the end-sequent there are no products in the characteristic term $\Theta(\varphi)$. Therefore the characteristic clause set $\mathrm{CL}(\varphi)$ contains just the union of all cut-ancestors in the axioms of φ; hence every $C \in \mathrm{CL}(\varphi)$ is of the form (1) $\vdash A$, (2) $A \vdash$, or (3) $A \vdash A$ for atoms A. Clauses of the form (3) are tautologies and can be deleted. We are left with a finite set of unit clauses which are contained in the Herbrand class. As $\mathrm{CL}(\varphi)$ is unsatisfiable $\mathrm{CL}(\varphi)$ contains two clauses of the form $C_1\colon \vdash A$ and $C_2\colon B \vdash$ s.t. $\{A, B\}$ is unifiable by some m.g.u. ϑ. Let ϑ' be a minimal ground unifier of $\{A, B\}$ (constructed as in Corollary 6.5.1). We consider the projections $\varphi[C_1\vartheta']$ and $\varphi[C_2\vartheta']$. Then the proof ψ:

$$\dfrac{\dfrac{\begin{array}{cc}\varphi[C_1\vartheta'] & \varphi[C_2\vartheta'] \\ \Gamma \vdash \Delta, A\vartheta' & A\vartheta', \Gamma \vdash \Delta\end{array}}{\Gamma, \Gamma \vdash \Delta, \Delta}\ cut}{\Gamma \vdash \Delta}\ c^*$$

is a CERES-normal form of φ. Note that the unification ϑ can lead to an exponential increase (in fact $\|A\vartheta\|$ can be exponential in $\|A\|$). Therefore there exists a number k s.t.

$$\|\varphi[C_i\vartheta]\| \leq 2^{k*\|\varphi\|} \text{ for } i = 1, 2.$$

Finally we obtain

$$\|\psi\| \leq 2^{r*\|\varphi\|} \text{ for some } r \in \mathbb{N}.$$

As there are no binary rule applications in the proofs $\varphi[C_1\vartheta]$ and $\varphi[C_1\vartheta]$, transforming ψ into a cut-free proof is only linear. Thus cut-elimination on **UIE** can be done in exponential time. □

Remark: The complexity cut-elimination on **UIE** is only linear if we do not measure the complexity by $\| \ \|$ but by proof length $l(\ )$. Clearly, cut-elimination on **UIE** is of linear symbolic complexity if only propositional proofs are considered. ◇

In [17] we have shown that cut-elimination on proofs with a single monotone cut is nonelementary. In a first step the cut can be transformed into negation normal norm. In a second one the negations in the cut formula (by the NNF-transformation they are immediately above atoms) can be eliminated by generalized disjunctions added at the left-hand side of the end-sequent; for details see [17]. We show now that a further restriction on the arity of inferences in the proofs leads to an elementary cut-elimination class.

Definition 8.1.4 A formula A is called *monotone* if the logical operators occurring in A are in $\{\wedge, \vee, \forall, \exists\}$. $\diamond$

Definition 8.1.5 **UILM** is the class of all skolemized **LK**-proofs φ from the standard axiom set, s.t. φ contains only one cut which is monotone, and all inferences in the left cut-derivation which go into the end-sequent are unary. $\diamond$

For cut-elimination via CERES on **UILM** we use a refinement of resolution, the so-called hyperresolution method. Hyperresolution is not only one of the most efficient refinements in theorem proving but is also a powerful tool for the decision problem of first-order clause classes [61]. For the representation of clauses we use normal forms which strongly reduce internal redundancy. One important normalization technique is condensation which was first used in resolution decision procedures (see [53]).

Definition 8.1.6 (condensation) Let C be a clause and D be a factor of C s.t. D is a proper subclause of C; then we say that D is obtained from C by *condensation.* A clause which does not admit condensations is called *condensed.* A condensation of a clause C is a clause D which is condensed and can be obtained by (iterated) condensation from C. $\diamond$

Example 8.1.1 Let $C = P(x), P(y) \vdash Q(y), Q(z)$.
Then $D\colon P(y) \vdash Q(y), Q(z)$ is obtained from C by condensation. A further condensation yields the clause $E\colon P(y) \vdash Q(y)$ which is condensed and E is the condensation of C. $\diamond$

Definition 8.1.7 Let C be a clause and D be a condensation of C. $N_c(C)$, *the condensation normal form*, is the clause which is obtained from D by renaming the variables to $\{x_1, \ldots, x_n, \ldots\}$ and by ordering the atoms in C_+ and in C_- by a total ordering. $\diamond$

Remark: It is easy to verify that, for all clauses C, $N_c(C)$ is logically equivalent to C; thus N_c is a sound normalization. $\diamond$

Example 8.1.2 Let C be the clause in Example 8.1.1. Then

$$N_c(C) = \{P(x_1) \vdash Q(x_1)\}.$$

$\diamond$

Definition 8.1.8 (hyperresolution) Let $\mathcal{C}$ be a set of clauses, $C_1, \ldots, C_n$ positive clauses in $\mathcal{C}$, and D be a nonpositive clause in $\mathcal{C}$. Then the sequence λ: $(D; C_1, \ldots, C_n)$ is called a *clash sequence*. We define

$E_0 = D$,
E_{i+1} is a PRF-resolvent of E_i and a renamed variant of C_{i+1} for $i < n$.

For the definition of a PRF-resolvent see Definition 3.3.10. If E_n exists then it is a positive clause and the normalization $N_c(E_n)$ is called a *hyperresolvent* of λ over $\mathcal{C}$. Let $\rho_H(\mathcal{C})$ be the set of all hyperresolvents over $\mathcal{C}$. We define

$$RH(\mathcal{C}) = \mathcal{C} \cup \rho_H(\mathcal{C}),$$
$$RH^{i+1}(\mathcal{C}) = RH(RH^i(\mathcal{C})) \text{ for } i \in \mathbb{N},\ RH^*(\mathcal{C}) = \bigcup_{i=0}^{\infty} RH^i(\mathcal{C}).$$

By $RH^*_+(\mathcal{C})$ we denote the set of positive clauses in $RH^*(\mathcal{C})$. $\diamond$

Note that $RH^*(\mathcal{C})$ is just the deductive closure of $\mathcal{C}$ under RH. RH is refutationally complete, i.e. for any unsatisfiable set of clauses $\mathcal{C}$ we have $\vdash \in RH^*(\mathcal{C})$ (for details see [61]).

Example 8.1.3 Let $\mathcal{C} = \{C_1, C_2, C_3, C_4, C_5\}$ for

$$\begin{aligned} C_1 &= \vdash P(x, f(x)), \\ C_2 &= \vdash P(f(x), x), \\ C_3 &= P(x, y) \vdash P(y, x), \\ C_4 &= P(x, y), P(y, z) \vdash P(x, z), \\ C_5 &= P(c, c) \vdash . \end{aligned}$$

We use the clash sequence $(C_3; C_1)$. We rename C_1 to $C_1'\colon \vdash P(u, f(u))$. We have $E_0 = C_3$ and E_1 is the (only) resolvent of C_3 and C_1 where

$$E_1 = \vdash P(f(u), u),\ N_c(E_1) = \vdash P(f(x_1), x_1).$$

$C_6\colon N_c(E_1)$ is a hyperresolvent resolvent of $(C_3; C_1)$.
Now we consider the clash sequence $(C_4; C_1, C_6)$. With appropriate renamings of C_1 and C_2 we obtain

$$\begin{aligned} E_1' &= P(f(u), z) \vdash P(u, z), \\ E_2' &= \vdash P(v, v). \end{aligned}$$

Therefore C_7 for $C_7 = N_c(E_2') = \vdash P(x_1, x_1)$ is a hyperresolvent of $(C_4; C_1, C_6)$. Finally we consider the clash sequence $(C_5; C_7)$ which gives the hyperresolvent $\vdash$.

In terms of the resolution operator RH we obtain

$$
\begin{aligned}
RH(\mathcal{C}) &= \mathcal{C} \cup \{\vdash P(f(x_1), x_1)\}, \\
RH^2(\mathcal{C}) &= \mathcal{C} \cup \{\vdash P(f(x_1), x_1);\ \vdash P(x_1, x_1)\}, \\
RH^3(\mathcal{C}) &= \mathcal{C} \cup \{\vdash P(f(x_1), x_1);\ \vdash P(x_1, x_1);\ \vdash\}.
\end{aligned}
$$

In particular $RH^*(\mathcal{C}) = RH^3(\mathcal{C})$. ◇

Theorem 8.1.1 CERES *is fast on* **UILM**. *Therefore cut-elimination on* **UILM** *is of elementary complexity.*

Proof: Let φ be a proof in **UILM** then φ is of the form $\varphi[\psi]_\nu$ where ψ is the only cut-derivation in φ (occurring at the node ν in the proof). Assume that $\psi =$

$$
\frac{\begin{matrix}(\psi_1) \\ \Gamma \vdash \Delta, A\end{matrix} \quad \begin{matrix}(\psi_2) \\ A, \Pi \vdash \Lambda\end{matrix}}{\Gamma, \Pi \vdash \Delta, \Lambda}\ cut
$$

As there is only this single cut in φ the end-sequent $S\colon \Gamma, \Pi \vdash \Delta, \Lambda$ of ψ is skolemized. Therefore we may apply CERES to ψ itself and replace ψ in φ by the obtained CERES-normal form ψ'.

We compute $\mathrm{CL}(\psi)$: first consider the ancestors of the cut in the axioms. Note that the cut is monotone. In ψ_1 the ancestors of the cut in the axioms are of the form $\vdash A_i$ $(i = 1, \ldots n)$ for some atoms A_i, in ψ_2 they are of the form $B_j \vdash$ $(j = 1, \ldots m)$ for some atoms B_j. As the left derivation ψ_1 does not contain binary inferences going into S, the characteristic term $\Theta(\psi)$ is of the form $t_1 \oplus t_2$ where t_1 does not contain products. Therefore the clause set $\mathrm{CL}(\psi)$ is of the form $\mathcal{C}_1 \cup \mathcal{C}_2$, where

$$
\begin{aligned}
\mathcal{C}_1 &= \{\vdash A_1;\ \ldots;\ \vdash A_n\},\ \text{and} \\
\mathcal{C}_2 &\subseteq \bigcup \{B_{j_1}, \ldots, B_{j_k} \vdash \mid \{j_1, \ldots, j_k\} \subseteq \{1, \ldots, m\},\ k \leq m\}.
\end{aligned}
$$

$\|\mathcal{C}_2\|$ is at most exponential in $\|\Theta(\psi)\|$, for which we have $\|\Theta(\psi)\| \leq \|\psi\|$.

Hence $\mathrm{CL}(\psi)$ is a set of Horn clauses consisting of positive unit clauses and negative clauses only. As $\mathrm{CL}(\psi)$ is unsatisfiable there exists a refutation ρ by hyperresolution of $\mathrm{CL}(\psi)$. Moreover there are no mixed clauses in $\mathrm{CL}(\psi)$, so ρ must consist of a single hyperresolvent, based on the clash sequence

$$
\gamma\colon (B_{j_1}, \ldots, B_{j_k} \vdash;\ \vdash A_{i_1}, \ldots, \vdash A_{i_k}),
$$

of the form

$$\cfrac{\vdash A_{i_k} \qquad \cfrac{\cfrac{\vdash A_{i_1} \quad B'_{j_1}, \ldots, B'_{j_k} \vdash}{\vdots}}{B'_{j_k} \vdash}}{\vdash}$$

Where the B'_{j_k} are instances of the B_{j_k}. Now construct a minimal ground projection ρ' of ρ and define $\psi' = \psi(\rho')$; then ψ' is a CERES-normal form of ψ. As the computation of a minimal ground projection can lead to an exponential blow up w.r.t. then size of the clash we obtain

$$\|\rho'\| \leq 2 * m * 2^{k*a}$$

where a is the maximal complexity of an atom in γ. Clearly $a \leq \|\psi\|$ and so

$$\|\rho'\| \leq 2^{r*\|\psi\|} \leq 2^{r*\|\varphi\|}$$

for some constant r; we see that the CERES-normal form ψ' of ψ is at most exponential in ψ.
Now $\varphi' = \varphi[\psi']_\nu$ is a proof of the same end-sequent with only atomic cuts and

$$\|\varphi'\| \leq \|\varphi\| + \|\psi'\| \leq \|\varphi\| + 2^{r*\|\varphi\|}.$$

Therefore CERES is fast on **UILM** and thus cut-elimination is elementary on **UILM**. □

Remark: Specific strategies of Gentzen's method can be shown to behave nonelementarily on **UILM**. The fully nondeterministic Gentzen method, however, is elementary on **UILM**. However it is very intransparent to use Gentzen's method as a method to *prove* fast cut-elimination for this class. ◇

Definition 8.1.9 **UIRM** is the class of all skolemized **LK**-proofs φ from the standard axiom set, s.t. φ contains only one cut which is monotone, and all inferences in the right cut-derivation which go into the end-sequent are unary. ◇

Theorem 8.1.2 *Cut-elimination is elementary on* **UIRM**.

Proof: Like for **UILM**.

Corollary 8.1.1 *Cut-elimination is elementary on* **UIRM** $\cup$ **UILM**

Proof: Obvious.

The hyperresolution refinement can also be used to prove fast cut-elimination for another class of proofs, which we call **AXDC**.

Definition 8.1.10 A proof $\varphi \in \Phi^s$ from the standard axiom set is in the class **AXDC** if different axioms in φ are variable-disjoint. $\diamond$

Theorem 8.1.3 CERES *is fast on* **AXDC**. *Therefore cut-elimination is elementary on* **AXDC**.

Proof: Let φ be in **AXDC** and let

$$\{\vdash A_1, \ldots, \vdash A_n,\ B_1 \vdash, \ldots, B_m \vdash\}$$

be the set of subsequents of axioms which are cut-ancestors (tautologies are omitted). Then no two different subsequences of axioms in the set share variables. Let $\mathcal{C} = \mathrm{CL}(\varphi)$. Then any clause $C \in \mathcal{C}$ is disconnected (two different atoms occurring in C do not share variables). We prove that, by using hyperresolution with condensing, we can find a refutation of $\mathcal{C}$ within exponential time relative to $\|\mathcal{C}\|$.

To this aim it is sufficient to show that the total size of derivable positive clauses $RH^*_+(\mathcal{C})$ is exponential in $\|\mathcal{C}\|$ (note that all clauses in $RH^*(\mathcal{C})\backslash\mathcal{C}$ are positive, and thus are contained in $RH^*_+(\mathcal{C})$). We will see below that there is no exponential increase in the size of atoms.

First we observe that all clauses in $RH^*_+(\mathcal{C})$ (i.e. the clauses which are actually derivable) are also variable disjoint. This is easy to see as all clauses are disconnected and have to be renamed prior to resolution; in fact, if a clause is used twice in a clash sequence, a renamed variant has to be constructed.

Now let, for $i = 1, \ldots, k$, be $\mathcal{A}_i$ the set of atoms occurring in the clause head of the i-th clause in $\mathcal{C}$.

By definition of hyperresolution the clauses in $RH^*_+(\mathcal{C})$ are "accumulations" of renamed instances of subsets of the $\mathcal{A}_i$. As all clauses are disconnected no unification substitution is ever stored in the resolvents and in the hyperresolvent. Moreover the hyperresolvents are condensed and normalized, i.e. they do not contain different variants of atoms. Therefore every hyperresolvent is of the form

$$\vdash A_1, \ldots, A_r$$

where the A_i are variants of atoms in $\mathcal{A}_1 \cup \ldots \cup \mathcal{A}_k$ and $r \leq h$ for $h = |\mathcal{A}_1| + \ldots + |\mathcal{A}_k|$. Therefore the number of possible hyperresolvents is $\leq 2^h$. But $2^h < 2^{\|\mathcal{C}\|}$.

So we have shown that computing the contradiction by RH is at most exponential in $\|\mathrm{CL}(\varphi)\|$.
Note that deciding unification is of linear complexity only, and computing the most general unifiers explicitly is not necessary for the computation of $RH^*_+(\mathcal{C})$.
As $\|\mathrm{CL}(\varphi)\|$ may be exponential in $\|\varphi\|$, computing the CERES normal form is at most double exponential. □

We used the refinement of hyperresolution to construct fast cut-elimination procedures by CERES for the classes **UILM**, **UIRM** and **AXDC**. For the analysis of the next class **MC** to be defined below we need another refinement, namely ordered resolution.
Let A be an atom; then $\tau(A)$ denotes the maximal term depth in A. For clauses C we define $\tau(C) = \max\{\tau(A) \mid A \text{ in } C\}$. $\tau_{\max}(x, A)$ denotes the maximal depth of the variable x in A. $V(A)$ defines the set of variables in A.

Definition 8.1.11 (depth ordering) Let A and B be atoms; we define $A <_d B$ if (1) $V(A) \subseteq V(B)$, (2) $\tau(A) < \tau(B)$ and (3) for all $x \in V(A)$: $\tau_{\max}(x, A) < \tau_{\max}(x, B)$. ◇

In [61] it is shown that $<_d$ is a so-called atom ordering.

Definition 8.1.12 (ordered resolution) Let C and D be clauses in a clause set $\mathcal{C}$ and E be a resolvent of C, D with resolved atom A. We define $N_c(E) \in \rho_{<_d}(\mathcal{C})$ iff there is no atom B in E s.t. $A <_d B$. The corresponding resolution operator is defined by:

$$R_{<_d}(\mathcal{C}) = \mathcal{C} \cup \rho_{<_d}(\mathcal{C}),\ R^*_{<_d}(\mathcal{C}) = \bigcup_{i=0}^{\infty} R^i_{<_d}(\mathcal{C}).$$

◇

$R_{<_d}$ is complete (see [61]), i.e. $\vdash \in R^*_{<_d}(\mathcal{C})$ if $\mathcal{C}$ is unsatisfiable.

Example 8.1.4 Let $\mathcal{C} = \{C_1, C_2, C_3\}$ for

$$\begin{array}{rcl} C_1 & = & \vdash P(a), \\ C_2 & = & P(x) \vdash P(f(x)), \\ C_3 & = & P(f(f(a)))\vdash . \end{array}$$

C_1 and C_2 have the resolvent $\vdash P(f(a))$, where the resolved atom is $P(a)$. Obviously $P(a) <_d P(f(a))$, thus by Definition 8.1.12 this resolution does not produce an ordered resolvent and $\vdash P(f(a)) \notin \rho_{<_d}(\mathcal{C})$. C_1 and C_3 cannot be resolved. It remains to resolve C_2 and C_3.
C_2 and C_3 define one resolvent, namely C_4: $P(f(a))\vdash$, the resolved atom being $P(f(f(a)))$. Here we have $P(f(a)) <_d P(f(f(a))$ and C_4 is admitted. Therefore we obtain

$$\rho_{<_d}(\mathcal{C}) = \{P(f(a))\vdash\}.$$

Continuing on the extended clause set $\mathcal{C} \cup \{C_4\}$ we get an ordered resolvent from C_4 and C_2 (this is the only new resolvent which can be obtained) and

$$\rho_{<_d}(\mathcal{C} \cup \{C_4\}) = \{C_4, C_5\} \text{ for } C_5 = P(a)\vdash.$$

In this resolution the resolved atom is $P(f(a))$. Obviously

$$\rho_{<_d}(\mathcal{C} \cup \{C_4, C_5\}) = \{C_4, C_5, \vdash\}.$$

For the operator $R_{<_d}$ we get

$$\begin{array}{rcl}
R_{<_d}(\mathcal{C}) & = & \mathcal{C} \cup \{C_4\}, \\
R^2_{<_d}(\mathcal{C}) & = & \mathcal{C} \cup \{C_4, C_5\}, \\
R^3_{<_d}(\mathcal{C}) & = & \mathcal{C} \cup \{C_4, C_5, \vdash\}, \\
R^*_{<_d}(\mathcal{C}) & = & R^3_{<_d}(\mathcal{C}).
\end{array}$$

◇

Definition 8.1.13 Let **MC** be the set of all skolemized **LK**-proofs φ over the axiom set of type $A \vdash A$ where A is quantifier-free, s.t. all function symbols occurring in φ are unary and all predicate symbols occurring in cut-formulas are also unary. ◇

Note that the end-sequents of proofs in **MC** may contain predicate symbols of arbitrary arity and thus define an undecidable class; therefore the class is nontrivial for cut-elimination (indeed the size of cut-free proofs is not bounded elementarily in the size of the end-sequents). Just consider the satisfiability problem of the prefix class $\forall\exists\forall$, which is undecidable (see [27]). As a consequence, the provability of sequents of the form

$$(\forall x)(\forall z)A(x, f(x), z) \vdash$$

(where $A(x, f(x), z)$ is a quantifier free matrix over the terms $x, f(x), z$) is undecidable too. Therefore there is no way to find cut-free proofs of elementary size within **MC** by exhaustive search.
In order to show that cut-elimination in **MC** is elementary we need some preparatory steps.

Definition 8.1.14 The class K is the set of all finite condensed sets of clauses $\mathcal{C}$ s.t. for all $C \in \mathcal{C}$: $|V(A)| \leq 1$ for all atoms A occurring in C. $\diamond$

Lemma 8.1.1 *$R^*_{<_d}(\mathcal{C})$ is finite for each $\mathcal{C} \in \mathrm{K}$ and $\tau(R^*_{<_d}(\mathcal{C})) \leq 2 * \tau(\mathcal{C})$.*

Proof: In [61], Theorem 5.2.1.

Definition 8.1.15 K_{mon} is the subclass of K containing only monadic predicate symbols and monadic function symbols. $\diamond$

Clearly Lemma 8.1.1 holds also for K_{mon}, but we may obtain sharper complexity bounds on the deductive closure.

Lemma 8.1.2 *Let $\mathcal{C} \in \mathrm{K}_{mon}$. Then*

(1) $|R^*_{<_d}(\mathcal{C})| \leq 2^{3r^2}$, *and*

(2) $\max\{\|C\| \mid C \in R^*_{<_d}(\mathcal{C})\} \leq 2r(\tau(\mathcal{C}) + 2)$

for $r = 2|\mathrm{PS}(\Sigma)||\mathrm{FS}(\Sigma)|^{2\tau(\mathcal{C})}(|\mathrm{CS}(\Sigma)| + 1)$ where Σ is the signature of $\mathcal{C}$.

Proof: Let $t = \tau(\mathcal{C})$ and Σ be the signature of $\mathcal{C}$. By Lemma 8.1.1 $R^*_{<_d}(\mathcal{C})$ is finite and $\tau(R^*_{<_d}(\mathcal{C})) \leq 2t$.
Now let A be an atom occurring in a clause in $R^*_{<_d}(\mathcal{C})$; then A is of the form $P(f_1 \ldots f_n s)$ where $s \in V \cup \mathrm{CS}(\Sigma)$, $P \in \mathrm{PS}(\Sigma)$, $f_i \in \mathrm{FS}(\Sigma)$, and $n \leq 2t$.
The number $g(2t, \Sigma)$, the number of ground atoms over Σ (or the number of atoms containing a fixed variable v in case $\mathrm{CS}(\Sigma) = \emptyset$) of depth $\leq 2t$ can be estimated by

$$g(2t, \Sigma) \leq |\mathrm{PS}(\Sigma)||\mathrm{FS}(\Sigma)|^{2t}(|\mathrm{CS}(\Sigma)| + 1).$$

As, by definition, all clauses in K_{mon} are condensed, the atoms $P(f_1 \ldots f_n c)$ and $P(f_1 \ldots f_n v)$ (for $c \in \mathrm{CS}(\sigma)$, $v \in V$) cannot appear in the same clause at the same side of the sequent sign: in fact, condensing would eliminate the atom $P(f_1 \ldots f_n v)$. The same situation holds for the atoms $P(f_1 \ldots f_n v_1)$

and $P(f_1 \dots f_n v_2)$ for different variables v_1, v_2. For this reason we have for every $C \in R^*_{<_d}(\mathcal{C})$

$$\max\{|C_+|, |C_-|\} \leq r$$

$$\text{for } r = |\mathrm{PS}(\Sigma)||\mathrm{FS}(\Sigma)|^{2t}(|\mathrm{CS}(\Sigma)| + 1).$$

and therefore

$$\max\{|C| \mid C \in R^*_{<_d}(\mathcal{C})\} \leq 2r.$$

As predicate symbols and function symbols are monadic we also have

$$\max\{|V(C)| \mid C \in R^*_{<_d}(\mathcal{C})\} \leq r.$$

Therefore, by standard renaming (enforced by the normalization operator N_c) the only variables which can occur in a clause in $R^*_{<_d}(\mathcal{C})$ are $x_1, \dots, x_r$. Therefore the number of possible atoms $a(R^*_{<_d}(\mathcal{C}))$ occurring in a clause in $R^*_{<_d}(\mathcal{C})$ is bounded by the number

$$a(R^*_{<_d}(\mathcal{C})) \leq |\mathrm{PS}(\Sigma)||\mathrm{FS}(\Sigma)|^{2t}(|\mathrm{CS}(\Sigma)| + r) \leq r(r+1).$$

As the clause length is at most r we obtain

$$|R^*_{<_d}(\mathcal{C})| \leq (r(r+1))^r \leq 2^{3r^2}.$$

This proves (1).

As the maximal number of atoms occurring in a clause is $2r$ and $\|A\| \leq 2 + \tau(\mathcal{C})$ for every atom occurring in $R^*_{<_d}(\mathcal{C})$ we have

$$\max\{\|C\| \mid C \in R^*_{<_d}(\mathcal{C})\} \leq 2r(2 + \tau(\mathcal{C})).$$

This proves (2). □

Theorem 8.1.4 CERES *is fast on* **MC**. *As a consequence, cut-elimination is elementary on* **MC**.

Proof: Let $\psi \in \mathbf{MC}$. We cannot apply CERES to ψ directly as ψ may contain nonatomic axioms. So we use the method defined in Lemma 4.1.1 and replace the axioms by their derivations from the standard axiom set; this way we obtain a proof $T(\psi)$ from the standard axiom set with $\|T(\psi)\| \leq k * \|\psi\|$ for a constant k independent of ψ. Now CERES can be applied to $\varphi: T(\psi)$. As the cut formulas contain only monadic function symbols and predicate symbols, the ancestors of the cuts in the axioms are of the form $\vdash A$ or $A \vdash$

where A is of the form $P(f_1 \dots f_n s)$ for $s \in \mathrm{CS} \cup V$. Note that, as always, we may omit tautologies in the construction of $\mathrm{CL}(\varphi)$. Therefore the clause set $\mathrm{CL}(\varphi)$ (defined by union and product) only consists of clauses built from atoms of this type. $\mathrm{CL}(\varphi)$ itself need not be in K_{mon}, but its condensation $\mathcal{C}$: $N_c(\mathrm{CL}(\varphi))$ is in K_{mon}. By Lemma 8.1.2 we get

$$|R^*_{<_d}(\mathcal{C})| \leq 2^{3r^2}$$

for $r = 2|\mathrm{PS}(\Sigma)||\mathrm{FS}(\Sigma)|^{2\tau(\mathcal{C})}(|\mathrm{CS}(\Sigma)| + 1)$ and $\Sigma = \Sigma(\mathcal{C})$.
So, as $\mathcal{C}$ is unsatisfiable, there exists a resolution refutation containing at most $\leq 2^{3r^2}$ different clauses. If the refutation is represented as a proof tree γ we obtain

$$l(\gamma) \leq r * 2^{2^{3r^2}}$$

Note that we also counted the applications of the condensations which are nothing else than repeated factors. The global unifier of γ which yields a propositional resolution refutation γ^* does not insert terms deeper than $2\tau(\mathcal{C})$ (this follows from the property that terms of depth greater than $\tau(\mathcal{C})$ are ground – see [61] Theorem 5.2.1). So also after global unification the clauses in γ^* are still of depth ≤ 2 and so

$$\|\gamma^*\| \leq \max\{\|C\| \mid C \in R^*_{<_d}(\mathcal{C})\} * r * 2^{2^{3r^2}}.$$

By Lemma 8.1.2 we get

$$\|\gamma^*\| \leq 2r(\tau(\mathcal{C}) + 2) * r * 2^{2^{3r^2}}.$$

Obviously $\tau(\mathcal{C}), |\mathrm{PS}(\Sigma)|, |\mathrm{FS}(\Sigma)|, |\mathrm{CS}(\Sigma)|$ are all bound by $\|\varphi\|$ and so $\|\gamma^*\|$ is elementary in $\|\varphi\|$. Eventually we obtain for the CERES normal form φ^* corresponding to γ^*

$$\|\varphi^*\| \leq k\|\gamma^*\|\|\varphi\|$$

for a constant k measuring additional contractions in the CERES normal form. So $\|\varphi^*\|$ is elementary in $\|\varphi\|$. □

The next theorem shows that all reductive methods based on $\mathcal{R}$ (see Definition 5.1.6) define only nonelementary cut-elimination sequences on **MC**. Therefore neither Gentzen's nor Tait's method can be used to prove that fast cut-elimination is possible on **MC**.

Theorem 8.1.5 *Cut-elimination based on $\mathcal{R}$ is nonelementary on* **MC**.

Proof: We prove that there exists no elementary bound on cut-elimination sequences based on $\mathcal{R}$ in terms of the size of the input proof. To this aim we choose the worst-case proof sequence $(\rho_n)_{n\in\mathbb{N}}$ of V.P. Orevkov [67]. The skolemization of $(\rho_n)_{n\in\mathbb{N}}$ yields a new proof sequence $(\xi_n)_{n\in\mathbb{N}}$ with only one unary function symbol f at the term level and only one ternary predicate symbol P at the level of atomic formulas. The end sequent of ξ_n is

$$\begin{array}{l}(\forall w)P(w,c,f(w)),\\ (\forall u,v,w)((\exists y)(P(y,c,u)\wedge(\exists z)(P(v,y,z)\wedge P(z,y,w)))\rightarrow P(v,u,w))\\ \quad\vdash\\ (\exists v_n)(P(c,c,v_n)\wedge(\exists v_{n-1})(P(c,v_n,v_{n-1})\wedge\ldots\wedge(\exists v_0)P(c,v_1,v_0)\ldots)),\end{array}$$

and the (only) cut formula $A_n(c)$ in ξ_n is defined inductively as

$$\begin{array}{l}A_0(\alpha)\equiv(\forall w_0)(\exists v_0)P(w_0,\alpha,v_0),\ \ \bar{A}_0(\alpha,\delta)\equiv(\exists v_0)P(\alpha,\delta,v_0),\\ \bar{A}_{i+1}(\alpha,\delta)\equiv(\exists v_{i+1})(A_i(v_{i+1})\wedge P(\alpha,\delta,v_{i+1})),\\ A_{i+1}(\alpha)\equiv(\forall w_{i+1})(A_i(w_{i+1})\rightarrow\bar{A}_{i+1}(w_{i+1},\alpha)).\end{array}$$

Also the new skolemized sequence $(\xi_n)_{n\in\mathbb{N}}$ is of nonelementary complexity for cut-elimination. Now we replace the predicate $\lambda x,y,z.P(x,y,z)$ by the following conjunction of new unary predicates:

$$\lambda x,y,z((Q_1(x)\wedge Q_2(y))\wedge Q_3(z)).$$

everywhere in the proof sequence and obtain a new sequence $(\varphi_n)_{n\in\mathbb{N}}$ of proofs belonging to the class **MC**. Therefore CERES is fast on $\{T(\varphi_n)\mid n\in\mathbb{N}\}$. But every $\mathcal{R}$-reduction step on φ_n is completely isomorphic to the corresponding step performed on ξ_n. Therefore all cut-elimination sequences on $(\varphi_n)_{n\in\mathbb{N}}$ are as long as those on the original sequence $(\xi_n)_{n\in\mathbb{N}}$. As a consequence the Gentzen procedure is of nonelementary complexity on $(\rho_n)_{n\in\mathbb{N}}$. □

Finally we want to illustrate the limitations of the CERES-method as a tool to prove fast cut-elimination.

Definition 8.1.16 The set of *quasi-monotone* formulas is defined inductively in the following way:

1. Atomic formulas and $\bot$ (representing falsum) are quasi-monotone.
2. If A and B are quasi-monotone then $(\forall x)A'$, $(\exists x)A'$ and $A\wedge B$ are quasi-monotone (where x is a bound variable and A' a variant of A containing x in place of a free variable)

3. If A is quasi-monotone and B is monotone then $B \rightarrow A$ is quasi-monotone.

A sequent $\Gamma \vdash \Delta$ is called a *QM-sequent* if Γ is quasi-monotone and Δ is monotone. $\diamond$

Definition 8.1.17 **LK**$\perp$ is **LK** with the standard axiom set and the axiom $\perp \vdash$. $\diamond$

Definition 8.1.18 $\mathcal{QMON}$ is the class of all **LK**$\perp$-proofs ω s.t. (1) the end sequent of ω is a QM-sequent, and (2) all cut formulas are monotone. $\diamond$

Theorem 8.1.6 *Cut-elimination is at most exponential on QMON.*

Proof: In [17, 65].

QMON is an essentially intuitionistic proof class, a feature which was used in the proof projection method defined in [17]. The CERES-method which is a method for classical logic does not distinguish between $\vee\colon l$ and $\rightarrow\colon l$ in the construction of the characteristic clause set, thus "eliminating" the intuitionistic character of the proof. In fact CERES does not yield characteristic clause sets which belong to well-known decidable clause classes. That does not mean that CERES is not fast on QMON; instead we do not know how to prove it by using only refutations of the characteristic clause sets in QMON.

8.2 CERES and the Interpolation Theorem

An interpolant for a valid formula $A \rightarrow B$ is a formula I in the language intersection of A and B such that $A \rightarrow I$ and $I \rightarrow B$ are valid. The existence of interpolants by Craig's lemma is one of the most fundamental properties of classical logic. It demonstrates that only contradictory formulas A and valid B admit the validity of $A \rightarrow B$ if A and B have nothing in common. Implicit definitions can be converted into explicit ones using Beth's theorem. On the other hand it shows a strong limitation of first order logic (contrary to higher order logic): concepts cannot be chosen provably notation invariant. One of the most significant properties of cut-free **LK**-derivations is that they allow by Maehara's lemma a direct construction of interpolants and, thereby, limit their complexity in terms of proof complexity (or even in the length of the proof, provided we first compute a general proof). In this chapter we develop an extension of Maehara's lemma (see [74]).

Definition 8.2.1 Let Γ be a sequence of formulas. We define $\Gamma \sim_p \Pi$ if Π is a permutation variant of Γ. $\diamond$

Note that, obviously, $\sim_p$ is an equivalence relation on sequences of formulas.

Definition 8.2.2 Let $S\colon \Gamma \vdash \Delta$ be a sequent, $\Gamma \sim_p \Gamma_1, \Gamma_2$ and $\Delta \sim_p \Delta_1, \Delta_2$. Then $\langle(\Gamma_1;\Delta_1),(\Gamma_2;\Delta_2)\rangle$ is called a *partition* of S. For two partitions $\mathcal{X}_1\colon \langle(\Gamma_1;\Delta_1),(\Gamma_2;\Delta_2)\rangle$ and $\mathcal{X}_2\colon \langle(\Gamma_1';\Delta_1'),(\Gamma_2';\Delta_2')\rangle$ of S we define $\mathcal{X}_1 = \mathcal{X}_2$ if

$$\Gamma_1 \sim_p \Gamma_1',\ \ \Delta_1 \sim_p \Delta_1',$$
$$\Gamma_2 \sim_p \Gamma_2',\ \ \Delta_2 \sim_p \Delta_2'.$$

$\diamond$

Note that, if $\langle(\Gamma_1;\Delta_1),(\Gamma_2;\Delta_2)\rangle$ is a partition of S, then $(\Gamma_1 \vdash \Delta_1) \circ (\Gamma_2 \vdash \Delta_2)$ is a permutation variant of S.

For technical reasons we extend the axiom set of **LK** by $\bot \vdash$ and $\vdash \top$ (representing false and true).

Definition 8.2.3 Let $\mathcal{A}_T$ be the standard axiom set from Definition 3.2.2. We define $\mathcal{A}_{\top\bot}$ as $\mathcal{A}_T \cup \{\vdash \top\} \cup \{\bot \vdash\}$. $\diamond$

Definition 8.2.4 A $\{\top,\bot\}$-formula is a first-order formula defined over a signature Σ for $\{\top,\bot\} \subseteq \Sigma(\mathrm{PS})$ s.t. $\top$ and $\bot$ are nullary predicate symbols (and thus also atomic formulas). $\diamond$

Definition 8.2.5 Let S be a sequent and $\mathcal{X}\colon \langle(\Gamma_1;\Delta_1),(\Gamma_2;\Delta_2)\rangle$ be a partition of S. A triple $\Phi\colon (C,\varphi_1,\varphi_2)$ is called an *interpolation* of S w.r.t. $\mathcal{X}$ if

(1) C is a $\{\top,\bot\}$-formula.

(2) φ_1 is a proof of $\Gamma_1 \vdash \Delta_1, C$ and φ_2 of $C, \Gamma_2 \vdash \Delta_2$ from $\mathcal{A}_{\top\bot}$.

(3) $\mathrm{PS}(C) \subseteq (\mathrm{PS}(\Gamma_1,\Delta_1) \cap \mathrm{PS}(\Gamma_2,\Delta_2)) \cup \{\top,\bot\}$.

(4) $V(C) \subseteq V(\Gamma_1,\Delta_1) \cap V(\Gamma_2,\Delta_2)$.

(5) $\mathrm{CS}(C) \subseteq \mathrm{CS}(\Gamma_1,\Delta_1) \cap \mathrm{CS}(\Gamma_2,\Delta_2)$.

(6) $\mathrm{FS}(C) \subseteq \mathrm{FS}(\Gamma_1,\Delta_1) \cap \mathrm{FS}(\Gamma_2,\Delta_2)$.

If only the conditions (1)–(3) hold then we call Φ a *weak* interpolation. The formula C in an interpolation $(C, \varphi_1, \varphi_2)$ is called an *interpolant* (and a weak interpolant for weak interpolations).
The pair of sequents $((\Gamma_1 \vdash \Delta_1, C), (C, \Gamma_2 \vdash \Delta_2))$ is called an *interpolation pair* of Φ. A proof of the form

$$\frac{\begin{array}{cc}(\varphi_1) & (\varphi_2)\\ \Gamma_1 \vdash \Delta_1, C & C, \Gamma_2 \vdash \Delta_2\end{array}}{\Gamma_1, \Gamma_2 \vdash \Delta_1, \Delta_2}\ cut$$

is called a (weak) *interpolation derivation* for S w.r.t. $\mathcal{X}$. ◇

Remark: Note that the sequent S in Definition 8.2.5 is always provable; indeed, a (weak) interpolation derivation for S w.r.t. a partition is a proof of a permutation variant of S. ◇

Interpolants of provable sequents S do not only exist, but can be constructed from cut-free proofs of S as shown in Maehara's lemma (see [74]). We prove the lemma (in fact a stronger version than that given in [74]) in two stages. In the first step we construct a weak interpolation and from this we construct a full one.

Lemma 8.2.1 *Let S be a sequent which is provable in* **LK** *from $\mathcal{A}_{\top\bot}$, and $\mathcal{X}$ be a partition of S. Then there exists a weak interpolation for S w.r.t. $\mathcal{X}$.*

Proof: We prove by induction on $l(\varphi)$ that, for a cut-free proof φ of S and for a partition $\mathcal{X}$ of S, there exists a weak interpolation derivation ψ for S w.r.t. $\mathcal{X}$. Note that, by the cut-elimination theorem, there exists always a cut-free proof of S.
Induction base $l(\varphi) = 1$.
Then φ is either of the form $A \vdash A$ for an atomic formula A, or $\vdash \top$ or $\bot \vdash$. We consider first axioms of type $A \vdash A$. We distinguish four partitions $\mathcal{X}$ of $A \vdash A$:

(1) $\mathcal{X} = \langle (;A), (A;) \rangle$. The corresponding weak interpolation derivation ψ is

$$\frac{\dfrac{A \vdash A}{\vdash A, \neg A}\ \neg\colon r \qquad \dfrac{A \vdash A}{\neg A, A \vdash}\ \neg\colon l}{A \vdash A}\ cut$$

(2) $\mathcal{X} = \langle (A;), (;A) \rangle$. Then $\psi =$

$$\frac{A \vdash A \quad A \vdash A}{A \vdash A}\ cut$$

(3) $\mathcal{X} = \langle (A; A), (;) \rangle$. We define $\psi =$

$$\cfrac{\cfrac{A \vdash A}{A \vdash A, \bot}\ w{:}\,r \qquad \bot \vdash}{A \vdash A}\ cut$$

(4) $\mathcal{X} = \langle (;), (A; A) \rangle$. Here $\psi =$

$$\cfrac{\vdash \top \qquad \cfrac{A \vdash A}{\top, A \vdash A}\ w{:}\,l}{A \vdash A}$$

In all cases above we see that the cut formula I in the interpolation derivations is indeed a weak interpolant, because

$$\mathrm{PS}(I) \subseteq (\mathrm{PS}(\Gamma_1; \Gamma_2) \cap \mathrm{PS}(\Delta_1, \Delta_2)) \cup \{\top, \bot\}$$

for all $\mathcal{X} = \langle (\Gamma_1; \Gamma_2), (\Delta_1; \Delta_2) \rangle$.

Now we consider the axioms $\vdash \top$ and $\bot \vdash$. Let the axiom be $\vdash \top$. We have to distinguish two partitions

(1) $\mathcal{X} = \langle (; \top), (;) \rangle$. We define $\psi =$

$$\cfrac{\cfrac{\vdash \top}{\vdash \top, \bot}\ w{:}\,r \qquad \bot \vdash}{\vdash \top}\ cut$$

(2) $\mathcal{X} = \langle (;), (; \top) \rangle$. We define $\psi =$

$$\cfrac{\vdash \top \qquad \cfrac{\vdash \top}{\top \vdash \top}\ w{:}\,l}{\vdash \top}\ cut$$

In both cases above the cut formula is obviously an interpolant.
The case of the axiom $\bot \vdash$ is completely analogous.

(IH) assume that for all S having a proof φ with $l(\varphi) \leq n$ and for all partitions $\mathcal{X}$ of S there exists an interpolation derivation for S w.r.t. $\mathcal{X}$.

Now let φ be a proof of S with $l(\varphi) = n + 1$. We distinguish several cases corresponding to the last inference in φ.

(I) The last inference in φ is a structural one.

- The last inference is weakening. We consider only $w{:}\,r$; the case of $w{:}\,l$ is analogous. So $S = \Gamma \vdash \Delta, A$ and $\varphi =$

$$\cfrac{\begin{matrix}(\varphi')\\ \Gamma \vdash \Delta\end{matrix}}{\Gamma \vdash \Delta, A}\ w{:}\,r$$

Let $\mathcal{X} = \langle(\Gamma_1; \Delta_1, A), (\Gamma_2; \Delta_2)\rangle$ be a partition of S. We define the following partition of $\Gamma \vdash \Delta$:

$$\mathcal{X}' = \langle(\Gamma_1; \Delta_1), (\Gamma_2; \Delta_2)\rangle$$

By (IH) there exists an interpolation derivation $\psi' =$

$$\dfrac{\begin{matrix}(\chi_1) \\ \Gamma_1 \vdash \Delta_1, I\end{matrix} \quad \begin{matrix}(\chi_2) \\ I, \Gamma_2 \vdash \Delta_2\end{matrix}}{\Gamma_1, \Gamma_2 \vdash \Delta_1, \Delta_2}\ cut$$

for $\Gamma \vdash \Delta$ w.r.t. $\mathcal{X}'$, where $\Gamma_1, \Gamma_2 \vdash \Delta_1, \Delta_2$ is a permutation variant of $\Gamma \vdash \Delta$. In particular we have

$$(\star)\ \mathrm{PS}(I) \subseteq (\mathrm{PS}(\Gamma_1, \Delta_1) \cap \mathrm{PS}(\Gamma_2, \Delta_2)) \cup \{\top, \bot\}.$$

We define $\psi =$

$$\dfrac{\dfrac{\dfrac{\begin{matrix}(\chi_1) \\ \Gamma_1 \vdash \Delta_1, I\end{matrix}}{\Gamma_1 \vdash \Delta_1, I, A}\ w{:}\,r}{\Gamma_1 \vdash \Delta_1, A, I}\ p{:}\,r \quad \begin{matrix}(\chi_2) \\ I, \Gamma_2 \vdash \Delta_2\end{matrix}}{\Gamma_1, \Gamma_2 \vdash \Delta_1, A, \Delta_2}\ cut$$

The sequent $\Gamma_1, \Gamma_2 \vdash \Delta_1, A, \Delta_2$ is a permutation variant of S. Moreover, by $(\star)$,

$$PS(I) \subseteq (\mathrm{PS}(\Gamma_1, \Delta_1, A) \cap \mathrm{PS}(\Gamma_2, \Delta_2)) \cup \{\top, \bot\}.$$

Therefore, ψ is a weak interpolation derivation for S w.r.t. $\mathcal{X}$.

The case of the partition $\mathcal{X} = \langle(\Gamma_1; \Delta_1), (\Gamma_2; \Delta_2, A)\rangle$ is completely analogous.

• The last inference is a permutation. We consider only the case of a permutation to the right; the other one is analogous. So let $\varphi =$

$$\dfrac{\begin{matrix}\varphi' \\ \Gamma \vdash \Delta'\end{matrix}}{\Gamma \vdash \Delta}\ p{:}\,r$$

and $\mathcal{X} = \langle(\Gamma_1; \Delta_1), (\Gamma_2; \Delta_2)\rangle$ be a partition of $\Gamma \vdash \Delta$. We take the same partition $\mathcal{X}$ for the sequent $\Gamma \vdash \Delta'$. By (IH) there exists an interpolation derivation ψ' of the form

$$\dfrac{\begin{matrix}(\chi_1) \\ \Gamma_1 \vdash \Delta_1, I\end{matrix} \quad \begin{matrix}(\chi_2) \\ I, \Gamma_2 \vdash \Delta_2\end{matrix}}{\Gamma_1, \Gamma_2 \vdash \Delta_1, \Delta_2}\ cut$$

We simply define $\psi = \psi'$ as ψ' itself is also a weak interpolation derivation for S w.r.t. $\mathcal{X}$ because $\Gamma_1, \Gamma_2 \vdash \Delta_1, \Delta_2$ is a permutation variant of $\Gamma \vdash \Delta$.

• The last inference in φ is a contraction. We consider only the case $c\colon r$. So let $\varphi =$

$$\frac{\begin{matrix}(\varphi')\\ \Gamma \vdash \Delta, A, A\end{matrix}}{\Gamma \vdash \Delta, A}\; c\colon r$$

and $\mathcal{X} = \langle(\Gamma_1; \Delta_1), (\Gamma_2; \Delta_2, A)\rangle$. We define the partition

$$\mathcal{X}' = \langle(\Gamma_1; \Delta_1), (\Gamma_2; \Delta_2, A, A)\rangle$$

of $\Gamma \vdash \Delta, A, A$. By (IH) there exists an interpolation derivation for $\Gamma \vdash \Delta, A, A$ w.r.t. $\mathcal{X}'$ of the form $\psi' =$

$$\frac{\begin{matrix}(\chi_1)\\ \Gamma_1 \vdash \Delta_1, I\end{matrix} \quad \begin{matrix}(\chi_2)\\ I, \Gamma_2 \vdash \Delta_2, A, A\end{matrix}}{\Gamma_1, \Gamma_2 \vdash \Delta_1, \Delta_2, A, A}\; cut$$

and

$$\mathrm{PS}(I) \subseteq (\mathrm{PS}(\Gamma_1, \Delta_1) \cap \mathrm{PS}(\Gamma_2, \Delta_2, A, A)) \cup \{\top, \bot\}.$$

We define $\psi =$

$$\frac{\begin{matrix}(\chi_1)\\ \Gamma_1 \vdash \Delta_1, I\end{matrix} \quad \dfrac{\begin{matrix}(\chi_2)\\ I, \Gamma_2 \vdash \Delta_2, A, A\end{matrix}}{I, \Gamma_2 \vdash \Delta_2, A}\; c\colon r}{\Gamma_1, \Gamma_2 \vdash \Delta_1, \Delta_2, A}\; cut$$

ψ is a weak interpolation derivation for $\Gamma \vdash \Delta, A$ w.r.t. $\mathcal{X}$ by

$$\mathrm{PS}(\Gamma_2, \Delta_2, A, A) = \mathrm{PS}(\Gamma_2, \Delta_2, A).$$

Now let $\mathcal{X} = \langle(\Gamma_1; \Delta_1, A), (\Gamma_2; \Delta_2)\rangle$. We define

$$\mathcal{X}' = \langle(\Gamma_1; \Delta_1, A, A), (\Gamma_2; \Delta_2)\rangle$$

as a partition of $\Gamma \vdash \Delta, A, A$. By (IH) there exists a weak interpolation derivation $\psi' =$

$$\frac{\begin{matrix}(\chi_1)\\ \Gamma_1 \vdash \Delta_1, A, A, I\end{matrix} \quad \begin{matrix}(\chi_2)\\ I, \Gamma_2 \vdash \Delta_2\end{matrix}}{\Gamma_1, \Gamma_2 \vdash \Delta_1, A, A, \Delta_2}\; cut$$

We define $\psi =$

$$\dfrac{\dfrac{\dfrac{\overset{(\chi_1)}{\Gamma_1 \vdash \Delta_1, A, A, I}}{\Gamma_1 \vdash \Delta_1, I, A}\ p{:}\,r + c{:}\,r}{\Gamma_1 \vdash \Delta_1, A, I}\ p{:}\,r \qquad \overset{(\chi_2)}{I, \Gamma_2 \vdash \Delta_2}}{\Gamma_1, \Gamma_2 \vdash \Delta_1, A, \Delta_2}\ cut$$

(II) The last inference is a logical one. We consider only the cases $\neg{:}\,l$, $\wedge : r$, $\forall{:}\,l$ and $\forall{:}\,r$. It turns out that, in case the last rule is a unary propositional rule, the weak interpolant is that of the induction hypothesis. In case of binary rules the interpolants are either conjunctions or disjunctions of weak interpolants. In case of a quantifier rule, abstraction of an interpolant may become necessary.

- The last rule of φ is $\neg{:}\,l$. Then φ is of the form

$$\dfrac{\overset{(\varphi')}{\Gamma \vdash \Delta, A}}{\neg A, \Gamma \vdash \Delta}\ \neg{:}\,l$$

Let $\mathcal{X} = \langle(\neg A, \Gamma_1; \Delta_1), (\Gamma_2; \Delta_2)\rangle$. We define the partition

$$\mathcal{X}' = \langle(\Gamma_1; \Delta_1, A), (\Gamma_2; \Delta_2)\rangle$$

of $\Gamma \vdash \Delta, A$. By (IH) there exists a weak interpolation derivation $\psi' =$

$$\dfrac{\overset{(\chi_1)}{\Gamma_1 \vdash \Delta_1, A, I} \quad \overset{(\chi_2)}{I, \Gamma_2 \vdash \Delta_2}}{\Gamma_1, \Gamma_2 \vdash \Delta_1, A, \Delta_2}\ cut$$

s.t.

$$\mathrm{PS}(I) \subseteq (\mathrm{PS}(\Gamma_1, \Delta_1, A) \cap \mathrm{PS}(\Gamma_2, \Delta_2)) \cup \{\top, \bot\}.$$

We define $\psi =$

$$\dfrac{\dfrac{\overset{(\chi_1)}{\Gamma_1 \vdash \Delta_1, A, I}}{\neg A, \Gamma_1 \vdash \Delta_1, I}\ p{:}\,r + \neg{:}\,l \qquad \overset{(\chi_2)}{I, \Gamma_2 \vdash \Delta_2}}{\neg A, \Gamma_1, \Gamma_2 \vdash \Delta_1, \Delta_2}\ cut$$

By $\mathrm{PS}(\Gamma_1, \Delta_1, A) = \mathrm{PS}(\neg A, \Gamma_1, \Delta_1)$ ψ is indeed a weak interpolation derivation for $\neg A, \Gamma \vdash \Delta$ w.r.t. $\mathcal{X}$.

Now let $\mathcal{X} = \langle(\Gamma_1; \Delta_1), (\neg A, \Gamma_2; \Delta_2)\rangle$. We define the partition

$$\mathcal{X}' = \langle(\Gamma_1; \Delta_1), (\Gamma_2; \Delta_2, A)\rangle$$

of $\Gamma \vdash \Delta, A$. By (IH) there exists a weak interpolation derivation $\psi' =$

$$\dfrac{\begin{matrix}(\chi_1) \\ \Gamma_1 \vdash \Delta_1, I\end{matrix} \quad \begin{matrix}(\chi_2) \\ I, \Gamma_2 \vdash \Delta_2, A\end{matrix}}{\Gamma_1, \Gamma_2 \vdash \Delta_1, \Delta_2, A}\ cut$$

s.t.

$$\mathrm{PS}(I) \subseteq (\mathrm{PS}(\Gamma_1, \Delta_1) \cap \mathrm{PS}(\Gamma_2, \Delta_2, A)) \cup \{\top, \bot\}.$$

We define $\psi =$

$$\dfrac{\begin{matrix}(\chi_1) \\ \Gamma_1 \vdash \Delta_1, I\end{matrix} \quad \dfrac{\begin{matrix}(\chi_2) \\ I, \Gamma_2 \vdash \Delta_2, A\end{matrix}}{I, \neg A, \Gamma_2 \vdash \Delta_2}\ \neg{:}\, l + p{:}\, l}{\Gamma_1, \neg A, \Gamma_2 \vdash \Delta_1, \Delta_2}\ cut$$

- The last rule is $\wedge{:}\, r$. Then φ is of the form

$$\dfrac{\begin{matrix}(\varphi_1) \\ \Gamma \vdash \Delta, A\end{matrix} \quad \begin{matrix}(\varphi_2) \\ \Gamma \vdash \Delta, B\end{matrix}}{\Gamma \vdash \Delta, A \wedge B}\ \wedge{:}\, r$$

Let $\mathcal{X} = \langle(\Gamma_1; \Delta_1), (\Gamma_2; \Delta_2, A \wedge B)\rangle$. We define the partitions

$$\begin{aligned} \mathcal{X}_1 &= \langle(\Gamma_1; \Delta_1), (\Gamma_2; \Delta_2, A)\rangle \text{ for } \Gamma \vdash \Delta, A \text{ and} \\ \mathcal{X}_2 &= \langle(\Gamma_1; \Delta_1), (\Gamma_2; \Delta_2, B)\rangle \text{ for } \Gamma \vdash \Delta, B. \end{aligned}$$

By (IH) there exist weak interpolation derivations ψ_1 for $\Gamma \vdash \Delta, A$ w.r.t. $\mathcal{X}_1$, and ψ_2 for $\Gamma \vdash \Delta, B$ w.r.t. $\mathcal{X}_2$ of the following form: $\psi_1 =$

$$\dfrac{\begin{matrix}(\psi_{1,1}) \\ \Gamma_1 \vdash \Delta_1, I\end{matrix} \quad \begin{matrix}(\psi_{1,2}) \\ I, \Gamma_2 \vdash \Delta_2, A\end{matrix}}{\Gamma_1, \Gamma_2 \vdash \Delta_1, \Delta_2, A}\ cut$$

and $\psi_2 =$

$$\dfrac{\begin{matrix}(\psi_{2,1}) \\ \Gamma_1 \vdash \Delta_1, J\end{matrix} \quad \begin{matrix}(\psi_{2,2}) \\ J, \Gamma_2 \vdash \Delta_2, B\end{matrix}}{\Gamma_1, \Gamma_2 \vdash \Delta_1, \Delta_2, B}\ cut$$

Moreover we have

$$\begin{aligned} \mathrm{PS}(I) &\subseteq (\mathrm{PS}(\Gamma_1, \Delta_1) \cap \mathrm{PS}(\Gamma_2, \Delta_2, A)) \cup \{\top, \bot\} \\ \mathrm{PS}(J) &\subseteq (\mathrm{PS}(\Gamma_1, \Delta_1) \cap \mathrm{PS}(\Gamma_2, \Delta_2, B)) \cup \{\top, \bot\}. \end{aligned}$$

We define the weak interpolation derivation ψ for $\Gamma \vdash \Delta, A \wedge B$ w.r.t. $\mathcal{X}$:

$$\dfrac{\dfrac{\stackrel{(\psi_{1,1})}{\Gamma_1 \vdash \Delta_1, I} \quad \stackrel{(\psi_{2,1})}{\Gamma_1 \vdash \Delta_1, J}}{\Gamma_1 \vdash \Delta_1, I \wedge J}\ \wedge: r \qquad \dfrac{\dfrac{\stackrel{(\psi_{1,2})}{I, \Gamma_2 \vdash \Delta_2, A}}{I \wedge J, \Gamma_2 \vdash \Delta_2, A}\ \wedge: l_2 \quad \dfrac{\stackrel{(\psi_{2,2})}{J, \Gamma_2 \vdash \Delta_2, B}}{I \wedge J, \Gamma_2 \vdash \Delta_2, B}\ \wedge: l_1}{I \wedge J, \Gamma_2 \vdash \Delta_2, A \wedge B}\ \wedge: r}{\Gamma_1, \Gamma_2 \vdash \Delta_1, \Delta_2, A \wedge B}\ cut$$

$I \wedge J$ is indeed a weak interpolant by

$$\begin{aligned} \mathrm{PS}(I \wedge J) &\subseteq (\mathrm{PS}(\Gamma_1, \Delta_1) \cap (\mathrm{PS}(\Gamma_2, \Delta_2, A) \cup \mathrm{PS}(\Gamma_2, \Delta_2, B))) \cup \{\top, \bot\} \\ &= (\mathrm{PS}(\Gamma_1, \Delta_1) \cap \mathrm{PS}(\Gamma_2, \Delta_2, A \wedge B)) \cup \{\top, \bot\}. \end{aligned}$$

Now let $\mathcal{X} = \langle (\Gamma_1; \Delta_1, A \wedge B), (\Gamma_2; \Delta_2) \rangle$.
We define the partitions of the premise sequents

$$\begin{aligned} \mathcal{X}_1 &= \langle (\Gamma_1; \Delta_1, A), (\Gamma_2; \Delta_2) \rangle \text{ for } \Gamma \vdash \Delta, A \text{ and} \\ \mathcal{X}_2 &= \langle (\Gamma_1; \Delta_1, B), (\Gamma_2; \Delta_2) \rangle \text{ for } \Gamma \vdash \Delta, B. \end{aligned}$$

By (IH) there exist weak interpolation derivations ψ_1 for $\Gamma \vdash \Delta, A$ w.r.t. $\mathcal{X}_1$, and ψ_2 for $\Gamma \vdash \Delta, B$ w.r.t. $\mathcal{X}_2$. We have $\psi_1 =$

$$\dfrac{\stackrel{(\psi_{1,1})}{\Gamma_1 \vdash \Delta_1, A, I} \quad \stackrel{(\psi_{1,2})}{I, \Gamma_2 \vdash \Delta_2}}{\Gamma_1, \Gamma_2 \vdash \Delta_1, A, \Delta_2}\ cut$$

and $\psi_2 =$

$$\dfrac{\stackrel{(\psi_{2,1})}{\Gamma_1 \vdash \Delta_1, B, J} \quad \stackrel{(\psi_{2,2})}{J, \Gamma_2 \vdash \Delta_2}}{\Gamma_1, \Gamma_2 \vdash \Delta_1, B, \Delta_2}\ cut$$

As I and J are weak interpolants we have

$$\begin{aligned} \mathrm{PS}(I) &\subseteq (\mathrm{PS}(\Gamma_1, \Delta_1, A) \cap \mathrm{PS}(\Gamma_2, \Delta_2)) \cup \{\top, \bot\}, \\ \mathrm{PS}(J) &\subseteq (\mathrm{PS}(\Gamma_1, \Delta_1, B) \cap \mathrm{PS}(\Gamma_2, \Delta_2)) \cup \{\top, \bot\}. \end{aligned}$$

We define $\psi =$

$$\dfrac{\dfrac{\dfrac{\dfrac{\dfrac{\stackrel{(\psi_{1,1})}{\Gamma_1 \vdash \Delta_1, A, I}}{\Gamma_1 \vdash \Delta_1, A, I \vee J}\ \vee: r_1}{\Gamma_1 \vdash \Delta_1, I \vee J, A}\ p: r \quad \dfrac{\dfrac{\stackrel{(\psi_{2,1})}{\Gamma_1 \vdash \Delta_1, B, J}}{\Gamma_1 \vdash \Delta_1, B, I \vee J}\ \vee: r_2}{\Gamma_1 \vdash \Delta_1, I \vee J, B}\ p: r}{\Gamma_1 \vdash \Delta_1, I \vee J, A \wedge B}\ \wedge: r}{\Gamma_1 \vdash \Delta_1, A \wedge B, I \vee J}\ p: r \qquad \dfrac{\stackrel{(\psi_{1,2})}{I, \Gamma_2 \vdash \Delta_2} \quad \stackrel{(\psi_{2,2})}{J, \Gamma_2 \vdash \Delta_2}}{I \vee J, \Gamma_2 \vdash \Delta_2}\ \vee: l}{\Gamma_1, \Gamma_2 \vdash \Delta_1, A \wedge B, \Delta_2}\ cut$$

ψ is indeed a weak interpolation derivation for $\Gamma \vdash \Delta, A \wedge B$ w.r.t. $\mathcal{X}$ by

$$\begin{aligned} \mathrm{PS}(I \vee J) &= \mathrm{PS}(I) \cup \mathrm{PS}(J) \\ &\subseteq ((\mathrm{PS}(\Gamma_1, \Delta_1, A) \cup \mathrm{PS}(\Gamma_1, \Delta_1, B)) \cap \mathrm{PS}(\Gamma_2, \Delta_2)) \cup \{\top, \bot\} \\ &= (\mathrm{PS}(\Gamma_1, \Delta_1, A \vee B) \cap \mathrm{PS}(\Gamma_2, \Delta_2)) \cup \{\top, \bot\}. \end{aligned}$$

- The last rule in φ is $\forall\colon l$. Then φ is of the form

$$\frac{\begin{matrix}(\varphi') \\ A\{x \leftarrow t\}, \Gamma \vdash \Delta\end{matrix}}{(\forall x)A, \Gamma \vdash \Delta}\ \forall\colon l$$

Let $\mathcal{X} = \langle(((\forall x)A, \Gamma_1; \Delta_1), (\Gamma_2; \Delta_2)\rangle$. We define

$$\mathcal{X}' = \langle(A\{x \leftarrow t\}, \Gamma_1; \Delta_1), (\Gamma_2; \Delta_2)\rangle$$

as partition of $A\{x \leftarrow t\}, \Gamma \vdash \Delta$. By (IH) there exists a weak interpolation derivation ψ' of the form

$$\frac{\begin{matrix}(\chi_1) \\ A\{x \leftarrow t\}, \Gamma_1 \vdash \Delta_1, I\end{matrix} \quad \begin{matrix}(\chi_2) \\ I, \Gamma_2 \vdash \Delta_2\end{matrix}}{A\{x \leftarrow t\}, \Gamma_1, \Gamma_2 \vdash \Delta_1, \Delta_2}\ cut$$

where

$$\mathrm{PS}(I) \subseteq (\mathrm{PS}(A\{x \leftarrow t\}, \Gamma_1, \Delta_1) \cap \mathrm{PS}(\Gamma_2, \Delta_2)) \cup \{\top, \bot\}.$$

We define $\psi =$

$$\frac{\dfrac{\begin{matrix}(\chi_1) \\ A\{x \leftarrow t\}, \Gamma_1 \vdash \Delta_1, I\end{matrix}}{(\forall x)A, \Gamma_1 \vdash \Delta_1, I}\ \forall\colon l \quad \begin{matrix}(\chi_2) \\ I, \Gamma_2 \vdash \Delta_2\end{matrix}}{(\forall x)A, \Gamma_1, \Gamma_2 \vdash \Delta_1, \Delta_2}\ cut$$

ψ is a weak interpolation derivation by $\mathrm{PS}((\forall x)A) = \mathrm{PS}(A\{x \leftarrow t\})$.
Note that, in general, I is not a (full) interpolant for $(\forall x)A, \Gamma \vdash \Delta$ w.r.t. $\mathcal{X}$ even if I is a (full) interpolant for $A\{x \leftarrow t\}, \Gamma \vdash \Delta$ w.r.t. $\mathcal{X}'$. Indeed, by the rule $(\forall\colon l)$, some function symbols, constants, or variables can be removed from one side which still occur in I.
The case of the partition $\langle((\Gamma_1; \Delta_1), ((\forall x)A, \Gamma_2; \Delta_2)\rangle$ is analogous.

- The last rule in φ is $\forall\colon r$. Then φ is of the form

$$\frac{\begin{matrix}(\varphi') \\ \Gamma \vdash \Delta, A\{x \leftarrow \alpha\}\end{matrix}}{\Gamma \vdash \Delta, (\forall x)A}\ \forall\colon r$$

where α does not occur in $\Gamma \vdash \Delta, (\forall x)A$. We consider the partition $\mathcal{X} = \langle(\Gamma_1;\Delta_1), (\Gamma_2;\Delta_2,(\forall x)A)\rangle$ of $\Gamma \vdash \Delta, (\forall x)A$. Let

$$\mathcal{X}' = \langle(\Gamma_1;\Delta_1), (\Gamma_2;\Delta_2, A\{x \leftarrow \alpha\})\rangle$$

be the corresponding partition of $S'\colon \Gamma \vdash \Delta, A\{x \leftarrow \alpha\}$. By (IH) there exists a weak interpolation derivation ψ' for S' w.r.t. $\mathcal{X}'$. ψ' is of the form

$$\frac{\begin{matrix}(\chi_1) \\ \Gamma_1 \vdash \Delta_1, I\end{matrix} \quad \begin{matrix}(\chi_2) \\ I, \Gamma_2 \vdash \Delta_2, A\{x \leftarrow \alpha\}\end{matrix}}{\Gamma_1, \Gamma_2 \vdash \Delta_1, \Delta_2, A\{x \leftarrow \alpha\}}\ cut$$

and

$$\mathrm{PS}(I) \subseteq (\mathrm{PS}(\Gamma_1 \vdash \Delta_1) \cap \mathrm{PS}(\Gamma_2 \vdash \Delta_2, A\{x \leftarrow \alpha\})) \cup \{\top, \bot\}.$$

For the weak interpolation derivation ψ we distinguish two cases:

(a) α does not occur in I. Then, as α is an eigenvariable, α does not occur in $I, \Gamma_2 \vdash \Delta_2$ either. Therefore we may define $\psi =$

$$\frac{\begin{matrix}(\chi_1) \\ \Gamma_1 \vdash \Delta_1, I\end{matrix} \quad \dfrac{\begin{matrix}(\chi_2) \\ I, \Gamma_2 \vdash \Delta_2, A\{x \leftarrow \alpha\}\end{matrix}}{I, \Gamma_2 \vdash \Delta_2, (\forall x)A}\ \forall\colon r}{\Gamma_1, \Gamma_2 \vdash \Delta_1, \Delta_2, (\forall x)A}\ cut$$

ψ is a weak interpolation derivation for $\Gamma \vdash \Delta, (\forall x)A$ w.r.t. $\mathcal{X}$ (note that $\mathrm{PS}((\forall x)A) = \mathrm{PS}(A\{x \leftarrow \alpha\})$).

(b) α occurs in I. As α is an eigenvariable it does not occur in $\Gamma_1, \Gamma_2 \vdash \Delta_1, \Delta_2$. So we may define $\psi =$

$$\frac{\dfrac{\begin{matrix}(\chi_1) \\ \Gamma_1 \vdash \Delta_1, I\end{matrix}}{\Gamma_1 \vdash \Delta_1, (\forall x)I\{\alpha \leftarrow x\}}\ \forall\colon r \quad \dfrac{\dfrac{\begin{matrix}(\chi_2) \\ I, \Gamma_2 \vdash \Delta_2, A\{x \leftarrow \alpha\}\end{matrix}}{(\forall x)I\{\alpha \leftarrow x\}, \Gamma_2 \vdash \Delta_2, A\{x \leftarrow \alpha\}}\ \forall\colon l}{(\forall x)I\{\alpha \leftarrow x\}, \Gamma_2 \vdash \Delta_2, (\forall x)A}\ \forall\colon r}{\Gamma_1, \Gamma_2 \vdash \Delta_1, \Delta_2, (\forall x)A}\ cut$$

Note that, by

$$\mathrm{PS}(I) = \mathrm{PS}((\forall x)I\{\alpha \leftarrow x\}),$$

the new interpolant is $(\forall x)I\{\alpha \leftarrow x\}$.
Now let $\mathcal{X} = \langle(\Gamma_1;\Delta_1,(\forall x)A), (\Gamma_2;\Delta_2)\rangle$. To $\mathcal{X}$ we define

$$\mathcal{X}' = \langle(\Gamma_1;\Delta_1, A\{x \leftarrow \alpha\}), (\Gamma_2;\Delta_2)\rangle.$$

By (IH) we have the following weak interpolation derivation ψ' w.r.t. $\mathcal{X}'$:

$$\dfrac{\begin{array}{cc}(\chi_1) & (\chi_2)\\ \Gamma_1 \vdash \Delta_1, A\{x \leftarrow \alpha\}, I & I, \Gamma_2 \vdash \Delta_2\end{array}}{\Gamma_1, \Gamma_2 \vdash \Delta_1, A\{x \leftarrow \alpha\}, \Delta_2}\ cut$$

again we distinguish two cases in the construction of ψ:

(a) α does not occur in I. Then we define $\psi =$

$$\dfrac{\dfrac{\dfrac{\dfrac{\dfrac{(\chi_1)}{\Gamma_1 \vdash \Delta_1, A\{x \leftarrow \alpha\}, I}}{\Gamma_1 \vdash \Delta_1, I, A\{x \leftarrow \alpha\}}\ p{:}\,r}{\Gamma_1 \vdash \Delta_1, I, (\forall x)A}\ \forall{:}\,r}{\Gamma_1 \vdash \Delta_1, (\forall x)A, I}\ p{:}\,r \quad \begin{array}{c}(\chi_2)\\ I, \Gamma_2 \vdash \Delta_2\end{array}}{\Gamma_1, \Gamma_2 \vdash \Delta_1, (\forall x)A, \Delta_2}\ cut$$

(b) α occurs in I. Here we define $\psi =$

$$\dfrac{\dfrac{\dfrac{\dfrac{\dfrac{\begin{array}{c}(\chi_1)\\ \Gamma_1 \vdash \Delta_1, A\{x \leftarrow \alpha\}, I\end{array}}{\Gamma_1 \vdash \Delta_1, A\{x \leftarrow \alpha\}, (\exists x)I\{\alpha \leftarrow x\}}\ \exists{:}\,r}{\Gamma_1 \vdash \Delta_1, (\exists x)I\{\alpha \leftarrow x\}, A\{x \leftarrow \alpha\}}\ p{:}\,r}{\Gamma_1 \vdash \Delta_1, (\exists x)I\{\alpha \leftarrow x\}, (\forall x)A}\ \forall{:}\,r}{\Gamma_1 \vdash \Delta_1, (\forall x)A, (\exists x)I\{\alpha \leftarrow x\}}\ p{:}\,r \quad \dfrac{\begin{array}{c}(\chi_2)\\ I, \Gamma_2 \vdash \Delta_2\end{array}}{(\exists x)I\{\alpha \leftarrow x\}, \Gamma_2 \vdash \Delta_2}\ \exists{:}\,l}{\Gamma_1, \Gamma_2 \vdash \Delta_1, (\forall x)A, \Delta_2}\ cut$$

The weak interpolant here is $(\exists x)I\{\alpha \leftarrow x\}$.

Note that, for $\forall{:}\,r$ being the last rule, the weak interpolant is either I, $(\forall x)I\{\alpha \leftarrow x\}$ or $(\exists x)I\{\alpha \leftarrow x\}$. □

Definition 8.2.6 Let A be a formula s.t. $A = A[t]_{\mathcal{M}}$ for a term t and a set of occurrences $\mathcal{M}$ of t in A. Let x be a bound variable which does not occur in A. Then the formulas $(\forall x)A[x]_{\mathcal{M}}$ and $(\exists x)A[x]_{\mathcal{M}}$ are called *abstractions* of A. We define abstraction to be closed under reflexivity and transitivity: A is an abstraction of A; if A is an abstraction of B and B is an abstraction of C then A is an abstraction of C. ◇

Definition 8.2.7 Let $\Phi\colon (C, \varphi_1, \varphi_2)$ be a weak interpolation of a sequent S w.r.t. a partition $\langle(\Gamma; \Delta), (\Pi; \Lambda)\rangle$ of S. A term t is called *critical* for Φ if either

- $t \in V(C)$ and $t \notin V(\Gamma,\Delta) \cap V(\Pi,\Lambda)$ or
- $t \in \mathrm{CS}(C)$ and $t \notin \mathrm{CS}(\Gamma,\Delta) \cap \mathrm{CS}(\Pi,\Lambda)$ or
- t is of the form $f(t_1,\ldots,t_m)$, $f \in \mathrm{FS}(C)$ and $f \notin \mathrm{FS}(\Gamma,\Delta) \cap \mathrm{FS}(\Pi,\Lambda)$.

$\diamond$

Remark: A term t is critical for an interpolation Φ w.r.t. $\mathcal{X}$ if it occurs in the interpolant and in one part of the partition $\mathcal{X}$, but not in the other one (or it does not occur at all in $\mathcal{X}$). By definition, interpolants (in contrast to weak interpolants) may not contain critical terms. $\diamond$

Lemma 8.2.2 *Let* $\Pi\colon (C,\varphi_1,\varphi_2)$ *be a weak interpolation of a sequent* S *w.r.t. a partition* $\mathcal{X}$*. Then there exists an interpolation* (D,ψ_1,ψ_2) *of* S *w.r.t.* $\mathcal{X}$ *s.t.* D *is an abstraction of* C*.*

Proof: Let $\mathcal{X} = \langle(\Gamma;\Delta),(\Pi;\Lambda)\rangle$ be the partition of S. We start with the weak interpolation derivation $\psi =$

$$\frac{\begin{array}{cc}(\varphi_1) & (\varphi_2)\\ \Gamma \vdash \Delta, C & C, \Pi \vdash \Lambda\end{array}}{\Gamma, \Pi \vdash \Delta, \Lambda}\ cut$$

and transform it into an interpolation derivation. We define $(\psi_i, C_i, \varphi_1^i, \varphi_2^i)$ for all i inductively.

$$\psi_0 = \psi,\ C_0 = C,\ \varphi_1^0 = \varphi_1,\ \varphi_2^0 = \varphi_2.$$

Assume that we have already defined $(\psi_i, C_i, \varphi_1^i, \varphi_2^i)$ and ψ_i is the weak interpolation derivation

$$\frac{\begin{array}{cc}(\varphi_1^i) & (\varphi_2^i)\\ \Gamma \vdash \Delta, C_i & C_i, \Pi \vdash \Lambda\end{array}}{\Gamma, \Pi \vdash \Delta, \Lambda}\ cut$$

of the interpolation $(C_i, \varphi_1^i, \varphi_2^i)$ of S w.r.t. $\mathcal{X}$.
If C_i does not contain critical terms we set

$$\psi_{i+1} = \psi_i,\ C_{i+1} = C_i,\ \varphi_1^{i+1} = \varphi_1^i,\ \varphi_2^{i+1} = \varphi_2^i.$$

Now let us assume that C_i contains critical terms. We select a critical term t s.t. $\|t\|$ is maximal. In particular $C_i = C_i[t]_{\mathcal{M}}$ where $\mathcal{M}$ are all occurrences of t in C_i. If $\|t\| = 1$ then $t \in V(S) \cup \mathrm{CS}(S)$. We consider first the more interesting case $\|t\| > 1$. Then $t = f(t_1, \ldots, t_m)$ for $f \in \mathrm{FS}(S)$. As t is critical we have

$$f \notin \mathrm{FS}(\Gamma, \Delta) \cap \mathrm{FS}(\Pi, \Lambda).$$

We distinguish three cases:

(1) $f \in \mathrm{FS}(\Gamma, \Delta)$ (and, clearly $f \notin \mathrm{FS}(\Pi, \Lambda)$). Consider the proof φ_i^2 of $C[t]_{\mathcal{M}}, \Pi \vdash \Lambda$. By replacing the term t by a free variable α we obtain the formula $C_i[\alpha]_{\mathcal{M}}$ and by replacing all occurrences of t in φ_2 by α we define the proof $\varphi_2^{i+1} =$

$$\cfrac{\begin{array}{c}(\varphi_i^2\{t/\alpha\})\\ C_i[\alpha]_{\mathcal{M}}, \Pi \vdash \Lambda\end{array}}{(\exists x_i)C_i[x_i]_{\mathcal{M}}, \Pi \vdash \Lambda}\ \exists\colon l$$

where x_i is a bound variable not occurring in C_i. Note that φ_2^{i+1} is indeed a proof. First of all $f(t_1, \ldots, t_m)$ does not occur in Π, Λ; so the replacement does not change Π, Λ. Moreover $\varphi_i^2\{t/\alpha\}$ is indeed a proof (no quantifier introduction rules can be damaged by this replacement) and all contraction rules are preserved. Finally the rule $\exists\colon l$ is sound as α does not occur in $(\exists x_i)C_i[x_i]_\lambda, \Pi \vdash \Lambda$.

$\varphi_1^{i+1} =$

$$\cfrac{\begin{array}{c}(\varphi_1^i)\\ \Gamma \vdash \Delta, C_i[t]_{\mathcal{M}}\end{array}}{\Gamma \vdash \Delta, (\exists x_i)C_i[x_i]_{\mathcal{M}}}\ \exists\colon r$$

Finally we define $C_{i+1} = (\exists x_i)C_i[x_i]_{\mathcal{M}}$ and $\psi_{i+1} =$

$$\cfrac{\begin{array}{cc}(\varphi_1^{i+1}) & (\varphi_2^{i+1})\\ \Gamma \vdash \Delta, (\exists x_i)C_i[x_i]_{\mathcal{M}} & (\exists x_i)C_i[x_i]_{\mathcal{M}}, \Pi \vdash \Lambda\end{array}}{\Gamma, \Pi \vdash \Delta, \Lambda}\ cut$$

Clearly ψ_{i+1} is the weak interpolation derivation of the weak interpolation $(C_{i+1}, \varphi_1^{i+1}, \varphi_2^{i+1})$ of S w.r.t. $\mathcal{X}$. Note that C_{i+1} and C_i contain the same predicate symbols!

(2) $f \in \mathrm{FS}(\Pi, \Lambda)$. Then the roles of the sides change and we define $\varphi_1^{i+1} =$

$$\cfrac{\begin{array}{c}(\varphi_1^i\{t/\alpha\})\\ \Gamma \vdash \Delta, C_i[\alpha]_{\mathcal{M}}\end{array}}{\Gamma \vdash \Delta, (\forall x_i)C_i[x_i]_{\mathcal{M}}}\ \forall\colon r$$

$\varphi_2^{i+1} =$

$$\frac{\begin{array}{c}(\varphi_2^i)\\ C_i[t]_{\mathcal{M}}, \Pi \vdash \Delta\end{array}}{(\forall x_i)C_i[x_i]_{\mathcal{M}}, \Pi \vdash \Delta}\ \forall\colon l$$

now we define $C_{i+1} = (\forall x_i)C_i[x_i]_{\mathcal{M}}$ and $\psi_{i+1} =$

$$\frac{\begin{array}{c}(\varphi_1^{i+1})\\ \Gamma \vdash \Delta, (\forall x_i)C_i[x_i]_{\mathcal{M}}\end{array} \quad \begin{array}{c}(\varphi_2^{i+1})\\ (\forall x_i)C_i[x_i]_{\mathcal{M}}, \Pi \vdash \Lambda\end{array}}{\Gamma, \Pi \vdash \Delta, \Lambda}\ cut$$

And, again, ψ_{i+1} is the weak interpolation derivation of the weak interpolation $(C_{i+1}, \varphi_1^{i+1}, \varphi_2^{i+1})$ of S w.r.t. $\mathcal{X}$.

(3) $f \notin \mathrm{FS}(\Gamma, \Pi, \Delta, \Lambda)$. Then we may take the interpolation derivation of (1) or of (2); the weak interpolant C_{i+1} can be chosen as $(\forall x_i)C_i[x_i]_{\mathcal{M}}$ or as $(\exists x_i)C_i[x_i]_{\mathcal{M}}$.

If $\|t\| = 1$ for the critical term with maximal size, then $t \in V \cup \mathrm{CS}$. The construction of the interpolation derivation proceeds exactly as for function terms t, only in the case $t \in V$ we can use t as eigenvariable directly. A constant symbol t is simply replaced by an eigenvariable α.

By the construction above $(C_i, \varphi_1^i, \varphi_2^i)$ is a weak interpolation of S w.r.t. $\mathcal{X}$. But in any step the number of occurrences of critical terms is strictly reduced until some C_i does not contain critical terms anymore. Let r be the number of occurrences of critical terms in C. Then $(C_r, \varphi_1^r, \varphi_2^r)$ is an *interpolation* of S w.r.t. $\mathcal{X}$.
For all i we get $C_{i+1} = (Q_i x_i)C_i[x_i]_{\mathcal{M}_i}$ for $Q_i \in \{\forall, \exists\}$ and $\mathcal{M}_i$ a set of positions in C_i. So by Definition 8.2.6 C_{i+1} is an abstraction of C_i. As the relation "abstraction of" is reflexive and transitive all C_i are abstractions of C; so C_r is an abstraction of C. □

Theorem 8.2.1 (interpolation theorem) *Let S be a sequent which is provable in* **LK** *from $\mathcal{A}_{\top\bot}$, and $\mathcal{X}$ be a partition of S. Then there exists an interpolation of S w.r.t. $\mathcal{X}$.*

Proof: By Lemma 8.2.1 there exists a weak interpolation Φ of S w.r.t. $\mathcal{X}$; by Lemma 8.2.2 we can transform Φ into a full interpolation of S w.r.t. $\mathcal{X}$. □

Lemma 8.2.3 *Let φ be an* **LK**-*proof from $\mathcal{A}_{\top\bot}$ of the form*

$$\frac{\begin{array}{cc}(\varphi_1) & (\varphi_2)\\ \Gamma \vdash \Delta, P(\bar{t}) & P(\bar{t}), \Pi \vdash \Lambda\end{array}}{\Gamma, \Pi \vdash \Delta, \Lambda}\ cut$$

where $P(\bar{t})$ is an atom. Let $\mathcal{X}$ be a partition of the end-sequent $S\colon \Gamma, \Pi \vdash \Delta, \Lambda$. Then there exists a weak interpolation (A, ψ_1, ψ_2) of S w.r.t. $\mathcal{X}$ s.t. either $A = I \wedge J$ or $A = I \vee J$ where I is an interpolant of $\Gamma \vdash \Delta, P(\bar{t})$ and J an interpolant of $P(\bar{t}), \Pi \vdash \Lambda$ (w.r.t. appropriate partitions).

Proof: Let $\mathcal{X} = \langle(\Gamma_1, \Pi_1; \Delta_1, \Lambda_1), (\Gamma_2, \Pi_2; \Delta_2, \Lambda_2)\rangle$ be a partition of S. We construct a weak interpolation derivation for $S\colon \Gamma, \Pi \vdash \Delta, \Lambda$ w.r.t. $\mathcal{X}$. We distinguish the following cases:

(a) $P \in \mathrm{PS}(\Gamma_2, \Pi_2, \Delta_2, \Lambda_2)$. We define the partitions

$$\begin{array}{rcl}\mathcal{X}_1 & = & \langle(\Gamma_1; \Delta_1), (\Gamma_2; \Delta_2, P(\bar{t}))\rangle \text{ of } \Gamma \vdash \Delta, P(\bar{t}),\\ \mathcal{X}_2 & = & \langle(\Pi_1; \Lambda_1), (P(\bar{t}), \Pi_2; \Lambda_2)\rangle \text{ of } P(\bar{t}), \Pi \vdash \Lambda.\end{array}$$

By Theorem 8.2.1 there exist interpolation derivations ψ_1' (w.r.t. $\mathcal{X}_1$) and ψ_2' (w.r.t. $\mathcal{X}_2$) of the following form: $\psi_1' =$

$$\frac{\begin{array}{cc}(\chi_{1,1}) & (\chi_{1,2})\\ \Gamma_1 \vdash \Delta_1, I & I, \Gamma_2 \vdash \Delta_2, P(\bar{t})\end{array}}{\Gamma_1, \Gamma_2 \vdash \Delta_1, \Delta_2, P(\bar{t})}\ cut$$

and $\psi_2' =$

$$\frac{\begin{array}{cc}(\chi_{2,1}) & (\chi_{2,2})\\ \Pi_1 \vdash \Lambda_1, J & J, P(\bar{t}), \Pi_2 \vdash \Lambda_2\end{array}}{\Pi, P(\bar{t}) \vdash \Lambda}\ cut$$

From the proofs $\chi_{i,j}$ we define a weak interpolation derivation ψ for S w.r.t. $\mathcal{X}$ for $\psi =$

$$\frac{\psi_1 \quad \psi_2}{S}\ cut$$

where $\psi_1 =$

$$\frac{\begin{array}{cc}(\chi_{1,1}) & (\chi_{2,1})\\ \Gamma_1 \vdash \Delta_1, I & \Pi_1 \vdash \Lambda_1, J\end{array}}{\Gamma_1, \Pi_1 \vdash \Delta_1, \Lambda_1, I \wedge J}\ \wedge\colon r$$

and $\psi_2 =$

$$\cfrac{\cfrac{\overset{(\chi_{1,2})}{I, \Gamma_2 \vdash \Delta_2, P(\bar{t})} \quad \overset{(\chi_{2,2})}{P(\bar{t}), J, \Pi_2 \vdash \Lambda_2}}{I, J, \Gamma_2, \Pi_2 \vdash \Delta_2, \Lambda_2}\ cut + s^*}{I \wedge J, \Gamma_2, \Pi_2 \vdash \Delta_2, \Lambda_2}\ *$$

Note that, by construction

$$\begin{array}{l} \mathrm{PS}(I \wedge J) \subseteq \mathrm{PS}(\Gamma_1, \Pi_1 \vdash \Delta_1, \Lambda_1) \text{ and} \\ \mathrm{PS}(I \wedge J) \subseteq \mathrm{PS}(\Gamma_2, \Pi_2 \vdash \Delta_2, \Lambda_2) \cup \{P\} = \mathrm{PS}(\Gamma_2, \Pi_2 \vdash \Delta_2, \Lambda_2). \end{array}$$

(b) $P \in \mathrm{PS}(\Gamma_1, \Pi_1 \vdash \Delta_1, \Lambda_1)$.

We define the partitions

$$\begin{array}{rcl} \mathcal{X}_1 & = & \langle(\Gamma_1; \Delta_1, P(\bar{t})), (\Gamma_2; \Delta_2)\rangle \text{ of } \Gamma \vdash \Delta, P(\bar{t}), \\ \mathcal{X}_2 & = & \langle(P(\bar{t}), \Pi_1; \Lambda_1), (\Pi_2; \Lambda_2)\rangle \text{ of } P(\bar{t}), \Pi \vdash \Lambda. \end{array}$$

By Theorem 8.2.1 there exists interpolation derivations ψ_1' and ψ_2' of the following form: $\psi_1' =$

$$\cfrac{\overset{(\chi_{1,1})}{\Gamma_1 \vdash \Delta_1, P(\bar{t}), I} \quad \overset{(\chi_{1,2})}{I, \Gamma_2 \vdash \Delta_2}}{\Gamma_1, \Gamma_2 \vdash \Delta_1, P(\bar{t}), \Delta_2}\ cut$$

and $\psi_2' =$

$$\cfrac{\overset{(\chi_{2,1})}{P(\bar{t}), \Pi_1 \vdash \Lambda_1, J} \quad \overset{(\chi_{2,2})}{J, \Pi_2 \vdash \Lambda_2}}{P(\bar{t}), \Pi_1 \vdash \Lambda_1, \Lambda_2}\ cut$$

From the proofs $\chi_{i,j}$ we define a weak interpolation derivation ψ of the form

$$\cfrac{\psi_1 \quad \psi_2}{S}\ cut$$

where $\psi_1 =$

$$\cfrac{\cfrac{\cfrac{\overset{(\chi_{1,1})}{\Gamma_1 \vdash \Delta_1, P(\bar{t}), I}}{\Gamma_1 \vdash \Delta_1, I, P(\bar{t})}\ p{:}\,r \quad \overset{(\chi_{2,1})}{P(\bar{t}), \Pi_1 \vdash \Lambda_1, J}}{\Gamma_1, \Pi_1 \vdash \Delta_1, I, \Lambda_1, J}\ cut}{\Gamma_1, \Pi_1 \vdash \Delta_1, \Lambda_1, I \vee J}\ *$$

and $\psi_2 =$

$$\cfrac{\begin{array}{cc}(\chi_{1,2}) & (\chi_{2,2}) \\ I, \Gamma_2 \vdash \Delta_2 & J, \Pi_2 \vdash \Lambda_2\end{array}}{I \vee J, \Gamma_2, \Pi_2 \vdash \Delta_2, \Lambda_2} \; \vee\colon l$$

By construction we have

$$\begin{array}{l} \mathrm{PS}(I \vee J) \subseteq \mathrm{PS}(\Gamma_2, \Pi_2 \vdash \Delta_2, \Lambda_2), \\ \mathrm{PS}(I \vee J) \subseteq \mathrm{PS}(\Gamma_1, \Pi_1 \vdash \Delta_1, \Lambda_1) \cup \{P\} = \mathrm{PS}(\Gamma_1, \Pi_1 \vdash \Delta_1, \Lambda_1). \end{array}$$

(c) $P \notin \mathrm{PS}(\Gamma, \Pi \vdash \Delta, \Lambda)$. Then the constructions in (a) and (b) both work; indeed neither I nor J contains P and thus $I \wedge J$ and $I \vee J$ do not contain P. □

Definition 8.2.8 Let S be a set of formulas. A formula F is said to be a $\{\wedge, \vee\}$-combination of S if either

- $F \in S$, or
- $F = F_1 \wedge F_2$ where F_1, F_2 are $\{\wedge, \vee\}$-combinations of S, or
- $F = F_1 \vee F_2$ where F_1, F_2 are $\{\wedge, \vee\}$-combinations of S.

◇

Definition 8.2.9 Let φ be a skolemized proof of the sequent S and $\mathcal{P}(\varphi)$ be the set of all projections w.r.t. $(\varphi, \mathrm{CL}(\varphi))$. We define the concept of a *projection-derivation* as follows:

- Every $\psi \in \mathcal{P}(\varphi)$ is a projection derivation of its end sequent from $\{\psi\}$.
- Let ψ_1, ψ_2 be projection derivations of sequents $S_1\colon S' \circ (\vdash P(\bar{t}))$ and $S_2\colon S'' \circ (P(\bar{t}) \vdash)$ from $\mathcal{P}_1$ and $\mathcal{P}_2$ respectively s.t. S is a subsequent of $S' \circ S''$. Then the derivation ψ:

$$\cfrac{\cfrac{\begin{array}{cc}\psi_1 & \psi_2 \\ S_1 & S_2\end{array}}{S' \circ S''} \; cut}{S^*} \; c^*$$

is a projection derivation of $S \circ C \circ D$ from $\mathcal{P}_1 \cup \mathcal{P}_2$. S^* is obtained from $S' \circ S''$ by an arbitrary sequence of contractions (left and right).

Proposition 8.2.1 *Let $\varphi \in \Phi^s$ be a proof of S. Then the projection derivations of S from $\mathcal{P}(\varphi)$ w.r.t. $(\varphi, \mathrm{CL}(\varphi))$ are just the* CERES *normal forms of φ.*

Proof: It is easy to see that, from any projection derivation ψ w.r.t. $(\varphi, \mathrm{CL}$ $(\varphi))$, we can extract a p-resolution refutation γ of $\mathrm{CL}(\varphi)$ s.t. $\psi = \gamma(\varphi)$. On the other hand, constructing $\gamma(\varphi)$ for a p-resolution refutation γ means to construct a projection derivation.

Theorem 8.2.2 *Let φ be a skolemized proof of S and $\mathcal{X}$ be a partition of S. Then there exists a weak interpolant I w.r.t. $\mathcal{X}$ which is an $\{\wedge, \vee\}$-combination of interpolants of* $\mathrm{PES}(\varphi)$ *(see Definition 6.4.6).*

Proof: Consider a CERES normal form ψ of φ; then ψ defines a projection derivation of S from $\mathrm{PES}(\varphi)$. We show by induction on the number $c(\tau)$ of cuts in a projection derivation τ of a sequent S' from $\mathcal{P}'$ (where $\mathcal{P}'$ is a subset of $\mathrm{PES}(\varphi)$), that – for any partition $\mathcal{X}'$ of S' – there exists an interpolant I' w.r.t. $\mathcal{X}'$ s.t. I' is an $\{\wedge, \vee\}$-combination of interpolants of sequents in $\mathcal{P}'$.

$c(\tau) = 0$.
Then τ is itself a projection. Clearly any interpolant of the end-sequent S' of τ is a Boolean combination of an interpolant of S' (as τ is a projection derivation from $\{S'\}$.

(IH) Assume that for all projection derivations τ s.t. $c(\tau) \leq n$ the assertion holds.

Let τ be a projection derivation with $c(\tau) = n + 1$. Then τ is of the form

$$\cfrac{\cfrac{\overset{\tau_1}{S_1 \circ (\vdash P(\bar{t}))} \quad \overset{\tau_2}{S_2 \circ (P(\bar{t}) \vdash)}}{S_1 \circ S_2}\ cut}{S^*}\ c^*$$

By definition τ_1 is a projection derivation of $S_1 \circ (\vdash P(\bar{t}))$ from $\mathcal{P}_1$ with $c(\tau_1) \leq n$, the same for τ_2. Then τ is a projection derivation of S^* from $\mathcal{P}_1 \cup \mathcal{P}_2$. By the induction hypothesis, for any partitions of the sequents S_1': $S_1 \circ (\vdash P(\bar{t}))$ and S_2': $S_2 \circ (P(\bar{t}) \vdash)$ there exist weak interpolants I_1 of S_1' and I_2 of S_2' which are $\{\wedge, \vee\}$-combination of interpolants of the set of end-sequences in $\mathcal{P}_1$ and $\mathcal{P}_2$, respectively. Now consider any partition $\mathcal{X}'$ of the sequent $S_1 \circ S_2$; then, by Lemma 8.2.3, there exists a weak interpolant I' w.r.t. $\mathcal{X}'$ which is of the form $C \wedge D$, or $C \vee D$, where C is a weak interpolant of S_1' and D is a weak interpolant of S_2'. We take C and D as the interpolants of the corresponding partitions which are $\{\wedge, \vee\}$-combinations of interpolants from end-sequents in $\mathcal{P}_1$ and $\mathcal{P}_2$. Then, clearly, I' is a weak interpolant which is an $\{\wedge, \vee\}$-combination of interpolants of sequents in $\mathcal{P}_1 \cup \mathcal{P}_2$. Finally we have to consider arbitrary partitions and interpolants in the sequent S^*; but these come directly from $S_1 \circ S_2$ via an obvious mapping. This concludes the induction proof. □

Corollary 8.2.1 *Let Let φ be a skolemized proof of a closed sequent S and $\mathcal{X}$ be a partition of S. Then there exists a interpolant I w.r.t. $\mathcal{X}$ which is an abstraction of an $\{\wedge, \vee\}$-combination of weak interpolants of* $\mathrm{PES}(\psi)$.

Proof: By Theorems 8.2.2 and 8.2.3.

Corollary 8.2.1 tells us that there are always interpolants which are built from interpolants of proof projections. These proof projections are parts of the *original proof φ prior to cut-elimination.* This specific form of an interpolant cannot be obtained when reductive cut-elimination is performed; in the latter case the ACNF of a proof φ does not contain any visible fragments from φ itself.

Still Corollary 8.2.1 only holds for skolemized end-sequents. The question remains whether the form of the interpolant is preserved when the original (non-skolemized) proof is considered. Below we give a positive answer:

Theorem 8.2.3 *Let φ be a proof of a sequent S and let $\mathcal{X}$ be a partition of S. Then there exists an interpolant I of S w.r.t. $\mathcal{X}$ s.t. I is an abstraction of an $\{\wedge, \vee\}$-combination of interpolants of* $\mathrm{PES}(sk(\varphi))$.

Proof: We prove that an interpolant of $sk(S)$ w.r.t. the partition $\mathcal{X}'$ (which is exactly the same partition as for S – only with skolemized formulas) is also the interpolant of S. Note that the interpolant I constructed in Corollary 8.2.1 does not contain Skolem symbols. So let $\mathcal{X} = \langle(\Gamma; \Delta), (\Pi; \Lambda)\rangle$ and $\mathcal{X}' = \langle(\Gamma'; \Delta'), (\Pi'; \Lambda')\rangle$ and I be the interpolant of $sk(S)$ w.r.t. $\mathcal{X}'$. Let A' be the formula representing $\Gamma' \vdash \Delta'$ and B' for $\Pi' \vdash \Lambda'$. Then, by soundness of **LK** the formulas

$$A' \rightarrow I \text{ and } I \rightarrow B'$$

are valid. As skolemization is validity-preserving also the formulas $A \rightarrow I$ and $I \rightarrow B$ are valid (we apply only partial skolemization on the formulas). As I is an interpolant of $sk(S)$ w.r.t. $\mathcal{X}'$ and I does not contain Skolem symbols, I is also an interpolant of S w.r.t. $\mathcal{X}$. □

8.3 Generalization of Proofs

Examples are of eminent importance for mathematics, although examples illustrating general facts are redundant. This hints at the capacity of good examples to represent arguments for universal statements in a compact manner. (In Babylonian mathematics, universal statements were taught and learned by examples only). In this chapter we provide a logical interpretation of the notion of a "good example". It is associated with a proof of a concrete fact which can be generalized.

Definition 8.3.1 (preproof, generalized proof) Let $\overset{(\Psi)}{S}$ be a cut-free **LK**-derivation from atomic $A \vdash A$, S contains weak quantifiers only, and $S'(t_1, \ldots, t_n) \equiv S$ for terms $t_1, \ldots, t_n$.
A preproof with respect to $\lambda x_1 \ldots x_n S'(x_1, \ldots, x_n)$ is defined inductively:

- Root: $S'(\alpha_1, \ldots, \alpha_n)$, where $\alpha_1, \ldots, \alpha_n$ are new variables.
- Inner inference: $\dfrac{S_1}{S}$ or $\dfrac{S_2 \quad S_3}{S}$
  - Propositional or structural: s.t. S^* is already constructed for S. S_1^* or S_2^*, S_3^* are induced from S^* by the form of the rule.
  - Quantificational:

    $$\frac{\Pi \vdash \Gamma, A(t)}{\Pi \vdash \Gamma, \exists x A(x)}, S^* = \Pi' \vdash \Gamma', \exists x A'(x), \text{ then } S_1^* \equiv \Pi' \vdash \Gamma', A(\beta)$$

    where β is a new variable.

    $$\frac{A(t), \Pi \vdash \Gamma}{\forall x A(x), \Pi \vdash \Gamma}, S^* = \forall x A(x), \Pi' \vdash \Gamma', \text{ then } S_1^* \equiv A(\beta), \Pi' \vdash \Gamma'$$

    where β is a new variable.
- Axiom positions: Do nothing

The generalized proof with respect to $\lambda x_1 \ldots x_n S'(x_1, \ldots, x_n)$ is obtained by unifying the two sides of the axiom positions by the unification algorithm UAL (see 3.1). $\diamond$

Example 8.3.1

$$\cfrac{\cfrac{P(ff0, ff0) \vdash P(ff0, ff0)}{\forall x P(fx, fx) \vdash P(ff0, ff0)}}{\forall x P(fx, fx) \vdash \exists y P(y, ff0)}$$

Preproof w.r.t. to $\lambda z[\forall x P(fx, fx) \vdash \exists y P(y, z)]$:

$$\dfrac{\dfrac{P(f\alpha, f\alpha) \vdash P(\beta, \gamma)}{\forall x P(fx, fx) \vdash P(\beta, \gamma)}}{\forall x P(fx, fx) \vdash \exists y P(y, \gamma)}$$

Generalized proof w.r.t. to $\lambda z[\forall x P(fx, fx) \vdash \exists y P(y, z)]$:

$$\dfrac{\dfrac{P(f\alpha, f\alpha) \vdash P(f\alpha, f\alpha)}{\forall x P(fx, fx) \vdash P(f\alpha, f\alpha)}}{\forall x P(fx, fx) \vdash \exists y P(y, f\alpha)}$$

◇

Proposition 8.3.1 *The generalized proof for* $\overset{(\Psi)}{S}$ *with respect to* $\lambda x_1 \dots x_n S'(x_1, \dots, x_n)$ *is an* **LK**-*derivation* $\overset{(\Psi^*)}{S^*(r_1, \dots, r_n)}$ *such that* $\Psi \equiv \Psi^* \sigma$ *and* $S \equiv S^*(r_1, \dots, r_n)\sigma$ *for some* σ.

Proof: By properties of the m.g.u. □

Definition 8.3.2 (generalized CERES normal form) Let $\overset{(\Psi)}{\Pi \vdash \Gamma}$ be a CERES normal form with $\Pi'(t_1, \dots, t_n) \vdash \Gamma'(t_1, \dots, t_n) \equiv \Pi \vdash \Gamma$ for some terms $t_1, \dots, t_n$. A generalized CERES normal form for $\overset{(\Psi)}{\Pi \vdash \Gamma}$ with respect to $\lambda x_1 \dots x_n[\Pi'(x_1, \dots, x_n) \vdash \Gamma'(x_1, \dots, x_n)]$ is constructed as follows: First calculate generalized proofs for the projections $\overset{(T)}{\Delta, \Pi \vdash \Gamma, \Psi}$ (where

$$\Delta \equiv P_{i_1}(t_{11}, \dots, t_{1k_1}), \dots, P_{i_l}(t_{l1}, \dots, t_{lk_l}) \equiv \Delta'(t_{11}, \dots, t_{lk_l})$$

and

$$\Psi \equiv Q_{j_1}(s_{11}, \dots, s_{1m_1}), \dots, Q_{j_r}(s_{r1}, \dots, s_{rm_r}) \equiv \Psi'(s_{11}, \dots, s_{rm_r}),$$

and $\Delta \vdash \Psi$ from the characteristic clause set) with respect to

$$\begin{array}{l}\lambda x_1 \dots x_n y_{11} \dots y_{lk_l} z_{11} \dots z_{rm_r}[\\ \Delta'(y_{11}, \dots, y_{lk_l}), \Pi'(x_1, \dots, x_n) \vdash \Gamma'(x_1, \dots, x_n), \Psi'(z_{11}, \dots, z_{rm_r})].\end{array}$$

Then calculate the general resolution refutation from the generalized clauses. ◇

Example 8.3.2

$$\cfrac{\cfrac{\cfrac{P(fa) \vdash P(fa)}{\forall x P(x) \vdash P(fa)}}{\forall x P(x) \vdash \forall x P(fx)} \quad \cfrac{P(ffa) \vdash P(ffa)}{\forall x P(fx) \vdash P(ffa)}}{\forall x P(x) \vdash P(ffa)}$$

Projections:

$$\cfrac{\cfrac{\cfrac{P(fa) \vdash P(fa)}{\forall x P(x) \vdash P(fa)}}{\forall x P(x) \vdash P(fa), P(ffa)}}{\forall x P(x) \vdash P(ffa), P(fa)}$$

and

$$\cfrac{\cfrac{P(ffa) \vdash P(ffa)}{\forall x P(x), P(ffa) \vdash P(ffa)}}{P(ffa), \forall x P(x) \vdash P(ffa)}$$

Generalization with respect to

$$\lambda uv[\forall x P(x) \vdash P(f(u)), P(v)] \qquad \lambda wu[P(w), \forall x P(x) \vdash P(f(u))]$$

$$\cfrac{\cfrac{\cfrac{P(\alpha) \vdash P(\alpha)}{\forall x P(x) \vdash P(\alpha)}}{\forall x P(x) \vdash P(\alpha), P(f\beta)}}{\forall x P(x) \vdash P(f\beta), P(\alpha)}$$

and

$$\cfrac{\cfrac{P(f\beta) \vdash P(f\beta)}{\forall x P(x), P(f\beta) \vdash P(f\beta)}}{P(f\beta), \forall x P(x) \vdash P(f\beta)}$$

Original clause form $\{\vdash P(f\alpha), P(ff\alpha) \vdash\}$.
Generalized clause form $\{\vdash P(\alpha), P(f\beta) \vdash\}$.

General ground resolution proof: $\cfrac{\vdash P(f\beta) \quad P(f\beta) \vdash}{\vdash}$

Apply $\sigma : f(\beta) \rightarrow \alpha$ to the projections and combine the proof parts as usual.
$\diamond$

Remark: In case all axioms are of the form $A \vdash A$ for A atomic, working with the generalized CERES normal forms can lead to an unbounded speed up with respect to usual CERES normal forms even with unchanged end sequents. $\diamond$

Definition 8.3.3 (term basis) A term basis for k and $\lambda x_1 \ldots x_n S(x_1, \ldots, x_n)$ is a set of tuples $\langle t_{11} \ldots t_{1n} \rangle, \ldots, \langle t_{l1} \ldots t_{ln} \rangle$ such that

1. $S(t_{i1}, \ldots, t_{in})$ is **LK**-derivable for $1 \leq i \leq l$
2. If $S(u_1, \ldots, u_n)$ is **LK**-derivable with depth $\leq k$, then $\langle u_1, \ldots, u_n \rangle = \langle t_{i1} \ldots t_{in} \rangle \sigma$ for some i and σ.

$\diamond$

Theorem 8.3.1 *A term basis exists for every k and $\lambda x_1 \ldots x_n S(x_1, \ldots, x_n)$.*

Proof: For every **LK**-derivation of depth k there is a CERES normal form of depth $\leq \Phi(k)$ by CERES. For any depth and $\lambda x_1 \ldots x_n S(x_1, \ldots, x_n)$ there are obviously only finitely many generalized CERES normal forms.

Corollary 8.3.1 *Let $O_0 \equiv 0$ and $O_{n+1} \equiv (O_n + 0)$. Let Π contain identity axioms and $\forall x(x + 0 = x)$. Let for all n $\Pi \vdash \Gamma, A(O_n)$ be derivable within a fixed depth. Then $\Pi \vdash \Gamma, \forall x A(x)$ is derivable.*

Corollary 8.3.2 *Let Π contain identity axioms and let*

$$\Pi \vdash \forall x(x = 0 \vee x = S(0) \vee \cdots \vee x = S^l(0) \vee \exists y x = S^{l+1}(y))$$

*be **LK**-derivable for all l. Then $\Pi \vdash \Gamma, A(S^n(0))$ is **LK**-derivable for all n within a fixed depth iff $\Pi \vdash \Gamma, \forall x A(x)$ is LK-derivable.*

Generalized Cut-free proofs can be calculated in the presence of strong quantifiers, see [55]. (For a general discussion of the topic, see [23, 24]). In presence of nonatomic logical axioms, a limitation of the size of axioms and cuts using Parikh's theorem has to be employed (cf. eg [55]). The results of this chapter can be extended to proofs with schematic quantifier free non-logical axioms, but not to proofs with (subst) as axiom. This follows from the undecidability of second order unification, cf. [55].

8.4 CERES and Herbrand Sequent Extraction

A further proof theoretic strength of CERES is illustrated by the possibility to demonstrate that a specific Herbrand sequent *cannot* be extracted from a given proof with cuts – even without eliminating them. In fact one can show that a Herbrand sequent is composed from the Herbrand sequents of the projections after deletion of the clause parts.

Example 8.4.1 Let

$$D_k^f = (\forall x)(P(f(x)) \rightarrow P(f(s^{2^k}(x)))),$$
$$D_k^g = (\forall x)(P(g(x)) \rightarrow P(g(s^{2^k}(x)))).$$

Moreover let ϕ_k^f be the obvious cut-free proof of $D_k^f \vdash D_{k+1}^f$ and χ_k^f be the cut-free **LK**-proof

$$\cfrac{\cfrac{\begin{matrix}\text{obvious proof}\\ D_n^f, P(f(0)) \vdash P(f(s^{2^n}(0)))\end{matrix}}{D_n^f, P(f(0)) \vdash P(f(s^{2^n}(0))) \vee P(g(s^{2^n}(0)))}\ \vee : r}{D_n^f, P(f(0)), P(g(0)), D_0^g \vdash P(f(s^{2^n}(0))) \vee P(g(s^{2^n}(0)))}\ w:^*$$

Now we combine the proofs $\phi_0^f, \ldots, \phi_{n-1}^f, \chi_n^f$ by cuts on the formulas $D_1^f, \ldots,$ D_n^f to a proof ψ_n of the end sequent

$$P(f(0)), P(g(0)), D_0^f, D_0^g \vdash P(f(s^{2^n}(0))) \vee P(g(s^{2^n}(0))).$$

We write Δ for $P(f(0)), P(g(0)), D_0^f, D_0^g$ and A for $P(f(s^{2^n}(0))) \vee P(g(s^{2^n}(0)))$. Note that only the proofs ϕ_0^f and χ_n^f have projections with nontautological clauses. Indeed, for ϕ_0^f we obtain a projection sequent of the form

$$P(f(x)), \Delta \vdash A, P(f(s(x)))$$

and for χ_n^f sequents of the form

$$\Delta \vdash A, P(f(0)), \quad P(f(s^{2^n}(0))), \Delta \vdash A.$$

After deletion of the clause parts and the pruning of weakenings the Herbrand sequents of the projections have the form

$$S_1\colon\ P(f(x)) \rightarrow P(f(s(x))) \vdash \text{ for } \phi_0^f,$$

and

$$S_2\colon\ P(f(0)) \vdash P(f(s^{2^n}(0))) \vee P(g(s^{2^n}(0))) \text{ for } \chi_n^f.$$

Therefore, the valid Herbrand sequent

$$P(f(0)), P(g(0)), P(g(0)) \rightarrow P(g(s(0))), \ldots, P(g(s^{2^n-1}(0))) \rightarrow P(g(s^{2^n}(0)))$$
$$\vdash$$
$$P(f(s^{2^n}(0))) \vee P(g(s^{2^n}(0)))$$

cannot be obtained from ψ_n by cut-elimination and is not composed of S_1 and S_2 using possibly weakening. ◇

8.5 Analysis of Mathematical Proofs

The analysis of mathematical proofs is one of the most prominent activities of mathematical research. Many important notions of mathematics originated from the generalization of existing proofs. The advantage of logical methods of the analysis of proofs lies in their systematic nature. Nowadays two main forms of logical analysis of proofs can be distinguished:

(1) methods which allow the constructions of completely new proofs from given ones; here the original proofs serve only as a tool and the relation to the old proof is only of minor interest. The most prominent method of this type is functional interpretation as in [54] and Herbrand analysis as, e.g., in the analysis of Roth's theorem by Luckhard [63].

(2) Methods which construct elementary proofs to give *interpretations* of the original ones; the most important example is Girard's transformation of the Fürstenberg–Weiss proof of van der Waerden's theorem into the original combinatorial proof of van der Waerden by means of cut-elimination [40].

CERES can be applied in both directions: concerning (2) CERES gives a reasonable overview of elementary proofs obtainable by the usual forms of cut-elimination in a nondeterministic sense. This makes the formulation of negative results possible, namely that a certain class of nonelementary proofs *cannot* be the origin of given cut-free proofs. For (1) the characteristic clause term can be analyzed mathematically without any regards to the original proofs to obtain new proofs and better bounds.

8.5.1 Proof Analysis by Cut-Elimination

In this book we selected two simple examples of mathematical proofs to illustrate the proof analysis by CERES. A more involved example, the analysis of the topological proof of the infinity of primes by H. Fürstenberg [1] can be found in [13]. The two examples in this book also demonstrate that a final interpretation of the output of CERES by a mathematician is necessary. CERES can thus be considered as an interactive tool for mathematicians in the interpretation of proofs.

8.5.2 The System `ceres`

The cut-elimination system `ceres` is written in ANSI-C++.[1] There are two main tasks. On the one hand to compute an unsatisfiable set of clauses $\mathcal{C}$ characterizing the cut formulas. This is done by automatically extracting the characteristic clause term and the computation of the resulting characteristic clause set. On the other hand to evaluate a resolution refutation of the characteristic clause set gained from an external theorem prover[2] and to compute the necessary projection schemes of the clauses actually used to refute the characteristic clause set. The properly instantiated projection schemes, such that every instantiated projection derives a clause instance of the refutation, are concatenated by using the resolution refutation as a skeleton of the cut-free proof yielding an ACNF. Equality rules appearing within the input proof are propagated to the projection schemes in the usual way (as arbitrary binary rules) and during theorem proving treated by means of paramodulation which applications within the final resolution refutation are transformed to appropriate **LK** equality rules again. The definition introductions introduced in Chapter 7 do not require any other special treatment within `ceres` than the ordinary unary rules.
Our system also performs proof skolemization on the input proofs (if necessary) since skolemized proofs are a crucial requirement for the CERES-method to be applied; the skolemization method is that of Andrews [2].
The system `ceres` expects an **LKDe**-proof φ and a set of atomic axioms as input; the output is a CERES normal of φ. Input and output are formatted using the well known data representation language XML,[3] which allows the use of arbitrary and well known utilities for editing, transformation and presentation and standardized programming libraries. To increase performance and to avoid redundancy most parts of the proofs are internally represented as directed acyclic graphs. This representation turns also out to be very handy for the internal unification algorithms.

The formal analysis of mathematical proofs (especially by the human mathematician as pre- and post-processor) relies on a suitable format for the input and output of proofs and on an appropriate aid in dealing with them. We developed an intermediary proof language (HLK) connecting the language

[1] The C++ Programming Language following the International Standard 14882:1998 approved as an American National Standard (see http://www.ansi.org).

[2] The current version of `ceres` uses the automated theorem prover Prover9 (see http://www.cs.unm.edu/ mccune/mace4/), but any refutational theorem prover capable of paramodulation may be used.

[3] See http://www.w3.org/XML/ for more information on the Extensible Markup Language specification.

of mathematical proofs with **LK** using equality and definitions. Furthermore we implemented a proof-viewer and proof-editor (ProofTool) with a graphical user interface. HLK and ProofTool make input and the analysis of the output of `ceres` much more comfortable. Thereby the usage of definitions as well as the integration of equality into the underlying calculus play an essential role for the overlooking, understanding and the analysis of complex mathematical proofs by humans. As the final proof is usually to long to be interpreted by humans, `ceres` also contains an algorithm for Herbrand sequent extraction which stronlgy reduces redundancy and makes the output intellegible. A more detailed description of the `ceres`-system can be found in [48] and on the webpage.[4] The application of Herbrand sequent extraction is described in [47, 79].

8.5.3 The Tape Proof

The example below (the tape proof) is taken from [78]; it was formalized in **LK** and analyzed by CERES in the papers [11] by the original version of CERES and [12] (by the extended version of CERES based on **LKDe**). In this section we use the extended version of CERES to give a simple formalization and a mathematical analysis of the tape proof. The end-sequent of the tape proof formalizes the statement: on a tape with infinitely many cells which are all labeled by 0 or by 1 there are two cells labeled by the same number. $f(x) = 0$ expresses that the cell nr. x is labeled by 0. Indexing of cells is done by number terms defined over $0, 1$ and $+$. The proof φ below uses two lemmas:

(1) *there are infinitely many cells labeled by* 0 *and*
(2) *there are infinitely many cells labeled by* 1.

These lemmas are eliminated by CERES and a more direct argument is obtained in the resulting proof φ'. In the text below the ancestors of the cuts in φ are indicated in boldface.

Let φ be the proof

$$\dfrac{\dfrac{\overset{(\tau)}{A \vdash \mathbf{I_0}, \mathbf{I_1}} \quad \overset{(\epsilon_0)}{\mathbf{I_0} \vdash (\exists p, q)(p \neq q \wedge f(p) = f(q))}}{A \vdash (\exists p, q)(p \neq q \wedge f(p) = f(q)), \mathbf{I_1}}\ cut \quad \overset{(\epsilon_1)}{\mathbf{I_1} \vdash (\exists p, q)(p \neq q \wedge f(p) = f(q))}}{A \vdash (\exists p, q)(p \neq q \wedge f(p) = f(q))}\ cut$$

[4] http://www.logic.at/ceres/

where $\tau =$

$$\begin{array}{c}
(\tau') \\
f(n_0+n_1)=0 \lor f(n_0+n_1)=1 \vdash \mathbf{f(n_0+n_1)=0, f(n_1+n_0)=1} \\
\hline
\forall x(f(x)=0 \lor f(x)=1) \vdash \mathbf{f(n_0+n_1)=0, f(n_1+n_0)=1} \quad \forall\colon l \\
\hline
A \vdash \mathbf{f(n_0+n_1)=0, f(n_1+n_0)=1} \quad \mathit{def}(A)\colon l \\
\hline
A \vdash \mathbf{f(n_0+n_1)=0, (\exists k) f(n_1+k)=1} \quad \exists\colon r \\
\hline
A \vdash \mathbf{(\exists k) f(n_0+k)=0, (\exists k) f(n_1+k)=1} \quad \exists\colon r \\
\hline
A \vdash \mathbf{(\exists k) f(n_0+k)=0, (\forall n)(\exists k) f(n+k)=1} \quad \forall\colon r \\
\hline
A \vdash \mathbf{(\forall n)(\exists k) f(n+k)=0, (\forall n)(\exists k) f(n+k)=1} \quad \forall\colon r \\
\hline
A \vdash \mathbf{I_0, (\forall n)(\exists k) f(n+k)=1} \quad \mathit{def}(I_0)\colon r \\
\hline
A \vdash \mathbf{I_0, I_1} \quad \mathit{def}(I_1)\colon r
\end{array}$$

For $\tau' =$

$$\dfrac{f(n_0+n_1)=0 \vdash \mathbf{f(n_0+n_1)=0} \qquad \dfrac{\overset{(\text{Axiom})}{\vdash n_1+n_0=n_0+n_1} \qquad f(n_1+n_0)=1 \vdash \mathbf{f(n_1+n_0)=1}}{f(n_0+n_1)=1 \vdash \mathbf{f(n_1+n_0)=1}}\ {=}\colon l}{f(n_0+n_1)=0 \lor f(n_0+n_1)=1 \vdash \mathbf{f(n_0+n_1)=0, f(n_1+n_0)=1}}\ \lor\colon l$$

And for $i = 1, 2$ we define the proofs $\epsilon_i =$

$$\begin{array}{c}
\psi \quad \eta_i \\
\hline
\mathbf{f(s)=i, f(t)=i} \vdash s \neq t \land f(s)=f(t) \quad \land\colon r \\
\hline
\mathbf{f(s)=i, f(t)=i} \vdash (\exists q)(s \neq q \land f(s)=f(q)) \quad \exists\colon r \\
\hline
\mathbf{f(s)=i, f(t)=i} \vdash (\exists p)(\exists q)(p \neq q \land f(p)=f(q)) \quad \exists\colon r \\
\hline
\mathbf{f(n_0+k_0)=i, (\exists k) f(((n_0+k_0)+1)+k)=i} \vdash (\exists p)(\exists q)(p \neq q \land f(p)=f(q)) \quad \exists\colon l \\
\hline
\mathbf{f(n_0+k_0)=i, (\forall n)(\exists k) f(n+k)=i} \vdash (\exists p)(\exists q)(p \neq q \land f(p)=f(q)) \quad \forall\colon l \\
\hline
\mathbf{(\exists k) f(n_0+k)=i, (\forall n)(\exists k) f(n+k)=i} \vdash (\exists p)(\exists q)(p \neq q \land f(p)=f(q)) \quad \exists\colon l \\
\hline
\mathbf{\forall n \exists k. f(n+k)=i, (\forall n)(\exists k) f(n+k)=i} \vdash (\exists p)(\exists q)(p \neq q \land f(p)=f(q)) \quad \forall\colon l \\
\hline
\mathbf{(\forall n)(\exists k) f(n+k)=i} \vdash (\exists p)(\exists q)(p \neq q \land f(p)=f(q)) \quad c\colon l \\
\hline
\mathbf{I_i} \vdash (\exists p)(\exists q)(p \neq q \land f(p)=f(q)) \quad \mathit{def}(I_i)\colon l
\end{array}$$

for $s = n_0 + k_0$, $t = ((n_0 + k_0) + 1) + k_1$, and the proofs
$\psi =$

$$\dfrac{\dfrac{\overset{(\text{axiom})}{\vdash (n_0+k_0)+(1+k_1) = ((n_0+k_0)+1)+k_1} \qquad \overset{(\text{axiom})}{n_0+k_0 = (n_0+k_0)+(1+k_1) \vdash}}{n_0+k_0 = ((n_0+k_0)+1)+k_1 \vdash}\ {=}\colon l1}{\vdash n_0+k_0 \neq ((n_0+k_0)+1)+k_1}\ \neg\colon r$$

and $\eta_i =$

$$\dfrac{\mathbf{f(s)} = \mathbf{i} \vdash f(s) = i \qquad \dfrac{\mathbf{f(t)} = \mathbf{i} \vdash f(t) = i \quad \overset{\text{(axiom)}}{\vdash i = i}}{\mathbf{f(t)} = \mathbf{i} \vdash i = f(t)} =: r2}{\mathbf{f(s)} = \mathbf{i}, \mathbf{f(t)} = \mathbf{i} \vdash f(s) = f(t)} =: r2$$

The characteristic clause set is (after variable renaming)

$$\begin{array}{rcl} \mathrm{CL}(\varphi) & = & \{\vdash f(x+y) = 0, f(y+x) = 1; \quad (C_1) \\ & & f(x+y) = 0, f(((x+y)+1)+z) = 0 \vdash; \quad (C_2) \\ & & f(x+y) = 1, f(((x+y)+1)+z) = 1 \vdash\} \quad (C_3). \end{array}$$

The axioms used for the proof are the standard axioms of type $A \vdash A$ and instances of $\vdash x = x$, of commutativity $\vdash x + y = y + x$, of associativity $\vdash (x + y) + z = x + (y + z)$, and of the axiom

$$x = x + (1 + y) \vdash,$$

expressing that $x + (1 + y) \neq x$ for all natural numbers x, y.
The comparison with the analysis of Urban's proof formulated in **LK** without equality [11] shows that this one is much more legible. In fact the set of characteristic clauses contains only 3 clauses (instead of 5), which are also simpler. This also facilitates the refutation of the clause set and makes the output proof simpler and more transparent. On the other hand, the analysis below shows that the mathematical argument obtained by cut-elimination is the same as in [11].

The program Otter found the following refutation of $\mathrm{CL}(\varphi)$ (based on hyperresolution only – without equality inference):
The first hyperesolvent, based on the clash sequence $(C_2; C_1, C_1)$, is

$$C_4 = \vdash f(y + x) = 1, \ f(z + ((x + y) + 1)) = 1,$$

with the intermediary clause

$$D_1 = f(((x + y) + 1) + z) = 0 \vdash f(y + x) = 1.$$

The next clash is sequence is $(C_3; C_4, C_4)$ which gives C_5 with intermediary clause D_2, where:

$$\begin{array}{rcl} C_5 & = & \vdash f(v' + u') = 1, \ f(v + u) = 1, \\ D_2 & = & f(x + y) = 1 \vdash f(v + u) = 1. \end{array}$$

Factoring C_5 gives C_6: $\vdash f(v+u) = 1$ (which roughly expresses that all fields are labelled by 1). The final clash sequence $(C_3; C_6, C_6)$ obviously results in the empty clause $\vdash$ with intermediary clause D_3: $f(((x+y)+1)+z) = 1 \vdash$. The hyperresolution proof ψ_3 in form of a tree can be obtained from the following resolution trees ψ_1 and ψ_2 defined below, where C' and ψ' stand for renamed variants of C and of ψ, respectively.

$\psi_1 =$

$$\dfrac{\vdash C_1\{x \leftarrow u', y \leftarrow v'\} \qquad \dfrac{\begin{array}{c} C_1 \\ \vdash f(x+y) = 0, f(y+x) = 1 \end{array} \quad \begin{array}{c} C_2\{u \leftarrow x, v \leftarrow y\} \\ f(u+v) = 0, f(t(u,v,z)) = 0 \vdash \end{array}}{f(t(x,y,z)) = 0 \vdash f(y+x) = 1 \ (D_1)}}{\vdash f(y+x) = 1,\ f(t'(x,y,z)) = 1 \ (C_4)}$$

for $t(x,y,z) = ((x+y)+1)+z$ and $t'(x,y,z) = z + ((x+y)+1)$. We give a mathematical interpretation of ψ_1. To this aim we first compute the global m.g.u. of ψ_1 which is

$$\sigma = \{u \leftarrow x,\ v \leftarrow y,\ u' \leftarrow (x+y)+1,\ v' \leftarrow z\}.$$

Moreover we use the properties of associativity and commutativity to simplify the terms. Then C_1 says that all cells $x+y$ are either labeled by 0 or by 1. C_2 expresses that not both cells $x+y$ and $t(x,y,z)$ are 0. Therefore, if cell $t(x,y,z)$ is 0 then cell $x+y$ must be different from 0 and thus 1 (clause D_1). An instance of C_1 tells that either cell $t(x,y,z)$ is 0 or it is 1. Hence either cell $x+y$ is 1 or cell $t(x,y,z)$ is one (this is the statement expressed by clause C_4).

$\psi_2 =$

$$\dfrac{\dfrac{\psi_1 \qquad \dfrac{\psi_1 \quad \begin{array}{c} C_3\{x \leftarrow u, y \leftarrow v, z \leftarrow w\} \\ f(u+v) = 1, f(((u+v)+1)+w) = 1 \vdash \end{array}}{f(u+v) = 1 \vdash f(y+x) = 1 \ (D_2')}}{\vdash f(y+x) = 1, f(v+u) = 1 \ (C_5')}}{\vdash f(v+u) = 1 \ (C_6)}$$

The most general unifier in the resolution of C_4 with C_3' in ψ_2 is

$$\sigma_1 = \{z \leftarrow (u+v)+1,\ w \leftarrow (x+y)+1\}.$$

Let $t = u+v+1+x+y+1$. Then the two clauses express:

- either cell $x+y$ is 1 or cell t is 1,
- not both cells $u+v$ and t are 1.

So, if cell $u+v$ is 1 then cell t is different from 1 and therefore cell $x+y$ is 1 (C_2'). Note that the clause C_2' represents the formula

$$(\forall u, v, x, y)(f(u+v) = 1 \rightarrow f(x+y) = 1)$$

which is equivalent (via quantifier shifting) to

$$(\exists u, v) f(u+v) = 1 \rightarrow (\forall x, y) f(y+x) = 1.$$

Therefore the existence of u, v s.t. cell $u+v$ is 1 implies that *all* cells $x+y$ are 1 (which means that all cells are 1). Now, again, we use ψ_1 which derives C_4, from which the existence of these u, v follows. We conclude that all cells $x+y$ are 1 (clause C_6).
Now we define $\psi_3 =$

$$\frac{\begin{array}{c}(\psi_2)\\ \vdash f(v+u)=1\end{array} \qquad \dfrac{\begin{array}{c}(\psi_2)\\ \vdash f(v+u)=1\end{array} \quad f(x+y)=1, f(((x+y)+1)+z)=1 \vdash \ (C_3)}{f(((x+y)+1)+z)=1 \vdash \ (D_3)}}{\vdash}$$

The refutation obtained in ψ_3 is simple. We know that, for all u, v, the cell $v+u$ is 1. But C_3 expresses that one of the cells $x+y$, $x+y+z+1$ must be different from 1. With the substitutions

$$\lambda_1 = \{v \leftarrow x, u \leftarrow y\},\ \lambda_2 = \{v \leftarrow (x+y)+1, u \leftarrow z\}$$

we obtain a contradiction.

Instantiation of ψ_3 by the uniform most general unifier σ of all resolutions gives a deduction tree $\psi_3\sigma$ in **LKDe**; indeed, after application of σ, resolution becomes cut and factoring becomes contraction. The proof $\psi_3\sigma$ is the skeleton of an **LKDe**-proof of the end-sequent with only atomic cuts. Then the leaves of the tree $\psi_3\sigma$ have to be replaced by the proof projections. E.g., the clause C_1 is replaced by the proof $\varphi[C_1]$, where $s = n_0 + n_1$ and $t = n_1 + n_0$:

$$\dfrac{\dfrac{\dfrac{\dfrac{f(s)=0 \vdash f(s)=0 \qquad \dfrac{\begin{array}{c}(\text{Axiom})\\ \vdash t = s\end{array} \quad f(t)=1 \vdash f(t)=1}{f(s)=1 \vdash f(t)=1}\ {=}{:}\,l}{f(s)=0 \vee f(s)=1 \vdash f(s)=0, f(t)=1}\ \vee{:}\,l}{(\forall x)(f(x)=0 \vee f(x)=1) \vdash f(s)=0, f(t)=1}\ \forall{:}\,l}{A \vdash f(s)=0, f(t)=1}\ def(A){:}\,l}{A \vdash (\exists p)(\exists q)(p \neq q \wedge f(p)=f(q)), f(s)=0, f(t)=1}\ w{:}\,r$$

Furthermore C_2 is replaced by the projection $\varphi[C_2]$ and C_3 by $\varphi[C_3]$, where (for $i = 0, 1$) $\varphi[C_{2+i}] =$

$$\cfrac{\cfrac{\cfrac{\cfrac{\psi \quad \eta_i}{f(s) = i, f(t) = i \vdash s \neq t \wedge f(s) = f(t)}\ \wedge: r}{f(s) = i, f(t) = i \vdash (\exists q)(s \neq q \wedge f(s) = f(q))}\ \exists: r}{f(s) = i, f(t) = i \vdash (\exists p)(\exists q)(p \neq q \wedge f(p) = f(q))}\ \exists: r}{f(s) = i, f(t) = i, A \vdash (\exists p)(\exists q)(p \neq q \wedge f(p) = f(q))}\ w: l$$

Note that ψ, η_0, η_1 are the same as in the definition of ϵ_0, ϵ_1 above.
By inserting the σ-instances of the projections into the resolution proof $\psi_3\sigma$ and performing some additional contractions, we eventually obtain the desired proof φ' of the end-sequent

$$A \vdash (\exists p)(\exists q)(p \neq q \wedge f(p) = f(q))$$

with only *atomic* cuts. φ' no longer uses the lemmas that infinitely many cells are labeled by 0 and by 1, respectively. The mathematical arguments in φ' are essentially those of the resolution refutation ψ_3', but transformed into a "direct" proof by inserting the projections. We see that already the resolution refutation of the characteristic clause set contains the essence of the mathematical argument.

8.5.4 The Lattice Proof

In this section, we demonstrate the usefulness of a Herbrand sequent for understanding a formal proof. We choose a simple example from lattice theory The analysis shown below largely follows this in [47]. There are several different, but equivalent, definitions of *lattice*. Usually, the equivalence of several statements is shown by proving a cycle of implications. While reducing the size of the proof, this practice has the drawback of not providing *direct* proofs between the statements. But by cut-elimination we can automatically generate a direct proof between any two of the equivalent statements. In this section, we will demonstrate how to apply cut-elimination with the system ceres followed by Herbrand sequent extraction for this purpose.
Definitions 8.5.2, 8.5.3 and 8.5.5 list different sets of properties that a 3-tuple $\langle L, \cap, \cup \rangle$ or a partially ordered set $\langle S, \leq \rangle$ must have in order to be considered a lattice.

Definition 8.5.1 (Semi-Lattice) A semi-lattice is a set L together with an operation $\circ$ which is

- commutative: $(\forall x)(\forall y)\ x \circ y = y \circ x$,
- associative: $(\forall x)(\forall y)(\forall z)\ (x \circ y) \circ z = x \circ (y \circ z)$ and
- idempotent: $(\forall x)\ x \circ x = x$.

◇

Definition 8.5.2 (Lattice: definition 1) A $L1$-lattice is a set L together with operations $\cap$ (*meet*) and $\cup$ (*join*) s.t. both $\langle L, \cap \rangle$ and $\langle L, \cup \rangle$ are semi-lattices and $\cap$ and $\cup$ are "inverse" in the sense that

$$(\forall x)(\forall y)(x \cap y = x \leftrightarrow x \cup y = y).$$

◇

Definition 8.5.3 (Lattice: definition 2) A $L2$-lattice is a set L together with operations $\cap$ and $\cup$ s.t. both $\langle L, \cap \rangle$ and $\langle L, \cup \rangle$ are semi-lattices and the absorption laws

$$(\forall x)(\forall y)\ (x \cap y) \cup x = x \quad \text{and} \quad (\forall x)(\forall y)\ (x \cup y) \cap x = x$$

hold. ◇

Definition 8.5.4 (Partial Order) A binary relation $\leq$ on a set S is called a *partial order* if it is

- reflexive (R): $(\forall x)\ x \leq x$,
- anti-symmetric (AS): $(\forall x)(\forall y)\ ((x \leq y \wedge y \leq x) \rightarrow x = y)$ and
- transitive (T): $(\forall x)(\forall y)(\forall z)\ ((x \leq y \wedge y \leq z) \rightarrow x \leq z)$.

◇

Definition 8.5.5 (Lattice: definition 3) A $L3$-lattice is a partially ordered set $\langle S, \leq \rangle$ s.t. for each two elements x, y of S there exist

- a greatest lower bound (GLB) $glb(x, y)$, i.e.

$$(\forall x)(\forall y)(glb(x,y) \leq x \wedge glb(x,y) \leq y \wedge (\forall z)((z \leq x \wedge z \leq y) \rightarrow z \leq glb(x,y)),$$

- and a least upper bound (LUB) $lub(x, y)$ i.e.

$$(\forall x)(\forall y)(x \leq lub(x,y) \wedge y \leq lub(x,y) \wedge (\forall z)((x \leq z \wedge y \leq z) \rightarrow lub(x,y) \leq z)).$$

The above three definitions of lattice are equivalent. We will formalize the following proofs of $L1 \to L3$ and $L3 \to L2$ in order to extract a direct proof of $L1 \to L2$, i.e. a proof which does not use the notion of partial order.

Proposition 8.5.1 *L1-lattices are L3-lattices.*

Proof: Let $\langle L, \cap, \cup \rangle$ be an L1-lattice. We define a relation $\leq$ by

$$x \leq y \leftrightarrow x \cap y = x.$$

By idempotence of $\cap$, $\leq$ is reflexive.
Anti-symmetry of $\leq$ follows from commutativity of $\cap$ as $(x \cap y = x \wedge y \cap x = y) \to x = y$.
To see that $\leq$ is transitive, assume (a) $x \cap y = x$ and (b) $y \cap z = y$ to derive

$$x \cap z =^{(a)} (x \cap y) \cap z =^{(\text{assoc.})} x \cap (y \cap z) =^{(b)} x \cap y =^{(a)} x$$

So $\leq$ is a partial order on L.
Now we prove that, for all elements $x, y \in L$, $x \cap y$ is a greatest lower bound w.r.t. $\leq$.
By associativity, commutativity and idempotence of $\cap$, we have $(x \cap y) \cap x = x \cap y$, i.e. $x \cap y \leq x$ and similarly $x \cap y \leq y$, so $\cap$ is a lower bound for $\leq$. To see that $\cap$ is also the greatest lower bound, assume there is a z with $z \leq x$ and $z \leq y$, i.e. $z \cap x = z$ and $z \cap y = z$. Then, by combining these two equations, $(z \cap y) \cap x = z$, and by associativity and commutativity of $\cap$, $z \leq x \cap y$.
Finally we prove that, for all elements $x, y \in L$, $x \cup y$ is a least upper bound of x, y w.r.t. $\leq$.
To see that $x \cup y$ is an upper bound, derive from the axioms of semi-lattices that

$$x \cup (x \cup y) = x \cup y$$

which, by the "inverse" condition of L1 gives $x \cap (x \cup y) = x$, i.e. $x \leq x \cup y$ and similarly for $y \leq x \cup y$.
Now assume there is a z with $x \leq z$ and $y \leq z$, i.e. $x \cap z = x$ and $y \cap z = z$ and by the "inverse" condition of L1: $x \cup z = z$ and $y \cup z = z$. From these two equations and the axioms of semi-lattices, derive $(x \cup y) \cup z = z$ which, by the "inverse" condition of $L1$, gives $(x \cup y) \cap z = x \cup y$, i.e. $x \cup y \leq z$. So $x \cup y$ is a least upper bound of x, y. □

Proposition 8.5.2 *L3-lattices are L2-lattices.*

Proof: Assume that $\langle L, \leq \rangle$ is an L3-lattice. For any two elements $x, y \in L$ we select a greatest lower bound $glb(x, y)$ of x, y and define $x \cap y = glb(x, y)$; similarly we define $x \cup y = lub(x, y)$.
We have to prove the absorbtion laws for $\cap$, $\cup$.
The first law is $(x \cap y) \cup x = x$. That $x \leq (x \cap y) \cup x$ follows immediately from $\cup$ being an upper bound. But $x \cap y \leq x$ because $\cap$ is a lower bound. Furthermore also $x \leq x$, so x is an upper bound of $x \cap y$ and x. But as $\cup$ is the lowest upper bound, we have $(x \cap y) \cup x \leq x$ which by anti-symmetry of $\leq$ proves $(x \cap y) \cup x = x$. The proof of the other absorption law $(x \cup y) \cap x = x$ is completely symmetric. □

By concatenation, the above two proofs show that all L1-lattices are L2-lattices. However, this proof is not a direct one, it uses the notion of partially ordered set which occurs neither in L1 nor in L2. By cut-elimination we obtain a direct formal proof automatically.
The analysis of the lattice proof, performed in CERES, followed the steps below (see also [47]):

1. *Formalization of the lattice proof in the sequent calculus* **LKDe**: semi-automated by HLK[5]. Firstly the proof was written in the language HandyLK, which can be considered as an intermediary language between informal mathematics and **LKDe**. Subsequently, HLK compiled it to an **LKDe**-proof ψ.

2. *Cut-Elimination of the formalized lattice proof*: fully automated by CERES, by applying the cut-elimination procedure based on resolution, sketched in Section 6.4, to ψ. First we obtain a characteristic clause set $\mathrm{CL}(\varphi)$, then a refutation of $\mathrm{CL}(\varphi)$ by resolution and paramodulation and, finally, a **LKDe**-proof φ in CERES normal form.

3. *Extraction of the Herbrand sequent of the ACNF*: fully automated by ceres, employing the algorithm described in Section 4.2.

4. *Use of the Herbrand sequent* to interpret and understand the proof φ^*, in order to obtain a new direct informal proof.

The full formal proof ψ has 260 rules (214 rules, if structural rules (except cut) are not counted). It is too large to be displayed here. Below we show only a part of it, which is close to the end-sequent and depicts the main structure of the proof, based on the cut-rule with $L3$ as the cut-formula.

[5] HLK Website: http://www.logic.at/hlk/

This cut divides the proof into two subproofs corresponding to Propositions 8.5.1 and 8.5.2. The full proofs, conveniently viewable with `ProofTool`,[6] are available in the website of CERES.

$$
\dfrac{\dfrac{\dfrac{\dfrac{\dfrac{[p_R]}{\vdash R} \quad \dfrac{\dfrac{[p_{AS}]}{\vdash AS} \quad \dfrac{[p_T]}{\vdash T}}{\vdash AS \wedge T}\;\wedge{:}\,r}{\vdash R \wedge (AS \wedge T)}\;\wedge{:}\,r}{\vdash POSET}\;d{:}\,r \quad \dfrac{[p_{GLB}] \quad [p_{LUB}]}{L1 \vdash GLB \wedge LUB}\;\wedge{:}\,r}{L1 \vdash POSET \wedge (GLB \wedge LUB)}\;\wedge{:}\,r}{L1 \vdash L3}\;d{:}\,r \quad \dfrac{[p_3^2]}{L3 \vdash L2}}{L1 \vdash L2}\;cut
$$

- $L1 \equiv (\forall x)(\forall y)(((x \cap y) = x \supset (x \cup y) = y) \wedge ((x \cup y) = y \supset (x \cap y) = x))$.
- $L2 \equiv (\forall x)(\forall y)((x \cap y) \cup x = x \wedge (x \cup y) \cap x = x)$.
- $L3 \equiv POSET \wedge (GLB \wedge LUB)$
- p_{AS}, p_T, p_R are proofs of, respectively, anti-symmetry (AS), transitivity (T) and reflexivity (R) of $\leq$ from the axioms of semi-lattices.
- p_3^2 is a proof that $L3$-lattices are $L2$-lattices, from the axioms of semi-lattices.

Remark: We formulated $L2$ in prenex form s.t. we can apply the algorithm for Herbrand sequent extraction defined in Section 4.2. In [47] $L2$ has been defined in the following non-prenex form:

$$(\forall x)(\forall y)((x \cap y) \cup x = x \wedge (\forall x)(\forall y)(x \cup y) \cap x = x).$$

The skolemized proof of our prenex version (obtained by applying quantifier distribution backwards) contains only two Skolem constants s_1, s_2, while in [47] four Skolem constants $s_1, \ldots, s_4$ have to be introduced. The algorithm for Herbrand extraction applied in [47] is more general than this presented in Section 4.2, but it coincides with our's in case of prenex end-sequents. $\diamond$

Prior to cut-elimination, the formalized proof is skolemized by `ceres`, resulting in a proof of the skolemized end-sequent $L1 \vdash (s_1 \cap s_2) \cup s_1 = s_1 \wedge (s_1 \cup s_2) \cap s_1 = s_1$, where s_1 and s_2 are skolem constants for the strongly quantified variables of $L2$. Then `ceres` eliminates cuts, producing a proof in atomic-cut normal (also available for visualization with `ProofTool` in the website of CERES).

[6] `ProofTool` Website: http://www.logic.at/prooftool/

The CERES normal form φ is still quite large (214 rules; 72 rules not counting structural rules (except cut)). It is interesting to note, however, that φ is smaller than the original proof ψ in this case, even though in the worst case cut-elimination can produce a non-elementary increase in the size of proofs as we have shown in Section 4.3.
Although φ itself is rather large, the extracted Herbrand sequent contains only 6 formulas. Therefore, the Herbrand sequent significantly reduces the amount of information that has to be analyzed in order to extract the direct mathematical argument contained in φ.
The Herbrand sequent of the CERES normal form , after set-normalization and removal of remaining sub-formulas introduced by weakening (or as the non-auxiliary formula of $\vee$ and $\wedge$ rules) in the ACNF, is:

$$\begin{array}{ll}
(A1) & s_1 \cup (s_1 \cup (s_1 \cap s_2)) = s_1 \cup (s_1 \cap s_2) \rightarrow s_1 \cap (s_1 \cup (s_1 \cap s_2)) = s_1, \\
(A2) & s_1 \cap s_1 = s_1 \rightarrow s_1 \cup s_1 = s_1, \\
(A3) & \underbrace{(s_1 \cap s_2) \cap s_1 = s_1 \cap s_2}_{(A3i)} \rightarrow (s_1 \cap s_2) \cup s_1 = s_1, \\
(A4) & (s_1 \cup (s_1 \cap s_2)) \cup s_1 = s_1 \rightarrow (s_1 \cup (s_1 \cap s_2)) \cap s_1 = s_1 \cup (s_1 \cap s_2), \\
(A5) & \underbrace{s_1 \cup (s_1 \cup s_2) = s_1 \cup s_2}_{(A5i)} \rightarrow s_1 \cap (s_1 \cup s_2) = s_1 \\
(C1) & \vdash \underbrace{(s_1 \cap s_2) \cup s_1 = s_1}_{(C1i)} \wedge \underbrace{(s_1 \cup s_2) \cap s_1 = s_1}_{(C1ii)}
\end{array}$$

After extracting a Herbrand sequent from the ACNF, the next step is to construct an informal, analytic proof of the theorem, based on the ACNF, but using only the information about the variable instantiations contained in its extracted Herbrand sequent. We want to stress that, in the analysis below, we are not performing syntactic manipulations of formulas of first-order logic, but instead we use the formulas from the Herbrand sequent of the CERES normal form φ as a *guide* to construct an analytical mathematical proof.

Theorem 8.5.1 *All $L1$-lattices $\langle L, \cap, \cup \rangle$ are $L2$-lattices.*

Proof: As both lattice definitions have associativity, commutativity and idempotence in common, it remains to show that the absorption laws hold for $\langle L, \cap, \cup \rangle$. We notice that, as expected, these properties coincide with the conjunction $(C1)$ for arbitrary s_1, s_2 on the right hand side of the Herbrand sequent and so we proceed by proving each conjunct for arbitrary $s_1, s_2 \in L$:

1. *We notice that $(A3i)+(A3)$ imply $(C1i)$. So we prove these properties:*

(a) *First we prove (A3i):*

$$s_1 \cap s_2 =^{(\text{idem.})} (s_1 \cap s_1) \cap s_2 =^{(\text{assoc.})} s_1 \cap (s_1 \cap s_2) =^{(\text{comm.})} s_1 \cap (s_2 \cap s_1) =^{(\text{assoc.})} (s_1 \cap s_2) \cap s_1$$

(b) *Assume* $(s_1 \cap s_2) \cap s_1 = s_1 \cap s_2$. *By definition of L1-lattices,* $(s_1 \cap s_2) \cup s_1 = s_1$. *Thus, we have proved (A3).*

2. *Again, we notice that (A5i) + (A5) + commutativity imply (C1ii) and use this fact:*

   (a) $s_1 \cup s_2 =^{(\text{idem.})} (s_1 \cup s_1) \cup s_2 =^{(\text{assoc.})} s_1 \cup (s_1 \cup s_2)$. *We have proved (A5i).*

   (b) *Assume* $s_1 \cup (s_1 \cup s_2) = s_1 \cup s_2$. *By definition of L1-lattices,* $s_1 \cap (s_1 \cup s_2) = s_1$. *This proves (A5).*

So we have shown that for arbitrary $s_1, s_2 \in L$, *we have* $(s_1 \cap s_2) \cup s_1 = s_1$ *and* $(s_1 \cup s_2) \cap s_1 = s_1$, *which completes the proof.*

Contrary to the proof in Section 8.5.4, we can now directly see the algebraic construction used to prove the theorem. This information was hidden in the synthetic argument that used the notion of partially ordered sets and was revealed by cut-elimination.
This example shows that the Herbrand sequent indeed contains the essential information of the ACNF, since an informal direct proof corresponding to the ACNF could be constructed by analyzing the extracted Herbrand sequent only.

Chapter 9

CERES in Nonclassical Logics

There are two main strategies to extend CERES to nonclassical logics: Nonclassical logics can sometimes be embedded into classical logics, i.e. their semantics can be classically formalized. Nonclassical proofs can thereby be translated into classical ones and CERES can be applied. This, however, changes the meaning of the information obtained from cut-free proofs (Herbrand disjuncts, interpolants, etc.). The second possibility is to adapt CERES to the logic in question. In this way CERES has been extended to a wide class of finitely-valued logics [19].

Considering the intended applications, intuitionistic logic and intermediate logics, i.e., logics over the standard language that are stronger than intuitionistic logic, but weaker than classical logic, are even more important targets for similar extensions. However, there are a number formidable obstacles to a straightforward generalization of CERES to this realm of logics:

- It is unclear whether and how classical resolution can be generalized, for the intended purpose, to intermediate logics.

- Gentzen's sequent format is too restrictive to obtain appropriate analytic calculi for many important intermediate logics.

- Skolemization, or rather the inverse de-Skolemization of proofs – an essential prerequisite for CERES – is not possible in general.

M. Baaz, A. Leitsch, *Methods of Cut-Elimination*, Trends in Logic 34,
DOI 10.1007/978-94-007-0320-9_9,

9.1 CERES in Finitely Valued Logics

The core of classical cut-elimination methods in the style of Gentzen [38] consists of the permutation of inferences and of the reduction of cuts to cuts on the immediate subformulas of the cut formulas. If we switch from two-valued to many-valued logic, the reduction steps become intrinsically tedious and opaque [10] in contrast to the extension of CERES to the many-valued case, which is straightforward.
We introduce CERES-m for correct (possible partial) calculi for m-valued first order logics based on m-valued connectives, distributive quantifiers [30] and arbitrary atomic initial sequents closed under substitution. We do not touch the completeness issue of these calculi, instead we derive clause terms from the proof representing the formulas which are ancestor formulas of the cut formulas, just as in Section 6.4. Like in the classical case the evaluation of these clause terms guarantees the existence of a resolution refutation as core of a proof with atomic cuts only. This resolution refutation is extended to a proof of the original end-sequent by adjoining cut-free parts of the original proof. Therefore, it is sufficient to refute the suitably assembled components of the initial sequents using a m-valued theorem prover [9]

9.1.1 Definitions

Definition 9.1.1 (language) The alphabet Σ consists of an infinite supply of variables, of infinite sets of n-ary function symbols and predicate symbols. Moreover, Σ contains a set W of truth symbols denoting the truth values of the logic, a finite number of connectives $\circ_1, \ldots, \circ_m$ of arity $n_1, \ldots, n_m$, and a finite number of quantifiers $Q_1, \ldots, Q_k$. ◇

Definition 9.1.2 (formula) An atomic formula is an expression of the form $P(t_1, \ldots, t_n)$ where P is an n-ary predicate symbol in Σ and $t_1, \ldots, t_n$ are terms over Σ. Atomic formulas are formulas.
If $\circ$ is an n-ary connective and $A_1, \ldots, A_n$ are formulas then $\circ(A_1, \ldots, A_n)$ is a formula.
If Q is quantifier in Σ and x is a variable then $(Qx)A$ is a formula. ◇

Definition 9.1.3 (signed formula) Let $w \in W$ and A be a formula. Then $w\colon A$ is called a signed formula. ◇

Definition 9.1.4 (sequent) A sequent is a finite sequence of signed formulas. The number of signed formulas occurring in a sequent S is called the *length* of S and is denoted by $l(S)$. $\hat{S}$ is called the *unsigned version* of S

if every signed formula $w{:}\,A$ in S is replaced by A. The length of unsigned versions is defined in the same way. A sequent S is called *atomic* if $\hat{S}$ is a sequence of atomic formulas. $\diamond$

Remark: Note that the classical sequent $(\forall x)P(x) \vdash Q(a)$ can be written as $\mathbf{f}{:}\,(\forall x)P(x), \mathbf{t}{:}\,Q(a)$. $\diamond$

m-valued sequents are sometimes written as m-sided sequents. We refrain from this notation, because it denotes a preferred order of truth values, which even in the two-valued case might induce unjustified conclusions.

Definition 9.1.5 (axiom set) A set $\mathcal{A}$ of atomic sequents is called an axiom set if $\mathcal{A}$ is closed under substitution. The definition is the same as for classical sequents (see Definition 3.2.1). $\diamond$

The calculus we are defining below is capable of formalizing any finitely valued logic. Concerning the quantifiers we assume them to be of distributive type [30]. Distribution quantifiers are functions from the non-empty sets of truth values to the set of truth values, where the domain represents the situation in the structure, i.e. the truth values actually taken.

Definition 9.1.6 Let $A(x)$ be a formula with free variable x. The *distribution Distr*$(A(x))$ of $A(x)$ is the set of all truth values in W to which $A(x)$ evaluates (for arbitrary assignments of domain elements to x). $\diamond$

Definition 9.1.7 Let q be a mapping $2^W \to W$. In interpreting the formula $(Qx)A(x)$ via q we first compute $\mathit{Distr}(A(x))$ and then $q(\mathit{Distr}(A(x)))$, which is the truth value of $(Qx)A(x)$ under the interpretation. $\diamond$

In the calculus defined below the distinction between quantifier introductions with (strong) and without eigenvariable conditions (weak) are vital.

Definition 9.1.8 A *strong quantifier* is a triple (V, w, w') (for $V \subseteq W$) s.t. $(Qx)A(x)$ evaluates to w if $\mathit{Distr}(A(x)) \subseteq V$ and to w' otherwise. A *weak quantifier* is a triple (u, w, w') s.t. $(Qx)A(x)$ evaluates to w if $u \in \mathit{Distr}(A(x))$, and to w' otherwise. $\diamond$

Remark: Strong and weak quantifiers are dual w.r.t. to set complementation. In fact to any strong quantifier there corresponds a weak one and vice versa. Like in classical logic we may speak about weak and strong occurrences of quantifiers in sequents and formulas. $\diamond$

Note that strong and weak quantifiers define merely a subclass of distribution quantifiers. Nevertheless the following property holds:

Proposition 9.1.1 *Any distributive quantifier can be expressed by strong and weak quantifiers and many valued associative, commutative and idempotent connectives (which are variants of conjunction and disjunction).*

Proof: In [9]. □

Definition 9.1.9 (LM-type calculi) We define an LM-type calculus **K**. The initial sequents are (arbitrary) atomic sequents of an axiom set $\mathcal{A}$. In the rules of **K** we always mark the auxiliary formulas (i.e. the formulas in the premise (premises) used for the inference) and the principal (i.e. the inferred) formula using different marking symbols. Thus, in our definition, classical $\wedge$-introduction to the right takes the form

$$\frac{\Gamma, \mathbf{t}\colon A^{+} \quad \Gamma, \mathbf{t}\colon B^{+}}{\Gamma, \mathbf{t}\colon A \wedge B^{*}}$$

If Γ, Δ is a sequent then Γ, Δ^{+} indicates that all signed formulas in Δ are auxiliary formulas of the defined inference. $\Gamma, \Delta, w\colon A^{*}$ indicates that $w\colon A$ is the principal formula (i.e. the inferred formula) of the inference.
Auxiliary formulas and the principal formula of an inference are always supposed to be rightmost. Therefore we usually avoid markings as the status of the formulas is clear from the notation.

Logical Rules:

Let $\circ$ be an n-nary connective. For any $w \in W$ we have an introduction rule $\circ\colon w$ of the form

$$\frac{\Gamma, \Delta_1^{+} \quad \ldots \quad \Gamma, \Delta_m^{+}}{\Gamma, w\colon \circ(\pi(\hat{\Delta}_1, \ldots, \hat{\Delta}_m, \hat{\Delta}))^{*}}\ \circ\colon w$$

where $l(\Delta_1, \ldots, \Delta_m, \Delta) = n$ (the Δ_i are sequences of signed formulas which are all auxiliary signed formulas of the inference) and $\pi(S)$ denotes a permutation of a sequent S.
Note that, for simplicity, we chose the additive version of all logical introduction rules.

In the introduction rules for quantifiers we distinguish *strong* and *weak* introduction rules. Any strong quantifier rule $Q\colon w$ (for a strong quantifier (V, w, w')) is of the form

$$\frac{\Gamma, u_1\colon A(\alpha_1)^{+}, \ldots, u_m\colon A(\alpha_m)^{+}}{\Gamma, w\colon (Qx)A(x)^{*}}\ Q\colon w$$

where the α_i are eigenvariables not occurring in Γ, and $V = \{u_1, \ldots, u_m\}$. Any weak quantifier rule (for a weak quantifier (u, w, w')) is of the form

$$\frac{\Gamma, u\colon A(t)^+}{\Gamma, w\colon (Qx)A(x)^*}\ Q\colon w$$

where t is a term containing no variables which are bound in $A(x)$. We say that t is *eliminated* by $Q\colon w$.

We have to define a special n-ary connective for every strong quantifier in order to carry out *Skolemization.* Indeed if we skip the introduction of a strong quantifier the m (possibly $m > 1$) auxiliary formulas must be contracted into a single one *after* the removal of the strong quantifier (see definition of Skolemization below). Thus for every rule

$$\frac{\Gamma, u_1\colon A(\alpha_1)^+, \ldots, u_m\colon A(\alpha_m)^+}{\Gamma, w\colon (Qx)A(x)^*}\ Q\colon w$$

we define a propositional rule

$$\frac{\Gamma, u_1\colon A(t)^+, \ldots, u_m\colon A(t)^+}{\Gamma, w\colon A(t)^*}\ c_Q\colon w$$

This new operator c_Q can be eliminated by the de-Skolemization procedure (to be defined below) afterwards.

Structural Rules:

The structural rule of weakening is defined like in **LK** (but we need only one weakening rule and may add more then one formula).

$$\frac{\Gamma}{\Gamma, \Delta}\ w$$

for sequents Γ and Δ.
To put the auxiliary formulas on the right positions we need permutation rules of the form

$$\frac{F_1, \ldots, F_n}{F_{\pi(1)}, \ldots, F_{\pi(n)}}\ \pi$$

where π is a permutation of $\{1, \ldots, n\}$ and the F_i are signed formulas .

Instead of the usual contraction rules we define an n-contraction rule for any $n \geq 2$ and $F_1 = \ldots = F_n = F$:

$$\frac{\Gamma, F_1, \ldots, F_n}{\Gamma, F}\ c:n$$

In contrast to **LK** we do not have a single cut rule, but instead rules $cut_{ww'}$ for any $w, w' \in W$ with $w \neq w'$. Any such rule is of the form

$$\frac{\Gamma, w\colon A \quad \Gamma', w'\colon A}{\Gamma, \Gamma'}\ cut_{ww'}$$

$\diamond$

Definition 9.1.10 (proof) A *proof* of a sequent S from an axiom set $\mathcal{A}$ is a directed labeled tree. The root is labeled by S, the leaves are labeled by elements of $\mathcal{A}$. The edges are defined according to the inference rules (in an n-ary rule the children of a node are labeled by the antecedents, the parent node is labeled by the consequent). Let N be a node in the proof φ then we write $\varphi.N$ for the corresponding subproof ending in N. For the number of nodes in φ we write $\|\varphi\|_l$ (compare to Definition 6.2.3). $\diamond$

Definition 9.1.11 Let **K** be an LM-type calculus. We define $\Phi[\mathbf{K}]$ as the set of all **K**-proofs. $\Phi^i[\mathbf{K}]$ is the subset of $\Phi[\mathbf{K}]$ consisting of all proofs with cut-complexity $\leq i$ ($\Phi^0[\mathbf{K}]$ is the set of proofs with at most atomic cuts). $\Phi^\emptyset[\mathbf{K}]$ is the subset of all cut-free proofs. $\diamond$

Example 9.1.1 We define $W = \{0, u, 1\}$ and the connectives as in the 3-valued Kleene logic, but introduce a new quantifier D ("D" for determined) which gives true iff all truth values are in $\{0, 1\}$. We only define the rules for $\vee$ and for D, as no other operators occur in the proof below.

$$\frac{0\colon A, 1\colon A \quad 0\colon B, 1\colon B \quad 1\colon A, 1\colon B}{1\colon A \vee B}\ \vee\colon 1$$

$$\frac{u\colon A, u\colon B}{u\colon A \vee B}\ \vee\colon u \qquad \frac{0\colon A \quad 0\colon B}{0\colon A \vee B}\ \vee\colon 0$$

$$\frac{0\colon A(\alpha), 1\colon A(\alpha)}{1\colon (Dx)A(x)}\ D\colon 1 \qquad \frac{u\colon A(t)}{0\colon (Dx)A(x)}\ D\colon 0$$

where α is an eigenvariable and t is a term containig no variables bound in $A(x)$. Note that $D\colon 1$ is a strong, and $D\colon 0$ a weak quantifier introduction. The formula $u\colon (Dx)A(x)$ can only be introduced via weakening.

For the notation of proofs we frequently abbreviate sequences of structural rules by $*$; thus $\pi^* + \vee\colon u$ means that $\vee\colon u$ is performed and permutations before and/or afterwards. This makes the proofs more legible and allows to focus on the logically relevant inferences. As in the definition of LM-type calculi we mark the auxiliary formulas of logical inferences and cut by $+$, the principle ones by $*$.
Let φ be the following proof

$$\dfrac{\varphi_1 \quad \varphi_2}{0\colon (Dx)((P(x) \vee Q(x)) \vee R(x)), 1\colon (Dx)P(x)}\ cut$$

where $\varphi_1 =$

$$\dfrac{\dfrac{\dfrac{\dfrac{\begin{matrix}(\psi')\\ 0\colon P(\alpha) \vee Q(\alpha),\ u\colon P(\alpha) \vee Q(\alpha),\ 1\colon P(\alpha) \vee Q(\alpha)\end{matrix}}{0\colon P(\alpha) \vee Q(\alpha),\ u\colon P(\alpha) \vee Q(\alpha),\ u\colon R(\alpha)^*,\ 1\colon P(\alpha) \vee Q(\alpha)}\ \pi^* + w}{0\colon A(\alpha) \vee Q(\alpha),\ u\colon (P(\alpha) \vee Q(\alpha)) \vee R(\alpha)^{+*},\ 1\colon P(\alpha) \vee Q(\alpha)}\ \pi^* + \vee\colon u}{0\colon (Dx)((P(x) \vee Q(x)) \vee R(x))^*,\ 0\colon P(\alpha) \vee Q(\alpha)^+,\ 1\colon P(\alpha) \vee Q(\alpha)^+}\ \pi^* + D\colon 0}{0\colon (Dx)((P(x) \vee Q(x)) \vee R(x)),\ 1\colon (Dx)(P(x) \vee Q(x))^*}\ D\colon 1$$

and $\varphi_2 =$

$$\dfrac{\dfrac{\dfrac{\dfrac{0\colon P(\beta),\ u\colon P(\beta),\ 1\colon P(\beta)}{0\colon P(\beta),\ 1\colon P(\beta),\ u\colon P(\beta)^+,\ u\colon Q(\beta)^{*+}}\ \pi^* + w}{0\colon P(\beta),\ u\colon P(\beta) \vee Q(\beta)^{*+},\ 1\colon P(\beta)}\ \pi^* + \vee\colon u}{0\colon (Dx)(P(x) \vee Q(x))^*,\ 0\colon P(\beta)^+,\ 1\colon P(\beta)^+}\ \pi^* + D\colon 0}{0\colon (Dx)(P(x) \vee Q(x)),\ 1\colon (Dx)P(x)^*}\ D\colon 1$$

we have to define ψ' as our axiom set must be atomic. We set

$$\psi' = \psi(A, B)\{A \leftarrow P(\alpha), A \leftarrow Q(\alpha)\}$$

and define
$\psi(A, B) =$

$$\dfrac{\dfrac{\psi_1(A, B) \quad \psi_2(A, B)}{0\colon A \vee B,\ u\colon A,\ u\colon B,\ 1\colon A \vee B}\ \pi^* + \vee\colon 0}{0\colon A \vee B,\ u\colon A \vee B,\ 1\colon A \vee B}\ \pi^* + \vee\colon u$$

and
$\psi_1(A, B) =$

$$\dfrac{0\colon A, u\colon A, u\colon B, 0\colon A, 1\colon A \quad 0\colon A, u\colon A, u\colon B, 0\colon B, 1\colon B \quad 0\colon A, u\colon A, u\colon B, 1\colon A, 1\colon B}{0\colon A,\ u\colon A,\ u\colon B,\ 1\colon A \vee B}\ \vee\colon 1$$

$\psi_2(A, B) =$

$$\frac{0{:}\,B, u{:}\,A, u{:}\,B, 0{:}\,A, 1{:}\,A \quad 0{:}\,B, u{:}\,A, u{:}\,B, 0{:}\,B, 1{:}\,B \quad 0{:}\,B, u{:}\,A, u{:}\,B, 1{:}\,A, 1{:}\,B}{0{:}\,B,\ u{:}\,A,\ u{:}\,B,\ 1{:}\,A \vee B}\ \vee{:}\,1$$

It is easy to see that the end sequent is valid as the axioms contain

$$0{:}\,A,\ u{:}\,A,\ 1{:}\,A \text{ and } 0{:}\,B,\ u{:}\,B,\ 1{:}\,B$$

as subsequents. ◇

Definition 9.1.12 (W-clause) A W-clause is an atomic sequent (where W is the set of truth symbols). The empty sequent is called empty clause and is denoted by $\Box$. ◇

Let S be an W-clause. S' is called a renamed variant of S if $S' = S\eta$ for a variable permutation η.

Definition 9.1.13 (W-resolution) We define a resolution calculus R_W which only depends on the set W (but not on the logical rules of **K**). R_W operates on W-clauses; its rules are:

1. $res_{ww'}$ for all $w, w' \in W$ and $w \neq w'$,
2. w-factoring for $w \in W$,
3. permutations.

Let $S{:}\,\Gamma, w{:}\,A$ and $S'{:}\,\Gamma', w'{:}\,A'$ (where $w \neq w'$) be two W-clauses and $S''{:}\,\Gamma'', w'{:}\,A''$ be a variant of S' s.t. S and S' are variable disjoint. Assume that $\{A, B'\}$ are unifiable by a most general unifier σ. Then the rule $res_{ww'}$ on S, S' generates a resolvent R for

$$R = \Gamma\sigma, \Gamma''\sigma.$$

Let $S{:}\,\Gamma, w{:}\,A_1, \ldots, w{:}\,A_m$ be a clause and σ be a most general unifier of $\{A_1, \ldots, A_m\}$. Then the clause

$$S'{:}\,\Gamma\sigma, w{:}\,A_1\sigma$$

is called a w-factor of S.

A W-resolution proof of a clause S from a set of clauses $\mathcal{S}$ is a directed labeled tree s.t. the root is labeled by S and the leaves are labeled by elements of $\mathcal{S}$. The edges correspond to the applications of w-factoring (unary), permutation (unary) and $res_{ww'}$ (binary). ◇

It is proved in [8] that W-resolution is complete. For the LM-type calculus we only require soundness w.r.t. the underlying logic. So from now on we assume that **K** is sound.

Note that we did not define clauses as sets of signed literals; therefore we need the permutation rule in order to "prepare" the clauses for resolution and factoring.

Definition 9.1.14 (ground projection) Let γ be a W-resolution proof and $\{x_1, \ldots, x_n\}$ be the variables occurring in the indexed clauses of γ. Then, for all ground terms $t_1, \ldots, t_n$, $\gamma\{x_1 \leftarrow t_1, \ldots, x_1 \leftarrow t_n\}$ is called a *ground projection* of γ (compare to Definition 3.3.14). ◇

Remark: Ground projections of resolution proofs are ordinary proofs in **K**; indeed factoring becomes n-contraction and resolution becomes cut. ◇

Definition 9.1.15 (ancestor relation) Let

$$\frac{S_1\colon \Gamma, \Delta_1^+ \quad \ldots \quad S_m\colon \Gamma, \Delta_m^+}{S\colon \Gamma, w\colon A^*}\, x$$

be an inference in a proof φ; let μ be the occurrence of the principal signed formula $w\colon A$ in S and ν_{ij} be the occurrence of the j-th auxiliary formula in S_i. Then all ν_{ij} are ancestors of μ.
The ancestor relation in φ is defined as the reflexive and transitive closure of the above relation.
By $S(N, \Omega)$ ($\bar{S}(N, \Omega)$) we denote the subsequent of S at the node N of φ consisting of all formulas which are (not) ancestors of a formula occurrence in Ω.

◇

Example 9.1.2 Let $\psi(A, B)$ as in Example 9.1.1:
$\psi(A, B) =$

$$\frac{\dfrac{\psi_1(A,B) \quad \psi_2(A,B)}{0\colon A \vee B^\dagger,\ u\colon A,\ u\colon B,\ 1\colon A \vee B}\, \pi^* + \vee\colon 0}{0\colon A \vee B^\dagger,\ u\colon A \vee B,\ 1\colon A \vee B}\, \pi^* + \vee\colon u$$

$\psi_1(A, B) =$

$$\frac{0\colon A^\dagger, u\colon A, u\colon B, 0\colon A, 1\colon A \quad 0\colon A^\dagger, u\colon A, u\colon B, 0\colon B, 1\colon B \quad 0\colon A^\dagger, u\colon A, u\colon B, 1\colon A, 1\colon B}{0\colon A^\dagger,\ u\colon A,\ u\colon B,\ 1\colon A \vee B}\, \vee\colon 1$$

$\psi_2(A, B) =$

$$\frac{0{:}\,B^\dagger, u{:}\,A, u{:}\,B, 0{:}\,A, 1{:}\,A \quad 0{:}\,B^\dagger, u{:}\,A, u{:}\,B, 0{:}\,B, 1{:}\,B \quad 0{:}\,B^\dagger, u{:}\,A, u{:}\,B, 1{:}\,A, 1{:}\,B}{0{:}\,B^\dagger,\ u{:}\,A,\ u{:}\,B,\ 1{:}\,A \vee B}\ \vee{:}\,1$$

Let N_0 be the root of $\psi(A, B)$ and μ be the occurrence of the first formula $(0{:}\,A \vee B)$ in N. The formula occurrences which are ancestors of μ are labeled with †. In the premises N_1, N_2 of the binary inference $\vee{:}\,0$ we have $S(N_1, \{\mu\}) = 0{:}\,A$ and $S(N_2, \{\mu\}) = 0{:}\,B$.

◇

9.1.2 Skolemization

As CERES-m (like CERES as defined in Section 6.4) augments a ground resolution proof with cut-free parts of the original proof related only to the end-sequent, eigenvariable conditions in these proof parts might be violated. To get rid of this problem, the endsequent of the proof and the formulas, which are ancestors of the end-sequent have to be Skolemized, i.e eigenvariables have to be replaced by suitable Skolem terms. To obtain a skolemization of the end-sequent, we have to represent (analyze) distributive quantifiers in terms of strong quantifiers (covering exclusively eigenvariables) and weak quantifiers (covering exclusively terms). This was the main motivation for the choice of our definition of quantifiers in Definition 9.1.9. The strong quantifiers are replaced by Skolem functions depending on the weakly quantified variables determined by the scope. Note that distributive quantifiers are in general mixed, i.e. they are neither weak nor strong, even in the two-valued case.

9.1.3 Skolemization of Proofs

Definition 9.1.16 (Skolemization) Let $\Delta{:}\,\Gamma, w{:}\,A$ be a sequent and $(Qx)B$ be a subformula of A at the position λ where Qx is a maximal strong quantifier in $w{:}\,A$. Let $y_1, \dots, y_m$ be free variables occurring in $(Qx)B$, then we define

$$sk(\Delta) = \Gamma, w{:}\,A[B\{x \leftarrow f(y_1, \dots, y_m)\}]_\lambda$$

where f is a function symbol not occurring in Δ.
If $w{:}\,A$ contains no strong quantifier then we define $sk(\Delta) = \Delta$.
A sequent S is in *Skolem form* if there exists no permutation S' of S s.t. $sk(S') \neq S'$. S' is called a Skolem form of S if S' is in Skolem form and can be obtained from S by permutations and the operator sk. ◇

The Skolemization of proofs can be defined in a way quite similar to the classical case (compare to Section 6.2).

Definition 9.1.17 (Skolemization of K-proofs) Let **K** be an LM-type calculus. We define a transformation of proofs which maps a proof φ of S from $\mathcal{A}$ into a proof $sk(\varphi)$ of S' from $\mathcal{A}'$ where S' is the Skolem form of S and $\mathcal{A}'$ is an instance of $\mathcal{A}$.
Locate an uppermost logical inference which introduces a signed formula $w\colon A$ which is not an ancestor of a cut and contains a strong quantifier.

(a) The formula is introduced by a strong quantifier inference:

$$\frac{\begin{array}{c}\psi[\alpha_1,\ldots,\alpha_m]\\ S'\colon\Gamma, u_1\colon A(\alpha_1)^+,\ldots,u_m\colon A(\alpha_m)^+\end{array}}{S\colon\Gamma, w\colon(Qx)A(x)^*}\ Q\colon w$$

in φ and N', N be the nodes in φ labeled by S', S. Let P be the path from the root to N', locate all weak quantifier inferences ξ_i ($i = 1,\ldots,n$) on P and all terms t_i eliminated by ξ_i. Then we delete the inference node N and replace the derivation ψ of N' by

$$\frac{\begin{array}{c}\psi[f(t_1,\ldots,t_n),\ldots,f(t_1,\ldots,t_n)]\\ S'\colon\Gamma, u_1\colon A(f(t_1,\ldots,t_n))^+,\ldots,u_m\colon A(f(t_1,\ldots,t_n))^+\end{array}}{S_0\colon\Gamma, w\colon A(f(t_1,\ldots,t_n))^*}\ c_Q : w$$

where f is a function symbol not occurring in φ and c_Q is the connective corresponding to Q. The sequents on P are adapted according to the inferences on P.

(b) The formula is inferred by a propositional inference or by weakening (within the principal formula $w\colon A$) then we replace $w\colon A$ by the Skolem form of $w\colon A$ where the Skolem function symbol does not occur in φ.

Furthermore the Skolemization works exactly as in Section 6.2: Let φ' be the proof after such a Skolemization step. We iterate the procedure until no occurrence of a strong quantifier is an ancestor of an occurrence in the end sequent. The resulting proof is called $sk(\varphi)$. Note that $sk(\varphi)$ is a proof from the *same* axiom set $\mathcal{A}$ as $\mathcal{A}$ is closed under substitution. ◇

Definition 9.1.18 A proof φ is called *Skolemized* if $sk(\varphi) = \varphi$. ◇

Note that Skolemized proofs may contain strong quantifiers, but these are ancestors of cut, in the end-sequent there are none.

Example 9.1.3 Let φ be the proof from Example 9.1.1:

$$\frac{\varphi_1 \quad \varphi_2}{0\colon (Dx)((P(x) \vee Q(x)) \vee R(x)), 1\colon (Dx)P(x)}\ cut$$

where $\varphi_1 =$

$$\cfrac{\cfrac{\cfrac{\cfrac{\begin{array}{c}(\psi')\\ 0\colon P(\alpha) \vee Q(\alpha),\ u\colon P(\alpha) \vee Q(\alpha),\ 1\colon P(\alpha) \vee Q(\alpha)\end{array}}{0\colon P(\alpha) \vee Q(\alpha),\ u\colon P(\alpha) \vee Q(\alpha),\ u\colon R(\alpha)^*,\ 1\colon P(\alpha) \vee Q(\alpha)}\ \pi^* + w}{0\colon P(\alpha) \vee Q(\alpha),\ u\colon (P(\alpha) \vee Q(\alpha)) \vee R(\alpha)^{+*},\ 1\colon P(\alpha) \vee Q(\alpha)}\ \pi^* + \vee\colon u}{0\colon (Dx)((P(x) \vee Q(x)) \vee R(x))^*,\ 0\colon P(\alpha) \vee Q(\alpha)^+,\ 1\colon P(\alpha) \vee Q(\alpha)^+}\ \pi^* + D\colon 0}{0\colon (Dx)((P(x) \vee Q(x)) \vee R(x)),\ 1\colon (Dx)(P(x) \vee Q(x))^*}\ D\colon 1$$

and $\varphi_2 =$

$$\cfrac{\cfrac{\cfrac{\cfrac{0\colon P(\beta),\ u\colon P(\beta),\ 1\colon P(\beta)}{0\colon P(\beta),\ 1\colon P(\beta),\ u\colon P(\beta)^+,\ u\colon Q(\beta)^{*+}}\ \pi^* + w}{0\colon P(\beta),\ u\colon P(\beta) \vee Q(\beta)^{*+},\ 1\colon P(\beta)}\ \pi^* + \vee\colon u}{0\colon (Dx)(P(x) \vee Q(x))^*,\ 0\colon P(\beta)^+,\ 1\colon P(\beta)^+}\ \pi^* + D\colon 0}{0\colon (Dx)(P(x) \vee Q(x)),\ 1\colon (Dx)P(x)^*}\ D\colon 1$$

The proof is not Skolemized as the endsequent contains a strong quantifier occurrence in the formula $1\colon (Dx)P(x)$. This formula comes from the proof φ_2. Thus we must Skolemize φ_2 and adapt the end sequent of φ. It is easy to verify that $sk(\varphi_2) =$

$$\cfrac{\cfrac{\cfrac{\cfrac{0\colon P(c),\ u\colon P(c),\ 1\colon P(c)}{0\colon P(c),\ 1\colon P(c),\ u\colon P(c)^+,\ u\colon Q(c)^{*+}}\ \pi^* + w}{0\colon P(c),\ u\colon P(c) \vee Q(c)^{*+},\ 1\colon P(c)}\ \pi^* + \vee\colon u}{0\colon (Dx)(P(x) \vee Q(x))^*,\ 0\colon P(c)^+,\ 1\colon P(c)^+}\ \pi^* + D\colon 0}{0\colon (Dx)(P(x) \vee Q(x)),\ 1\colon P(c)^*}\ c_{D-1}$$

Then $sk(\varphi) =$

$$\frac{\varphi_1 \quad sk(\varphi_2)}{0\colon (Dx)((P(x) \vee Q(x)) \vee R(x)), 1\colon P(c)}\ cut$$

Note that φ_1 cannot be Skolemized as the strong quantifiers in φ_1 are ancestors of the cut in φ. $\diamond$

Skolem functions can be replaced by the original structure of (strong and weak) quantifiers by the following straightforward algorithm at most exponential in the maximal size of the original proof and of the CERES-m proof of the Skolemized end sequent: Order the Skolem terms (terms, whose outermost function symbol is a Skolem function) by inclusion. The size of the proof resulting from CERES-m together with the number of inferences in the original proof limits the number of relevant Skolem terms. Always replace a maximal Skolem term by a fresh variable, and determine the formula F in the proof, for which the corresponding strong quantifier should be introduced. In re-introducing the quantifier we eliminate the newly introduced connectives c_Q. As the eigenvariable condition might be violated at the lowest possible position, where the quantifier can be introduced (because e.g. the quantified formula has to become part of a larger formula by an inference) suppress all inferences on F such that F occurs as side formula besides the original end-sequent. Then perform all inferences on F. This at most triples the size of the proof (a copy of the proof together with suitable contractions might be necessary).

The distributive quantifiers are by now represented by a combination of strong quantifiers, weak quantifiers and connectives. A simple permutation of inferences in the proof leads to the immediate derivation in several steps of the representation of the distributive quantifier from the premises of the distributive quantifier inference. The replacement of the representation by the distributive quantifier is then simple.

9.1.4 CERES-m

As in the classical case (see Chapter 6) we restrict cut-elimination to Skolemized proofs. After cut-elimination the obtained proof can be de-Skolemized, i.e. it can be transformed into a derivation of the original (un-Skolemized) end-sequent.

Definition 9.1.19 Let $\mathbf{K}$ be an LM-type calculus. We define $\Phi^s[\mathbf{K}]$ as the set of all Skolemized proofs in $\mathbf{K}$. $\Phi^s_\emptyset[\mathbf{K}]$ is the set of all cut-free proofs in $\Phi^s[\mathbf{K}]$ and, for all $i \geq 0$, $\Phi^s_i[\mathbf{K}]$ is the subset of $\Phi^s[\mathbf{K}]$ containing all proofs with cut-formulas of formula complexity $\leq i$. $\diamond$

Our goal is to transform a derivation in $\Phi^s[\mathbf{K}]$ into a derivation in $\Phi^s_0[\mathbf{K}]$ (i.e. we reduce all cuts to atomic ones). Like in the classical case, the first step in the corresponding procedure consists in the definition of a clause term corresponding to the sub-derivations of an $\mathbf{K}$-proof ending in a cut. In particular we focus on derivations of the cut formulas themselves, i.e. on the derivation of formulas having no successors in the end-sequent. Below we will see that this analysis of proofs, first defined for **LK**, is quite general and can easily be generalized to LM-type calculi. The definition of clause terms by binary operators $\oplus$ and $\otimes$ and their semantics is the same as in Definitions 7.1.1 and 7.1.2. Also the concepts of characteristic clause term and characteristic clause set can be defined exactly as in the Definitions 7.1.3 and 7.1.4

Example 9.1.4 Let φ' be the Skolemized proof defined in Example 9.1.3. It is easy to verify that the characteristic clause set $\mathrm{CL}(\varphi')$ is

$$\begin{array}{l}
\{u\colon P(c),\\
0\colon P(\alpha),\ 0\colon P(\alpha),\ 1\colon P(\alpha)\\
0\colon P(\alpha),\ 0\colon Q(\alpha),\ 1\colon Q(\alpha)\\
0\colon P(\alpha),\ 1\colon P(\alpha),\ 1\colon Q(\alpha)\\
0\colon Q(\alpha),\ 0\colon P(\alpha),\ 1\colon P(\alpha)\\
0\colon Q(\alpha),\ 0\colon Q(\alpha),\ 1\colon Q(\alpha)\\
0\colon Q(\alpha),\ 1\colon P(\alpha),\ 1\colon Q(\alpha)\}.
\end{array}$$

The set $\mathrm{CL}(\varphi')$ can be refuted via W-resolution for $W = \{0, u, 1\}$. A W-resolution refutation is ($0f$ stands for 0-factoring) $\gamma =$

$$\dfrac{\dfrac{\dfrac{0\colon P(\alpha), 0\colon P(\alpha), 1\colon P(\alpha) \quad u\colon P(c)}{0\colon P(c),\ 0\colon P(c)}\ res_{1u}}{0\colon P(c)}\ 0f \qquad u\colon P(c)}{\Box}\ res_{0u}$$

A ground projection of γ (even the only one) is $\gamma' = \gamma\{\alpha \leftarrow c\} =$

$$\dfrac{\dfrac{\dfrac{0\colon P(c), 0\colon P(c), 1\colon P(c) \quad u\colon P(c)}{0\colon P(c),\ 0\colon P(c)}\ cut_{1u}}{0\colon P(c)}\ c \qquad u\colon P(c)}{\Box}\ cut_{0u}$$

Obviously γ' is a proof in **K**.

$\diamond$

In Example 9.1.4 we have seen that the characteristic clause set of a proof is refutable by W-resolution. Like in the classical case this is a general principle as will be shown below.

Definition 9.1.20 From now on we write Ω for the set of all occurrences of cut-formulas in φ. So, for any node N in φ $S(N,\Omega)$ is the subsequent of S containing the ancestors of a cut. $\bar{S}(N,\Omega)$ denotes the subsequent of S containing all non-ancestors of a cut. $\diamond$

Remark: Note that for any sequent S occurring at a node N of φ, S is a permutation variant of $S(N,\Omega), \bar{S}(N,\Omega)$. $\diamond$

Theorem 9.1.1 *Let φ be a proof in an* LM*-calculus* $\mathbf{K}$*. Then there exists a W-resolution refutation of* $\mathrm{CL}(\varphi)$*.*

Proof: According to Definition 7.1.3 we have to show that

(∗) for all nodes N in φ there exists a proof of $S(N,\Omega)$ from $\mathcal{S}_N$,

where $\mathcal{S}_N$ is defined as $|\Theta(\varphi)/N|$ (i.e. the set of clauses corresponding to N, see Definition 7.1.3). If N_0 is the root node of φ labeled by S then, clearly, no ancestor of a cut exists in S and so $S(N_0,\Omega) = \Box$. But by definition $\mathcal{S}_{N_0} = \mathrm{CL}(\varphi)$. So we obtain a proof of $\Box$ from $\mathrm{CL}(\varphi)$ in $\mathbf{K}$. By the completeness of W-resolution there exists a W-resolution refutation of $\mathrm{CL}(\varphi)$.

It remains to prove (∗):
Let N be a leaf node in φ. Then by definition of $\mathrm{CL}(\varphi)$ $\mathcal{S}_N = \{S(N,\Omega)\}$. So $S(N,\Omega)$ itself is the required proof of $S(N,\Omega)$ from $\mathcal{S}_N$.

(IH):
Now assume inductively that for all nodes N of depth $\leq n$ in φ there exists a proof ψ_N of $S(N,\Omega)$ from $\mathcal{S}_N$.

So let N be a node of depth $n+1$ in φ. We distinguish the following cases:

(a) N is the consequent of M, i.e. N is the result of a unary inference in φ. That means $\varphi.N =$

$$\frac{\varphi.M}{S(N)}\ x$$

By (IH) there exists a proof ψ_M of $S(M,\Omega)$ from $\mathcal{S}_M$. By Definition 7.1.3 $\mathcal{S}_N = \mathcal{S}_M$. If the auxiliary formula of the last inference is in $S(M,\Omega)$ we define $\psi_N =$

$$\frac{\psi_M}{S'}\ x$$

Obviously S' is just $S(N,\Omega)$.

If the auxiliary formula of the last inference in $\varphi.N$ is not in $S(M,\Omega)$ we simply drop the inference and define $\psi_N = \psi.M$. As the ancestors of cut did not change ψ_N is just a proof of $S(N,\Omega)$ from $\mathcal{S}_N$.

(b) N is the consequent of an n-ary inference for $n \geq 2$, i.e. $\varphi.N =$

$$\frac{\varphi.M_1 \quad \ldots \quad \varphi.M_n}{S(N)}\ x$$

By (IH) there exist proofs ψ_{M_i} of $S(M_i,\Omega)$ from $\mathcal{S}_{M_i}$.

(b1) The auxiliary formulas of the last inference in $\varphi.N$ are in $S(M_i,\Omega)$, i.e. the inference yields an ancestor of a cut. Then, by Definition 7.1.3

$$\mathcal{S}_N \;=\; \mathcal{S}_{M_1} \cup \ldots \cup \mathcal{S}_{M_n}.$$

Then clearly the proof ψ_N:

$$\frac{\psi_{M_1} \quad \ldots \quad \psi_{M_n}}{S'}\ x$$

is a proof of S' from $\mathcal{S}_N$ and $S' = S(N,\Omega)$.

(b2) The auxiliary formulas of the last inference in $\varphi.N$ are not in $S(M_i,\Omega)$, i.e. the principal formula of the inference is not an ancestor of a cut. Then, by Definition 7.1.3

$$\mathcal{S}_N \;=\; \odot(\mathcal{S}_{M_1},\ldots,\mathcal{S}_{M_n}).$$

We write $\mathcal{S}_i$ for $\mathcal{S}_{M_i}$ and ψ_i for ψ_{M_i}, Γ_i for $S(M_i,\Omega)$ and define

$$\begin{aligned}\mathcal{D}_i &= \odot(\mathcal{S}_1,\ldots,\mathcal{S}_i),\\ \Delta_i &= \Gamma_1,\ldots,\Gamma_i,\end{aligned}$$

for $i = 1,\ldots,n$. Our aim is to define a proof ψ_N of $S(N,\Omega)$ from $\mathcal{S}_N$ where $\mathcal{S}_N = \mathcal{D}_n$.

We proceed inductively and define proofs χ_i of Δ_i from $\mathcal{D}_i$. Note that for $i = n$ we obtain a proof χ_n of $S(M_1,\Omega),\ldots,S(M_n,\Omega)$ from $\mathcal{S}_N$, and $S(N,\Omega) = S(M_1,\Omega),\ldots,S(M_n,\Omega)$. This is just what we want.

For $i = 1$ we define $\chi_1 = \psi_1$.

Assume that $i < n$ and we already have a proof χ_i of Δ_i from $\mathcal{D}_i$. For every $D \in \mathcal{S}_{i+1}$ we define a proof $\chi_i[D]$:

Replace all axioms C in χ_i by the derivation

$$\frac{C, D}{D, C}\ \pi$$

and simulate χ_i on the extended axioms (the clause D remains passive). The result is a proof $\chi'[D]$ of the sequent

$$D, \ldots, D, \Delta_i.$$

Note that the propagation of D through the proof is possible as no eigenvariable conditions can be violated, as we assume the original proof to be regular (if not then we may transform the ψ_i into proofs with mutually disjoint sets of eigenvariables). Then we define $\chi_i[D]$ as

$$\frac{\chi'[D]}{\Delta_i, D}\ c^* + \pi$$

Next we replace every axiom D in the derivation ψ_{i+1} by the proof $\chi_i[D]$ and (again) simulate ψ_{i+1} on the end-sequents of the $\chi_i[D]$ where the Δ_i remain passive. Again we can be sure that no eigenvariable condition is violated and we obtain a proof ρ of

$$\Delta_i, \ldots, \Delta_i, \Gamma_{i+1}.$$

from the clause set $\odot(\mathcal{D}_i, \mathcal{S}_{i+1})$ which is $\mathcal{D}_{i+1}$. Finally we define $\chi_{i+1} =$

$$\frac{\rho}{\Delta_i, \Gamma_{i+1}}\ \pi^* + c^*$$

Indeed, χ_{i+1} is a proof of Δ_{i+1} from $\mathcal{D}_{i+1}$. □

Like in the classical case we define projections of the proof φ relative to clauses C in $\mathrm{CL}(\varphi)$. We drop all inferences which infer ancestors of a cut formula; the result is a cut-free proof of the end sequent extended by the clause C.

Lemma 9.1.1 *Let φ be a deduction in $\Phi^s[\mathbf{K}]$ of a sequent S. Let C be a clause in $\mathrm{CL}(\varphi)$. Then there exists a deduction $\varphi[C]$ of C, S s.t. $\varphi[C]$ is cut-free (in particular $\varphi(C) \in \Phi^s_\emptyset[\mathbf{K}]$) and $\|\varphi[C]\|_l \leq 2 * \|\varphi\|_l$.*

Proof: Let $\mathcal{S}_N$ be $|\Theta(\varphi)/N|$ (like in the proof of Theorem 9.1.1). We prove that

($\star$) for every node N in φ and for every $C \in \mathcal{S}_N$ there exists a proof $T(\varphi, N, C)$ of $C, \bar{S}(N, \Omega)$ s.t.

$$\|T(\varphi, N, C)\|_l \leq 2\|\varphi.N\|_l.$$

Indeed, it is sufficient to prove ($\star$): for the root node N_0 we have $S = \bar{S}(N_0, \Omega)$ (no signed formula of the end sequent is an ancestor of Ω), $\varphi.N_0 = \varphi$ and $\mathrm{CL}(\varphi) = \mathcal{S}_{N_0}$; so at the end we just define $\varphi[C] = T(\varphi, N_0, C)$ for every $C \in \mathrm{CL}(\varphi)$.

We prove $\star$ by induction on the depth of a node N in φ.

(IB) N is a leaf in φ.

Then, by definition of $\mathcal{S}_N$ we have $\mathcal{S} = \{S(N, \Omega)\}$ and $C\colon S(N, \Omega)$ is the only clause in $\mathcal{S}_N$. Let $\Gamma = \bar{S}(N, \Omega)$. Then $S(N)$ (the sequent labeling the node N) is a permutation variant of C, Γ and we define $T(\varphi, N, C) =$

$$\frac{S(N)}{C, \Gamma}\ \pi$$

If no permutation is necessary we just define $T(\varphi, N, C) = S(N)$. In both cases

$$\|T(\varphi, N, C)\|_l \leq 2 = 2\|\varphi.N\|_l.$$

(IH) Assume ($\star$) holds for all nodes of depth $\leq k$.

Let N be a node of depth $k + 1$. We distinguish the following cases:

(1) N is inferred from M via a unary inference x. By Definition of the clause term we have $\mathcal{S}_N = \mathcal{S}_M$. So any clause in $\mathcal{S}_N$ is already in $\mathcal{S}_M$.

(1a) The auxiliary formula of x is an ancestor of Ω. Then clearly $\bar{S}(N, \Omega) = \bar{S}(M, \Omega)$ and we define $T(\varphi, N, C) = T(\varphi, M, C)$. Clearly

$$\|T(\varphi, N, C)\|_l = \|T(\varphi, M, C)\|_l \leq_{(IH)} 2\|\varphi.M\|_l < 2\|\varphi.N\|_l.$$

(1b) The auxiliary formula of x is not an ancestor of Ω. Let $\Gamma = \bar{S}(M, \Omega), \Gamma' = \bar{S}(N, \Omega)$; thus the auxiliary formula of x is in Γ.

By (IH) there exists a proof $\psi: T(\varphi, M, C)$ of C, Γ and $\|\psi\|_l \leq 2\|\varphi.M\|_l$. We define $T(\varphi, N, C) =$

$$\cfrac{\begin{matrix}(\psi)\\ C, \Gamma\end{matrix}}{C, \Gamma'}\ x$$

Note that x cannot be a strong quantifier inference as the proof φ is Skolemized and there are no strong quantifiers in the end sequent. Thus $T(\varphi, N, C)$ is well-defined. Moreover

$$\|T(\varphi, N, C)\|_l = \|T(\varphi, M, C)\|_l + 1 \leq_{(IH)} 2\|\varphi.M\|_l + 1 < 2\|\varphi.N\|_l.$$

(2) N is inferred from $M_1, \ldots, M_n$ via the inference x for $n \geq 2$. By (IH) there are proofs $T(\varphi, M_i, C_i)$ for $i = 1, \ldots, n$ and $C_i \in \mathcal{S}_{M_i}$. Let $\bar{S}(M_i, \Omega) = \Gamma_i$ and $\bar{S}(N, \Omega) = \Gamma'_1, \ldots, \Gamma'_n$. We abbreviate $T(\varphi, M_i, C_i)$ by ψ_i.

(2a) The auxiliary formulas of x are in $\Gamma_1, \ldots, \Gamma_n$. Let C be a clause in $\mathcal{S}_N$. Then, by definition of the characteristic clause set, $C = C_1, \ldots, C_n$ for $C_i \in \mathcal{S}_{M_i}$ ($\mathcal{S}_N$ is defined by merge). We define $T(\varphi, N, C)$ as

$$\cfrac{\begin{matrix}(\psi_1) & & (\psi_n)\\ C_1, \Gamma_1 & \ldots & C_n, \Gamma_n\end{matrix}}{C_1, \ldots, C_n, \Gamma'_1, \ldots, \Gamma'_n}\ x$$

By definition of $\| \, \|_l$ we have

$$\begin{aligned}\|\varphi.N\|_l &= 1 + \sum_{i=1}^{n} \|\varphi.M_i\|_l,\\ \|\psi_i\|_l &\leq 2\|\varphi.M_i\|_l \text{ by (IH)}\end{aligned}$$

Therefore

$$\|T(\varphi, N, C)\|_l = 1 + \sum_{i=1}^{n} \|\psi_i\|_l \leq 1 + 2\sum_{i=1}^{n} \|\varphi.M_i\|_l < 2\|\varphi.N\|_l.$$

(2b) The auxiliary formulas of x are not in $\Gamma_1, \ldots, \Gamma_n$. Let C by a clause in $\mathcal{S}_N$. Then x operates on ancestors of cuts and $\mathcal{S}_N = \bigcup_{i=1}^{n} \mathcal{S}_{M_i}$, thus $C \in \mathcal{S}_{M_i}$ for some $i \in \{1, \ldots, n\}$. Moreover $\Gamma'_i = \Gamma_i$ for $i = 1, \ldots, n$. We define $T(\varphi, N, C)$ as

$$\cfrac{\cfrac{\begin{matrix}(\psi_i)\\ C, \Gamma_i\end{matrix}}{C, \Gamma_i, \Gamma_1, \ldots, \Gamma_{i-1}, \Gamma_{i+1}, \ldots, \Gamma_n}\ w}{C, \Gamma_1, \ldots, \Gamma_n}\ \pi$$

Then

$$\|T(\varphi, N, C)\|_l \leq \|\psi_i\|_l + 2 < 2\|\varphi.N\|_l.$$

This concludes the induction proof. □

Example 9.1.5 Let φ' be the proof from Example 9.1.3. We have computed the set $\mathrm{CL}(\varphi')$ in Example 9.1.4. We select the clause

$$C{:}\, 0{:}\, P(\alpha),\ 0{:}\, P(\alpha),\ 1{:}\, P(\alpha)$$

and compute the projection $\varphi'[C]$:

$$\begin{array}{cl}
0{:}\, P(\alpha),\ u{:}\, P(\alpha),\ u{:}\, Q(\alpha),\ 0{:}\, P(\alpha),\ 1{:}\, P(\alpha) & \\
\hline
0{:}\, P(\alpha),\ 0{:}\, P(\alpha), 1{:}\, P(\alpha),\ u{:}\, P(\alpha),\ u{:}\, Q(\alpha) & \pi \\
\hline
0{:}\, P(\alpha),\ 0{:}\, P(\alpha), 1{:}\, P(\alpha),\ u{:}\, P(\alpha) \vee Q(\alpha) & \vee{:}\, u \\
\hline
0{:}\, P(\alpha),\ 0{:}\, P(\alpha), 1{:}\, P(\alpha),\ u{:}\, P(\alpha) \vee Q(\alpha),\ u{:}\, R(\alpha) & w \\
\hline
0{:}\, P(\alpha),\ 0{:}\, P(\alpha), 1{:}\, P(\alpha),\ u{:}\, (P(\alpha) \vee Q(\alpha)) \vee R(\alpha) & \vee{:}\, u \\
\hline
0{:}\, P(\alpha),\ 0{:}\, P(\alpha), 1{:}\, P(\alpha),\ 0{:}\, (Dx)((P(x) \vee Q(x)) \vee R(x)) & D{:}\, 0 \\
\hline
0{:}\, P(\alpha),\ 0{:}\, P(\alpha), 1{:}\, P(\alpha),\ 0{:}\, (Dx)((P(x) \vee Q(x)) \vee R(x)),\ 1{:}\, P(c) & w
\end{array}$$

◇

Let φ be a proof of S s.t. $\varphi \in \Phi^s[\mathbf{K}]$ and let γ be a W-resolution refutation of $\mathrm{CL}(\varphi)$. We define a ground projection γ' of γ which is a $\mathbf{K}$-proof of □ from instances of $\mathrm{CL}(\varphi)$. This proof γ' can be transformed into a proof $\gamma'[\varphi]$ of S from the axiom set $\mathcal{A}$ s.t. $\gamma'[\varphi] \in \Phi^s_0[\mathbf{K}]$ ($\gamma'[\varphi]$ is a proof with atomic cuts). Indeed, γ' is the skeleton of the proof of S with atomic cuts and the real core of the end result; $\gamma'[\varphi]$ can be considered as an application of γ' to (the projections of) φ.

Theorem 9.1.2 *Let φ be a proof of S from $\mathcal{A}$ in $\Phi^s[\mathbf{K}]$ and let γ' be a ground projection of a W-refutation of* $\mathrm{CL}(\varphi)$. *Then there exists a proof $\gamma'[\varphi]$ of S with $\gamma'[\varphi] \in \Phi^s_0[\mathbf{K}]$ and*

$$\|\gamma'[\varphi]\|_l \quad \leq \quad \|\gamma'\|_l(2 * \|\varphi\|_l + l(S) + 2).$$

Proof: We construct $\gamma'[\varphi]$:

(1) Replace every axiom C in γ' by the projection $\varphi[C]$. Then instead of C we obtain the proof $\varphi[C]$ of C, S. For every occurrence of an axiom C in γ we obtain a proof of length $\leq 2 * \|\varphi\|_l$ (by Lemma 9.1.1).

(2) Apply the permutation rule to all end sequents of the $\varphi[C]$ and infer S, C. The result is a proof $\psi[C]$ with $\|\psi[C]\|_l \leq 2 * \|\varphi\|_l + 1$.

(3) Simulate γ' on the extended sequents S, C, where the left part S remains passive (note that, according to our definition, inferences take place on the right). The result is a proof χ of a sequent $S, \ldots, S$ from $\mathcal{A}$ s.t.

$$\|\chi\|_l \leq \|\gamma'\|_l * (2 * \|\varphi\|_l + 1) + \|\gamma\|_l.$$

Note that χ is indeed a **K**-proof as all inferences in γ' are also inferences of **K**.

(4) Apply one permutation and contractions to the end sequent of χ for obtaining the end sequent S. The resulting proof is $\gamma'[\varphi]$, the proof we are searching for. As the number of occurrences of S in the end sequent is $\leq \|\gamma'\|_l$ the additional number of inferences is $\leq 1 + l(S) * \|\gamma'\|_l$. By putting things together we obtain

$$\|\gamma'[\varphi]\|_l \leq \|\gamma'\|_l(2 * \|\varphi\|_l + l(S) + 2).$$

□

Looking at the estimation in Theorem 9.1.2 we see that the main source of complexity is the length of the W-resolution proof γ'. Indeed, γ (and thus γ') can be considered as the characteristic part of $\gamma'[\varphi]$ representing the essence of cut-elimination. To sum up the procedure CERES-m for cut-elimination in any LM-type logic **K** can be defined as:

Definition 9.1.21 (CERES-m)

input :$\varphi \in \Phi[\mathbf{K}]$.
construct a Skolem form φ' of φ.
compute CL(φ').
construct a W-refutation γ of CL(φ').
compute a ground projection γ' of γ.
compute $\gamma'[\varphi']$ ($\gamma'[\varphi'] \in \Phi_0^s[\mathbf{K}]$).
de-Skolemize $\gamma'[\varphi']$ to φ'' ($\varphi'' \in \Phi^0[\mathbf{K}]$).

◇

Example 9.1.6 The proof φ from Example 9.1.1 has been Skolemized to a proof φ' in Example 9.1.3. In Example 9.1.4 we computed the characteristic clause set $\mathrm{CL}(\varphi')$ and gave a refutation γ of $\mathrm{CL}(\varphi')$ and a ground projection $\gamma' : \gamma\{\alpha \leftarrow c\}$. Recall γ':

$$\dfrac{\dfrac{\dfrac{0{:}\,P(c),\ 0{:}\,P(c),\ 1{:}\,P(c) \quad u{:}\,P(c)}{0{:}\,P(c),\ 0{:}\,P(c)}\,cut_{1u}}{0{:}\,P(c)}\,c \qquad u{:}\,P(c)}{\Box}\,cut_{0u}$$

and the instances $C_1' = u{:}\,P(c)$ and $C_2' = 0{:}\,P(c),\ 0{:}\,P(c),\ 1{:}\,P(c)$ of two signed clauses in $\mathrm{CL}(\varphi')$ which defined the axioms of γ'. We obtain $\gamma'[\varphi']$ by substituting the axioms C_1', C_2' by the projections $\varphi[C_1'], \varphi[C_2']$ ($\varphi[C_2']$ is an instance of the projection computed in Example 9.1.5). The end sequent of φ' is

$$S{:}\ 0{:}\,(Dx)((P(x) \vee Q(x)) \vee R(x)), 1{:}\,P(c)$$

So we obtain $\gamma'[\varphi'] =$

$$\dfrac{\dfrac{\dfrac{\dfrac{\dfrac{(\varphi'[C_2'])}{0{:}\,P(c),\ 0{:}\,P(c),\ 1{:}\,P(c),\ S}}{S,\ 0{:}\,P(c),\ 0{:}\,P(c),\ 1{:}\,P(c)}\,\pi \quad \dfrac{\dfrac{(\varphi[C_1'])}{u{:}\,P(c),\ S}}{S,\ u{:}\,P(c)}\,\pi}{S,\ S,\ 0{:}\,P(c),\ 0{:}\,P(c)}\,cut_{1u}}{S,\ S,\ 0{:}\,P(c)}\,c \qquad \dfrac{\dfrac{(\varphi[C_1'])}{u{:}\,P(c),\ S}}{S,\ u{:}\,P(c)}\,\pi}{\dfrac{S,\ S,\ S}{S}\,c^*}\,cut_{0u}$$

◇

9.2 CERES in Gödel Logic

Following [7] we single out a prominent intermediate logic, namely Gödel logic **G** (also called Gödel–Dummett logic), which is also one of the main formalizations of fuzzy logic (see, e.g., [42]) and therefore sometimes called intuitionistic fuzzy logic [75]. We show that essential features of CERES can be adapted to the calculus **HG** [3, 25] for **G** that uses *hypersequents*, a generalization of Gentzen's sequents to multisets of sequents. This adaption is far from trivial and, among other novel features, entails a new concept of "resolution": hyperclause resolution, which combines most general unification and cuts on atomic hypersequents. It also provides clues to a better understanding of resolution based cut elimination for sequent and hypersequent calculi, in general.

Due to the incorrectness of general de-Skolemization we will deal with HG-proofs with (arbitrary cut-formulas, but) end-hypersequents that contain either only weak quantifier occurrences or only prenex formulas. For the latter case we show that the corresponding class of proofs admits de-Skolemization. For simplicity we refrain in this chapter from the development of an analogue of clause terms and derive characteristic hyper-clause sets parallel to projections. In addition we restrict to atomic logical axioms. For symmetry we consider in this chapter both sequents and hyper-sequents as multisets. Our results can also be seen as a first step towards automatizing cut-elimination and proof analysis for intuitionistic and other intermediate logics.

9.2.1 First Order Gödel Logic and Hypersequents

Definition 9.2.1 (Semantics of Gödel logic) An *interpretation* $\mathcal{I}$ into $[0,1]$ consists of

1. a nonempty set $U = U^{\mathcal{I}}$, the "universe" of $\mathcal{I}$,
2. for each k-ary predicate symbol P, a function $P^{\mathcal{I}} : U^k \to [0,1]$,
3. for each k-ary function symbol f, a function $f^{\mathcal{I}} : U^k \to U$.
4. for each variable v, a value $v^{\mathcal{I}} \in U$.

Given an interpretation $\mathcal{I}$, we can naturally define a value $t^{\mathcal{I}}$ for any term t and a truth value $\mathcal{I}(A)$ for any formula A of $\mathcal{L}^{\mathcal{U}}$. For a terms $t = f(u_1, \ldots, u_k)$ we define $\mathcal{I}(t) = f^{\mathcal{I}}(u_1^{\mathcal{I}}, \ldots, u_k^{\mathcal{I}})$. For atomic formulas $A \equiv P(t_1, \ldots, t_n)$, we define $\mathcal{I}(A) = P^{\mathcal{I}}(t_1^{\mathcal{I}}, \ldots, t_n^{\mathcal{I}})$. For composite formulas A we define $\mathcal{I}(A)$ by:

$$
\begin{aligned}
\mathcal{I}(\bot) &= 0 \\
\mathcal{I}(A \wedge B) &= \min(\mathcal{I}(A), \mathcal{I}(B)) \\
\mathcal{I}(A \vee B) &= \max(\mathcal{I}(A), \mathcal{I}(B)) \\
\mathcal{I}(A \to B) &= 1 \text{ if } \mathcal{I}(A) \leq \mathcal{I}(B) \\
&= \mathcal{I}(B) \text{ otherwise}
\end{aligned}
$$

Let $\mathcal{I}_u^x$ be the interpretation which differs from $\mathcal{I}$ at most on x and $x^{I_u^x} = u$. Then we define the semantics of quantifiers in the following way:

$$
\begin{aligned}
\mathcal{I}(\forall x\, A(x)) &= \inf\{\mathcal{I}_u^x(A(x)) \mid u \in U\} \\
\mathcal{I}(\exists x\, A(x)) &= \sup\{\mathcal{I}_u^x(A(x)) \mid u \in U\}
\end{aligned}
$$

If $\mathcal{I}(A) = 1$, we say that $\mathcal{I}$ *satisfies* A, and write $\mathcal{I} \models A$.
We define **G** as the set of all formulas of $\mathcal{L}$ such that $\mathcal{I} \models A$ for all interpretations $\mathcal{I}$. $\diamond$

Syntactically, **G** arises from intuitionistic logic by adding the axiom of linearity $(A \to B) \vee (B \to A)$ and the quantifier shifting axiom $\forall x(A(x) \vee C) \to [(\forall x A(x)) \vee C]$, where the x does not occur free in C [22].
The importance of the logic is also indicated by the fact that it can alternatively be seen as a fuzzy logic, the logic characterized semantically by the class of all rooted linearly ordered Kripke frames with constant domains [26], but also as a temporal logic [21].
Hypersequent calculi [4] are simple and natural generalizations of Gentzen's sequent calculi. In our context, a hypersequent is a finite multiset of single-conclusioned ('intuitionistic') sequents, called *components*, written as

$$\Gamma_1 \vdash \Delta_1 \mid \cdots \mid \Gamma_n \vdash \Delta_n$$

where, for $i \in \{1, \ldots, n\}$, Γ_i is a multiset of formulas, and Δ_i is either empty or a single formula. The intended interpretation of the symbol "|" is disjunction at the meta-level.
A hypersequent calculus for propositional Gödel logic has been introduced by Avron [3, 4] and extended to first-order in [25]. The logical rules and internal structural rules of this calculus are essentially the same as those in Gentzen's sequent calculus **LJ** for intuitionistic logic; the only difference being the presence of contexts $\mathcal{H}$ representing (possibly empty) *side hypersequents.* In addition we have *external* contraction and weakening, and the so-called *communication rule.* We present an equivalent version **HG** of the calculi in [3, 25] with multiplicative logical rules (see, e.g., [77] for this terminology).

Definition 9.2.2 (HG)
Axioms: $\bot \vdash$, $\quad A \vdash A$, for *atomic* formulas A.

In the following rules, Δ is either empty or a single formula.

Internal Structural Rules:

$$\frac{\mathcal{H} \mid \Gamma \vdash \Delta}{\mathcal{H} \mid A, \Gamma \vdash \Delta}\ (iw\text{-}l) \qquad \frac{\mathcal{H} \mid \Gamma \vdash}{\mathcal{H} \mid \Gamma \vdash A}\ (iw\text{-}r) \qquad \frac{\mathcal{H} \mid A, A, \Gamma \vdash \Delta}{\mathcal{H} \mid A, \Gamma \vdash \Delta}\ (ic\text{-}l)$$

External Structural Rules:

$$\frac{\mathcal{H}}{\mathcal{H} \mid \Gamma \vdash \Delta}\ (ew) \qquad \frac{\mathcal{H} \mid \Gamma \vdash \Delta \mid \Gamma \vdash \Delta}{\mathcal{H} \mid \Gamma \vdash \Delta}\ (ec)$$

Logical Rules:

$$\frac{\mathcal{H} \mid A_1, \Gamma_1 \vdash \Delta \qquad \mathcal{H}' \mid A_2, \Gamma_2 \vdash \Delta}{\mathcal{H} \mid \mathcal{H}' \mid A_1 \vee A_2, \Gamma_1, \Gamma_2 \vdash \Delta}\ (\vee\text{-}l) \qquad \frac{\mathcal{H} \mid \Gamma \vdash A_i}{\mathcal{H} \mid \Gamma \vdash A_1 \vee A_2}\ (\vee_i\text{-}r)_{i \in \{1,2\}}$$

$$\frac{\mathcal{H} \mid A_i, \Gamma \vdash \Delta}{\mathcal{H} \mid A_1 \wedge A_2, \Gamma \vdash \Delta}\ (\wedge_i\text{-}l)_{i \in \{1,2\}} \qquad \frac{\mathcal{H} \mid \Gamma_1 \vdash A \qquad \mathcal{H}' \mid \Gamma_2 \vdash B}{\mathcal{H} \mid \mathcal{H}' \mid \Gamma_1, \Gamma_2 \vdash A \wedge B}\ (\wedge\text{-}r)$$

$$\frac{\mathcal{H} \mid \Gamma_1 \vdash A \qquad \mathcal{H}' \mid B, \Gamma_2 \vdash \Delta}{\mathcal{H} \mid \mathcal{H}' \mid A \rightarrow B, \Gamma_1, \Gamma_2 \vdash \Delta}\ (\rightarrow\text{-}l) \qquad \frac{\mathcal{H} \mid A, \Gamma \vdash B}{\mathcal{H} \mid \Gamma \vdash A \rightarrow B}\ (\rightarrow\text{-}r)$$

In the following quantifier rules t denotes an arbitrary term, and y denotes an *eigenvariable*, i.e., y does not occur in the lower hypersequent:

$$\frac{\mathcal{H} \mid A(t), \Gamma \vdash \Delta}{\mathcal{H} \mid (\forall x)A(x), \Gamma \vdash \Delta}\ (\forall\text{-}l) \qquad \frac{\mathcal{H} \mid \Gamma \vdash A(y)}{\mathcal{H} \mid \Gamma \vdash (\forall x)A(x)}\ (\forall\text{-}r)$$

$$\frac{\mathcal{H} \mid A(y), \Gamma \vdash \Delta}{\mathcal{H} \mid (\exists x)A(x), \Gamma \vdash \Delta}\ (\exists\text{-}l) \qquad \frac{\mathcal{H} \mid \Gamma \vdash A(t)}{\mathcal{H} \mid \Gamma \vdash (\exists x)A(x)}\ (\exists\text{-}r)$$

Like in [77] we call the exhibited formula in the lower hypersequent of each of these rules the *main formula*, and the corresponding subformulas exhibited in the upper hypersequents the *active formulas* of the inference.
The following **communication** rule of **HG** is specific to the logic **G**:

$$\frac{\mathcal{H} \mid \Gamma_1, \Gamma_2 \vdash \Delta_1 \quad \mathcal{H}' \mid \Gamma_1, \Gamma_2 \vdash \Delta_2}{\mathcal{H} \mid \mathcal{H}' \mid \Gamma_1 \vdash \Delta_1 \mid \Gamma_2 \vdash \Delta_2}\ (com)$$

This version of the communication is equivalent to the one introduced in [3] (see [4]).
Finally we have **cut**, where A is called the *cut-formula* of the inference:

$$\frac{\mathcal{H} \mid \Gamma_1 \vdash A \qquad \mathcal{H}' \mid A, \Gamma_2 \vdash \Delta}{\mathcal{H} \mid \mathcal{H}' \mid \Gamma_1, \Gamma_2 \vdash \Delta}\ (cut)$$

If A is atomic we speak of an *atomic cut*.

Remark: Note the absence of negation from our calculus: $\neg A$ is just an abbreviation of $A \rightarrow \bot$. (See, e.g., [77] for similar sytems for intuitionistic logic.)
Communication allows us to derive the following additional "distribution rule" which we will use in Section 9.2.5:

$$\frac{\mathcal{H} \mid \Gamma \vdash A \vee B}{\mathcal{H} \mid \Gamma \vdash A \mid \Gamma \vdash B} \; (distr)$$

◇

A derivation ρ using the rules of **HG** is viewed as an upward rooted tree. The root of ρ is called its *end-hypersequent*, which we will denote by $\mathcal{H}_\rho$. The leaf nodes are called *initial hypersequents.* A *proof* σ of a hypersequent $\mathcal{H}$ is a derivation with $\mathcal{H}_\sigma = \mathcal{H}$, where all initial hypersequents are axioms. The *ancestors paths* of a formula occurrence in a derivation is traced as for LK upwards to the initial hypersequents in the obvious way. That is, active formulas are immediate ancestors of the main formula of an inference. The other formula occurrences in the premises (i.e., upper hypersequents) are immediate ancestors of the corresponding formula occurrences in the lower hypersequent. (This includes also internal and external contraction: here, a formula in the lower hypersequent may have *two* corresponding occurrences, i.e. immediate ancestors, in the premises.)
The sub-hypersequent consisting of all ancestors of cut-formulas of an hypersequent $\mathcal{H}$ in a derivation is called the *cut-relevant part* of $\mathcal{H}$. The complementary sub-hypersequent of $\mathcal{H}$ consisting of all formula occurrences that are not ancestors of cut-formulas is the *cut-irrelevant part* of $\mathcal{H}$. An inference is called *cut-relevant* if its main formula is an ancestor of a cut-formula, and is called *cut-irrelevant* otherwise.
The hypersequent $\Gamma_1 \vdash \Delta_1 \mid \ldots \mid \Gamma_n \vdash \Delta_n$ is called *valid* if its translation $\bigvee_{1 \leq i \leq n}(\bigwedge_{A \in \Gamma_i} A \rightarrow [\Delta_i])$ is valid in **G**, where $[\Delta_i]$ is $\bot$ if Δ_i is empty, and the indicated implications collapse to Δ_i whenever Γ_i is empty. A set of hypersequents is called *unsatisfiable* if their translations entail $\bot$ in **G**. (Different but equivalent ways of defining validity and entailment in **G** have been indicated at the beginning of this section.)

Theorem 9.2.1 *A hypersequent $\mathcal{H}$ is provable in* **HG** *without cuts iff $\mathcal{H}$ is valid.*

Proof: In [5, 25]. □

Remark: It might surprise the reader that we rely on the *cut-free* completeness of **HG** in a paper dealing with *cut elimination.* However, this just emphasizes the fact that we are interested in a *particular* transformation of proofs with cuts (i.e., "lemmas") into cut-free proofs, that is adequate for automatization and proof analysis (compare [13, 20]).

9.2.2 The Method hyperCERES

Before presenting the details of our transformation of appropriate **HG**-proofs into cut-free proofs, which we call *hyperCERES*, we will assist the orientation of the reader and describe the overall procedure on a more abstract level using keywords that will be explained in the following sections.
The end-hypersequent $\mathcal{H}_\sigma$ of the **HG**-proof σ that forms the input of hyperCERES can be of two forms: either it contains only weak quantifier occurrences or it consists of prenex formulas only. (While in classical logic all formulas can be translated into equivalent prenex formulas, this does not hold for **G**.) In the latter case we have to Skolemize the proof first (step 1) and de-Skolemize it after cut elimination (step 7):

1. if necessary, construct a *Skolemized form* $\hat{\sigma}$ of σ, otherwise $\hat{\sigma} = \sigma$ (Section 9.2.3)

2. compute a *characteristic set of pairs* $\{\langle R_1(\hat{\sigma}), D_1\rangle, \ldots \langle R_n(\hat{\sigma}), D_n\rangle\}$, where $\Sigma_d(\hat{\sigma}) = \{D_1, \ldots, D_n\}$ is the *characteristic set of d-hyperclauses* – coding the cut formulas of $\hat{\sigma}$ – and each *reduced proof* $R_i(\hat{\sigma})$ is a cut-free proof of a cut-irrelevant sub-hypersequent of $\mathcal{H}_{\hat{\sigma}}$ augmented by D_i (Section 9.2.4)

3. translate $\Sigma_d(\hat{\sigma})$ into an equivalent set of hyperclauses $\Sigma(\hat{\sigma})$ and construct a (*hyperclause*) resolution refutation γ of $\Sigma(\hat{\sigma})$ (Section 9.2.5)

4. compute a *ground instantiation* γ' of γ using a ground substitution θ (Section 9.2.5)

5. apply θ to the reduced proofs $R_1(\hat{\sigma})$, ..., $R_n(\hat{\sigma})$, and assemble them into a single proof $\gamma'[\hat{\sigma}]$ using the atomic cuts and contractions that come from γ (Section 9.2.6)

6. eliminate the atomic cuts in $\gamma'[\hat{\sigma}]$ in the usual way (As is known, atomic cuts in **HG**-proofs can be moved upwards to the axioms, where they become redundant (see, e.g., [3, 5]).)

7. if necessary, *de-Skolemize* the proof $\gamma'[\hat{\sigma}]$ and apply final contractions and weakenings to obtain a cut-free proof of $\mathcal{H}_\sigma$ (Section 9.2.3)

It is well known (see, e.g., [67, 72]) that there is no elementary bound on the size of shortest cut-free proofs relative to the size of proofs with cuts of the same end-(hyper)sequent. While the non-elementary upper bound on the complexity of cut elimination obviously also applies to hyperCERES it should be pointed out that the global (hyperclause) resolution based method presented here is considerably faster in general, and never essentially slower, than traditional Gentzen- or Schütte–Tait-style cut elimination procedures [3, 5]. Moreover, the reliance on most general unification and atomic cuts, i.e., on resolution for the computational kernel of the procedure implies that hyperCERES is a potentially essential ingredient of (semi-)automated analysis of appropriately formalized proofs.

9.2.3 Skolemization and de-Skolemization

Like in the original CERES-method [18, 20], step 5 of hyperCERES is sound only if end-(hyper)sequents do not contain strong quantifier occurrences. The reason for this is that, in general, the eigenvariable condition might be violated when the reduced proofs (constructed in step 2) are combined with the resolution refutation (constructed in step 3) to replace the original cuts with atomic cuts. Consequently, like in CERES, we first *Skolemize* the proof; i.e., we replace all strong quantifier occurrences with appropriate Skolem terms. (Obviously this is necessary only if there are strong quantifier occurrences at all.) While this transformation is always sound (in fact also for **LJ**-proofs), the inverse *de-Skolemization*, i.e., the re-introduction of strong quantifier occurrences according to the information coded in the Skolem terms, is unsound in general. (For example, $\forall x(A(x) \vee B) \vdash A(c) \vee B$ is provable in **LJ** while its de-Skolemized version $\forall x(A(x) \vee B) \vdash \forall x A(x) \vee B$ is not.) However, as we will show below, de-Skolemization is possible for **HG**-proofs of *prenex* hypersequents (step 7).

By a *prenex hypersequent* we mean a hypersequent in which all formulas are in prenex form, i.e., all formulas begin with a (possibly empty) *prefix* of quantifier occurrences, followed by a quantifier-free formula. If $\Gamma \vdash \Delta$ is a component of a prenex hypersequent, then all existential quantifiers occurring in Γ and all universal quantifiers occurring in Δ are called *strong*. The other quantifier occurrences are called *weak*.

The *Skolemization* $\mathcal{H}^S$ of a prenex hypersequent $\mathcal{H}$ is obtained as follows. In every component $\Gamma \vdash \Delta$ of $\mathcal{H}$, delete each strong quantifier occurrence $\mathsf{Q}x$ and replace all corresponding occurrences of x by the *Skolem term* $f(\overline{y})$, where f is a new function symbol and $\overline{y}$ are the variables of the weak quantifier occurrences in the scope of which $\mathsf{Q}x$ occurs. (If $\mathsf{Q}x$ is not the scope of any weak quantifier then f is a constant symbol.)

Given an **HG**-proof σ of $\mathcal{H}$ its *Skolemization* $\hat{\sigma}$ is constructed in stages:

1. Replace the end-hypersequent $\mathcal{H}$ of σ by $\mathcal{H}^S$. Recall that this means that every occurrence of a strongly quantified variable x in $\mathcal{H}$ is replaced by a corresponding Skolem term $f(\overline{y})$.

2. Trace the indicated occurrences of x and of the eigenvariable y corresponding to its introduction throughout σ and replace all these occurrences by $f(\overline{y})$, too.

3. Delete the (now) spurious strong quantifiers and remove the corresponding inferences that introduce these quantifiers in σ.

4. For any inference in σ introducing a weakly quantified variable y by replacing $A(t)$ with $\mathsf{Q}yA(y)$, replace all corresponding occurrences of y in Skolem terms $f(\overline{y})$ by t.

It is straightforward to check that $\hat{\sigma}$ is an **HG**-proof of $\mathcal{H}^S$. (Note that strong quantifier occurrences in ancestors of cut formulas remain untouched by our Skolemization.)
It is shown in [6] that prenex formulas of **G** allow for de-Skolemization. We generalize this result to *proofs* of prenex hypersequents. Our main tool is the following result from [25].

Theorem 9.2.2 (mid-hypersequent theorem) *Any cut-free* **HG**-*proof* σ *of a prenex hypersequent* $\mathcal{H}$ *can be stepwise transformed into one in which no propositional rule is applied below any application of a quantifier rule.*

We call a hypersequent $\overline{\mathcal{H}_S}$ a *linked Skolem instance* of $\mathcal{H}$ if each formula A in $\overline{\mathcal{H}_S}$ is an instance of a Skolemized formula A^S that occurs in $\mathcal{H}^S$ on the same side (left or right) of a component as A. Moreover we link A to A^S. As we will see in Section 9.2.6, we obtain (cut-free proofs of) linked Skolem instances from step 5 (and 6) of hyperCERES.

Theorem 9.2.3 (De-Skolemization) *Given a cut-free* **HG**-*proof* $\hat{\rho}$ *of a linked Skolem instance* $\overline{\mathcal{H}_S}$ *of a prenex hypersequent* $\mathcal{H}$*, we can find a* **HG**-*proof* ρ *of* $\mathcal{H}$*.*

Proof: We construct ρ in stages as follows:

1. By applying Theorem 9.2.2 to $\hat{\rho}$ we obtain a proof ρ' of the following form:

$$\begin{array}{ccccc} \rho_1^p & \cdots & \rho_i^p & \cdots & \rho_n^p \\ \mathcal{G}_1 & & \mathcal{G}_i & & \mathcal{G}_n \\ & \ddots & \rho^{\mathsf{Q}} & \swarrow & \\ & & \overline{\mathcal{H}_S} & & \end{array}$$

where the mid-hypersequents $\mathcal{G}_1, \ldots, \mathcal{G}_n$ separate ρ' into a part ρ^{Q} containing only (weak) quantifier introductions and applications of structural rules and parts $\rho_1^p, \ldots, \rho_n^p$ containing only propositional and structural inferences.

2. Applications of the weakening rules, $(iw\text{-}l)$ and (ew), can be shifted upwards to the axioms in the usual manner, while applications of $(iw\text{-}r)$ can be safely deleted by replacing each axiom $\bot \vdash$ in the proof by $\bot \vdash \Delta$ for suitable Δ.

   Consequently, ρ^{Q} does not contain weakenings after this transformation step.

3. Note that – in contrast to **LK** – Theorem 9.2.2 induces many and not just one mid-hypersequents, in general. The reason for this is the possible presence of the *binary* structural rule (com) in ρ^{Q}. To obtain a proof ρ'' with a single mid-hypersequent, we have to move 'communications' upwards in ρ^{Q}; i.e., we have to permute applications of (com) with applications of (ic), (ec), $(\forall\text{-}l)$, and $(\exists\text{-}r)$, respectively. The only non-trivial case is $(\forall\text{-}l)$. Disregarding side-hypersequents, the corresponding transformation consists in replacing

$$\dfrac{\dfrac{\Gamma, P(x), \Sigma \vdash \Delta}{\Gamma, \forall x P(x), \Sigma \vdash \Delta}\,(\forall\text{-}l) \qquad \Gamma, \forall x P(x), \Sigma \vdash \Delta'}{\Gamma, \forall x P(x) \vdash \Delta \mid \Sigma \vdash \Delta'}\,(com)$$

   by

$$\dfrac{\dfrac{\dfrac{\dfrac{\dfrac{\Gamma, P(x), \Sigma \vdash \Delta}{\Gamma, P(x), \Sigma, \Gamma, \forall x P(x) \vdash \Delta}\,(iw)^* \qquad \dfrac{\Gamma, \forall x P(x), \Sigma \vdash \Delta'}{\Gamma, P(x), \Sigma, \Gamma, \forall x P(x) \vdash \Delta'}\,(iw)^*}{\Gamma, P(x), \Sigma \vdash \Delta' \mid \Gamma, \forall x P(x) \vdash \Delta}\,(com) \qquad \Gamma, P(x), \Sigma \vdash \Delta}{\Gamma, P(x) \vdash \Delta \mid \Sigma \vdash \Delta' \mid \Gamma, \forall x P(x) \vdash \Delta}\,(com)}{\Gamma, \forall x P(x) \vdash \Delta \mid \Sigma \vdash \Delta' \mid \Gamma, \forall x P(x) \vdash \Delta}\,(\forall\text{-}l)}{\Gamma, \forall x P(x) \vdash \Delta \mid \Sigma \vdash \Delta'}\,(ec)$$

4. For the final step we proceed like in [6], where the soundness of reintroducing strong quantifier occurrences for corresponding Skolem terms is shown: we ignore ρ'' and, given $\mathcal{H}$ and the links to its formulas, apply appropriate inferences to the mid-hypersequent as follows.

   (a) Infer all weak quantifier occurrences, which can be introduced at this stage according to the quantifier prefixes in $\mathcal{H}$.

   (b) Apply all possible internal and external contractions.

   (c) Among the strong quantifiers that immediately precede the already introduced quantifiers we pick one linked to an instance of a Skolem term, that is maximal with respect to the subterm ordering. This term is replaced everywhere by the eigenvariable of the corresponding strong quantifier inference.

   These three steps are iterated until the original hypersequent $\mathcal{H}$ is restored. □

9.2.4 Characteristic Hyperclauses and Reduced Proofs

All information of the original **HG**-proof σ that goes into the cut-formulas is collected in a set $\Sigma_d(\hat{\sigma})$, consisting of hypersequents whose components only contain atomic formulas on the left hand sides and a (possibly empty) disjunction of atomic formulas, on the right hand side. We will call hypersequents of this latter form *d-hyperclauses*. In the proof of Theorem 9.2.4 we will construct *characteristic d-hyperclauses* D_i together with corresponding *reduced proofs* $R_i(\hat{\sigma})$ which combine the cut-irrelevant part of the Skolemized proof $\hat{\sigma}$ with D_i. The pairs $\langle R_i(\hat{\sigma}), D_i\rangle$ provide the information needed to construct corresponding proofs containing only atomic cuts.
To assist concise argumentation we assume that the components of all hypersequents in a proof are labeled with unique sets of identifiers. More precisely, a derivation σ is *labeled* if there is a function from all components of hypersequents occurring in σ into the powerset of a set of *identifiers*, satisfying the following conditions: (We will put the label above the corresponding sequent arrow.)

- All components occurring in initial hypersequents of σ are assigned pairwise different singleton sets of identifiers.

- In all unary inferences the labels are transferred from the upper hypersequent to the lower hypersequent in the obvious way. In external weakening (ew) a fresh singleton set is assigned to the new component in the lower hypersequent. In external contraction (ec), if $\Gamma \overset{M}{\vdash} \Delta$ and $\Gamma \overset{N}{\vdash} \Delta$ are the two contracted components of the upper hypersequent, then $\Gamma \overset{M \cup N}{\vdash} \Delta$ is the corresponding component in the lower hypersequent.

- In all binary logical inferences the labels in the side-hypersequents are transferred in the obvious way, and the label of the component containing the main formula is the union of the labels of the components containing the active formulas.
- In (cut) the labels of the components containing the cut formulas are merged, like above, to obtain the label of the exhibited component of the lower hypersequent.
- In (com) the labels of all components are transferred from the premises to the lower hypersequent simply in the same sequence as exhibited in the statement of the rule.

Let $\mathcal{H}$ and $\mathcal{G}$ denote the labeled hypersequents

$$\Gamma_1 \overset{K_1}{\vdash} \Delta_1 \mid \cdots \mid \Gamma_k \overset{K_k}{\vdash} \Delta_k \mid \mathcal{H}' \quad \text{and} \quad \Gamma_1' \overset{K_1}{\vdash} \Delta_1' \mid \cdots \mid \Gamma_k' \overset{K_k}{\vdash} \Delta_k' \mid \mathcal{G}'$$

respectively, where the labels in $\mathcal{H}'$ and $\mathcal{G}'$ are pairwise different and also different from the labels $K_1, \ldots, K_k$. Then $\mathcal{H} \odot \mathcal{G}$ denotes the *merged* hypersequent

$$\Gamma_1, \Gamma_1' \overset{K_1}{\vdash} \Delta_1 \vee \Delta_1' \mid \cdots \mid \Gamma_k, \Gamma_k' \overset{K_k}{\vdash} \Delta_k \vee \Delta_k' \mid \mathcal{H}' \mid \mathcal{G}'$$

where $\Delta_i \vee \Delta_i'$ is Δ_i if Δ_i' is empty and is Δ_i' if Δ_i is empty (and thus $\Delta_i \vee \Delta_i'$ is empty if both are empty).

Theorem 9.2.4 *Given a Skolemized and labeled* $\mathbf{HG}$*-proof* $\hat{\sigma}$ *of* $\mathcal{H}_{\hat{\sigma}}$ *one can construct a* characteristic set of pairs $\{\langle R_1(\hat{\sigma}), D_1\rangle, \ldots \langle R_n(\hat{\sigma}), D_n\rangle\}$, *where, for all* $i \in \{1, \ldots, n\}$, D_i *is a labeled d-hyperclause and* $R_i(\hat{\sigma})$ *is a labeled* "(reduced)" *cut-free* $\mathbf{HG}$*-proof with the following properties:*

(1) *the end-hypersequent of* $R_i(\hat{\sigma})$ *is* $\mathcal{H}_{\hat{\sigma}}' \odot D_i$, *for some sub-hypersequent* $\mathcal{H}_{\hat{\sigma}}'$ *of* $\mathcal{H}_{\hat{\sigma}}$,

(2) *the characteristic d-hyperclause set* $\Sigma_d(\hat{\sigma}) = \{D_1, \ldots, D_n\}$ *is unsatisfiable.*

Proof: To show (1) and (2) we use the following induction hypotheses:

(1′) A characteristic set of pairs $\langle R_i(\hat{\sigma}'), D'_i\rangle$ exists for every sub-proof $\hat{\sigma}'$ of $\hat{\sigma}$, where $R_i(\hat{\sigma}')$ proves $\mathcal{H}'_{\hat{\sigma}'} \odot D'_i$ for some sub-hypersequent $\mathcal{H}'_{\hat{\sigma}'}$ of $\mathcal{H}_{\hat{\sigma}'}$ which is cut-irrelevant with respect to the original cuts in $\hat{\sigma}$. Moreover, the right hand sides in $\mathcal{H}'_{\hat{\sigma}'} \odot D'_i$ are formulas in either $\mathcal{H}'_{\hat{\sigma}'}$ or in D'_i.

(2′) There is a derivation of the cut-relevant part of $\mathcal{H}_{\hat{\sigma}'}$ from the set $\{D'_1, \dots, D'_m\}$ of d-hyperclauses constructed for $\hat{\sigma}'$.

Note that (2) follows from (2′) as the cut-relevant part of $\mathcal{H}_{\hat{\sigma}}$ is an empty hypersequent by definition. The proof proceeds by induction on the length of $\hat{\sigma}'$.
If $\hat{\sigma}'$ consists just of an axiom $A \overset{M}{\vdash} A$ then there is only one pair $\langle R(\hat{\sigma}'), D\rangle$ in the corresponding characteristic set. $R(\hat{\sigma}')$ is the axiom itself and D is the cut-relevant part of $A \overset{M}{\vdash} A$ (which might be the empty hypersequent). (1′) and (2′) trivially hold. Axioms of the form $\bot\ \vdash$ are handled in the same way.
If $\hat{\sigma}'$ is not an axiom we distinguish cases according to the last inference in $\hat{\sigma}'$.
($\vee$-l): $\hat{\sigma}'$ ends with the inference

$$\dfrac{\dfrac{\vdots\ \hat{\rho}}{\mathcal{H} \mid A_1, \Gamma_1 \overset{M}{\vdash} \Delta} \qquad \dfrac{\vdots\ \hat{\tau}}{\mathcal{H}' \mid A_2, \Gamma_2 \overset{N}{\vdash} \Delta}}{\mathcal{H} \mid \mathcal{H}' \mid A_1 \vee A_2, \Gamma_1, \Gamma_2 \overset{M \cup N}{\vdash} \Delta}\ (\vee\text{-}l)$$

By induction hypothesis (1′) there are characteristic sets of pairs $S_1 = \{\langle R_1(\hat{\rho}), E_1\rangle, \dots, \langle R_m(\hat{\rho}), E_m\rangle\}$ and $S_2 = \{\langle R_1(\hat{\tau}), F_1\rangle, \dots, \langle R_n(\hat{\tau}), F_n\rangle\}$, where the reduced proofs $R_i(\hat{\rho})$ and $R_j(\hat{\tau})$ end in $\mathcal{H}_{R_i(\hat{\rho})} = \mathcal{G}_i \odot E_i$ and in $\mathcal{H}_{R_i(\hat{\tau})} = \mathcal{G}'_j \odot F_j$, respectively, where $\mathcal{G}_i$ and $\mathcal{G}'_j$ are sub-hypersequents of the cut-irrelevant parts of $\mathcal{H} \mid A_1, \Gamma_1 \overset{M}{\vdash} \Delta$ and $\mathcal{H}' \mid A_2, \Gamma_2 \overset{N}{\vdash} \Delta$, respectively. Moreover, by (2′), there are derivations ρ_C and τ_C of the cut-relevant parts of the just mentioned hypersequents from $\{E_1, \dots, E_m\}$ and $\{F_1, \dots, F_n\}$, respectively.
Two cases can occur:

(a) If the inference is *cut-relevant*, then the characteristic set S of pairs corresponding to $\hat{\sigma}'$ is just $S_1 \cup S_2$. Condition (1′) trivially remains satisfied. Also (2′) is maintained because we obtain a derivation of the cut-relevant part of $\mathcal{H} \mid \mathcal{H}' \mid A_1 \vee A_2, \Gamma_1, \Gamma_2 \overset{M \cup N}{\vdash} \Delta$ by joining ρ_C and τ_C with the indicated application of $(\vee\text{-}l)$.

(b) If the inference is *cut-irrelevant*, then we obtain the set S corresponding to $\hat{\sigma}'$ by

$$S = \{\langle R_{ij}(\hat{\rho} \bowtie_{\vee\text{-}l} \hat{\tau}), E_i \bowtie_{ij} F_j\rangle : 1 \leq i \leq m, 1 \leq j \leq n\},$$

where $R_{ij}(\hat{\rho} \bowtie_{\vee\text{-}l} \hat{\tau})$ and $E_i \bowtie_{ij} F_J$ are defined as follows.

1. If A_1 does not occur at the indicated position in $\mathcal{H}_{R_i(\hat{\rho})}$ then $R_{ij}(\hat{\rho} \bowtie_{\vee\text{-}l} \hat{\tau})$ is $R_i(\hat{\rho})$ and $E_i \bowtie_{ij} F_j$ is E_i.
2. If A_2 does not occur at the indicated position in $\mathcal{H}_{R_j(\hat{\tau})}$ then $R_{ij}(\hat{\rho} \bowtie_{\vee\text{-}l} \hat{\tau})$ is $R_j(\hat{\tau})$ and $E_i \bowtie_{ij} F_j$ is F_j.
3. If neither A_1 nor A_2 occur as indicated in the reduced proofs, then $R_{ij}(\hat{\rho} \bowtie_{\vee\text{-}l} \hat{\tau})$ can be non-deterministically chosen to be either $R_i(\hat{\rho})$ or $R_j(\hat{\tau})$ and $E_i \bowtie_{ij} F_j$ is either E_i or F_j, accordingly.
4. If both A_1 and A_2 occur at the indicated positions, then $E_i \bowtie_{ij} F_j$ is $E_i' \odot F_j'$, where E_i' (F_j') is like E_i (F_j), except for changing the label M (N) to $M \cup N$.

   Note that our labeling mechanism guarantees that the appropriate components are identified in merging hypersequents.

   The corresponding reduced proof $R_{ij}(\hat{\rho} \bowtie_{\vee\text{-}l} \hat{\tau})$ is constructed as follows. Since A_1 and A_2 occur as exhibited in the end-hypersequents $\mathcal{G}_i \odot E_i$ and $\mathcal{G}_j' \odot F_j$ of $R_i(\hat{\rho})$ and $R_j(\hat{\tau})$, respectively, we want to join them by introducing $A_1 \vee A_2$ using $(\vee\text{-}l)$ like in $\hat{\sigma}'$. However, $(\vee\text{-}l)$ is only applicable if the right hand sides of the two relevant components in the premises are identical. To achieve this, we might first have to apply $(\vee\text{-}r)$ or $(iw\text{-}r)$ to the mentioned end-hypersequents. The resulting new end-hypersequent might still contain different components transferred from E_i and F_j, respectively, that need to be merged with other components. This can be achieved by first applying internal weakenings to make the relevant components identical, and then applying external contraction (ec) to remove redundant copies of identical components.

Note that in all four cases $(1')$ remains satisfied by definition of $R_{ij}(\hat{\rho} \bowtie_{\vee\text{-}l} \hat{\tau})$ and of $E_i \bowtie_{ij} F_j$. For cases 1, 2, and 3 also $(2')$ trivially still holds. To obtain $(2')$ for case 4, we proceed in two steps. First we merge the occurrences of clauses $E_1, \dots, E_m$ in the derivation ρ_C of the cut-relevant part $\mathcal{H}_c^{\hat{\rho}}$ of $\mathcal{H}_{\hat{\rho}}$ with clauses in $\{F_1, \dots, F_n\}$ to obtain a derivation $\rho_C(F_i)$ of $\mathcal{H}_c^{\hat{\rho}} \odot F_i$ for each $i \in \{1, \dots, n\}$. In a second step, each initial hypersequent F_i in the derivation τ_C of the cut-relevant part of $\mathcal{H}_{\hat{\tau}}$ is replaced by $\rho_C(F_i)$. By merging also the inner nodes of τ_C with $\mathcal{H}_c^{\hat{\rho}}$ we arrive at a derivation of the cut-relevant part of $\mathcal{H}_{\hat{\sigma}'}$. (Actually, as the rules of **HG** are multiplicative, redundant copies of identical formulas might arise, that are to be removed by finally applying corresponding contractions.)

$(\wedge_i\text{-}l)$, $(\rightarrow\text{-}r)$, $(\vee\text{-}r)$, $(\forall\text{-}l)$, $(\forall\text{-}r)$, $(\exists\text{-}l)$, $(\exists\text{-}r)$, $(ic\text{-}l)$: If the indicated last (unary) inference is *cut-relevant*, then the characteristic set of pairs remains the same as for the sub-proof ending with the premise of this inference.
If the inference is *cut-irrelevant*, then the hyperclauses $E_1, \dots, E_m$ of the pairs in characteristic set $\{\langle R_1(\hat{\rho}), E_1\rangle, \dots \langle R_m(\hat{\rho}), E_m\rangle\}$ for $\hat{\rho}$ remain unchanged. Each reduced proof $R_i(\hat{\rho})$ is augmented by the corresponding inference if its active formula occurs in the end-hypersequent $\mathcal{H}_{R_i(\hat{\rho})}$. If this is not the case then also $R_i(\hat{\rho})$ remains unchanged.
In any of these cases, $(1')$ and $(2')$ clearly remain satisfied.
(ew), $(iw\text{-}l)$, $(iw\text{-}r)$: The characteristic set of pairs remains unchanged and consequently $(1')$ still holds. Also $(2')$ trivially remains valid if the inference is cut-irrelevant. If a cut-relevant formula is introduced by weakening, then the derivation required for $(2')$ is obtained from the induction hypothesis by adding a corresponding application of a weakening rule.
$(\wedge\text{-}r)$, $(\rightarrow\text{-}l)$, (cut), (com): These cases are analogous to the one for $(\vee\text{-}l)$.
□

Example 9.2.1 Consider the labeled proof σ in Figure 9.1.
The cut-relevant parts of σ and the names of all corresponding cut-relevant inferences are underlined. The initial pair for the $\{1\}$-labeled axiom is $\langle \rho_1, \overset{\{1\}}{\vdash} Q\rangle$, where ρ_1 is $Q \overset{\{1\}}{\vdash} Q$. Since the succeeding inference $(\vee\text{-}r)$ is unary and cut-relevant, the pair remains unchanged in that step.
For the middle part of the proof let us look at the subproof σ' ending with an application of (com) yielding $Q \overset{\{2\}}{\vdash} \exists y P(y) \mid P(c) \overset{\{3\}}{\vdash} Q$. Since there are no cut-ancestors in the $\{2\}$-labeled axiom, the corresponding d-hyperclause is the empty $\overset{\{2\}}{\vdash}$. This is retained for the right premise of (com). The corresponding reduced derivation consists only of the first inference $(\exists\text{-}r)$ as the succeeding application of $(iw\text{-}l)$ is cut-relevant. For the left premise of the communication we obtain the d-hyperclause $Q \overset{\{3\}}{\vdash}$, which is then merged and

$$\dfrac{P(c) \overset{\{2\}}{\vdash} P(c)}{P(c) \overset{\{2\}}{\vdash} \exists y P(y)}\,(\exists\text{-}r)$$

$$\dfrac{P(c) \overset{\{2\}}{\vdash} \exists y P(y)}{\underline{Q}, P(c) \overset{\{2\}}{\vdash} \exists y P(y)}\,\underline{(iw\text{-}l)} \qquad \dfrac{\underline{Q} \overset{\{3\}}{\vdash} Q}{\underline{Q}, P(c) \overset{\{3\}}{\vdash} Q}\,(iw\text{-}l)$$

$$\dfrac{\underline{Q}, P(c) \overset{\{2\}}{\vdash} \exists y P(y) \qquad \underline{Q}, P(c) \overset{\{3\}}{\vdash} Q}{\underline{Q} \overset{\{2\}}{\vdash} \exists y P(y) \mid P(c) \overset{\{3\}}{\vdash} Q}\,(com) \qquad \dfrac{\underline{P(x)} \overset{\{4\}}{\vdash} P(x)}{\underline{P(x)} \overset{\{4\}}{\vdash} \exists y P(y)}\,(\exists\text{-}r)$$

$$\dfrac{\underline{Q} \overset{\{2\}}{\vdash} \exists y P(y) \mid P(c) \overset{\{3\}}{\vdash} Q \qquad \underline{P(x)} \overset{\{4\}}{\vdash} \exists y P(y)}{\underline{P(x) \vee Q} \overset{\{2,4\}}{\vdash} \exists y P(y) \mid P(c) \overset{\{3\}}{\vdash} Q}\,\underline{(\vee\text{-}l)}$$

$$\dfrac{\underline{P(x) \vee Q} \overset{\{2,4\}}{\vdash} \exists y P(y) \mid P(c) \overset{\{3\}}{\vdash} Q \qquad Q \overset{\{5\}}{\vdash} Q}{\underline{P(x) \vee Q} \overset{\{2,4\}}{\vdash} \exists y P(y) \mid P(c) \vee Q \overset{\{3,5\}}{\vdash} Q}\,(\vee\text{-}l)$$

$$\dfrac{Q \overset{\{1\}}{\vdash} \underline{Q}}{Q \overset{\{1\}}{\vdash} \underline{P(x) \vee Q}}\,\underline{(\vee\text{-}r)}$$

$$\dfrac{Q \overset{\{1\}}{\vdash} \underline{P(x) \vee Q} \qquad \underline{P(x) \vee Q} \overset{\{2,4\}}{\vdash} \exists y P(y) \mid P(c) \vee Q \overset{\{3,5\}}{\vdash} Q}{Q \overset{\{1,2,4\}}{\vdash} \exists y P(y) \mid P(c) \vee Q \overset{\{3,5\}}{\vdash} Q}\,\underline{(cut)}$$

Figure 9.1: Labelled proof σ with underlined cut-relevant part.

"communicated" with $\overset{\{2\}}{\vdash}$ to obtain for σ' the d-hyperclause $Q \overset{\{2\}}{\vdash} \mid \overset{\{3\}}{\vdash}$. This forms a pair with the reduced derivation $R(\sigma')$, which, in this case, is identical with σ'. (Note that neither the cut-relevant application of $(iw\text{-}l)$ nor Q appears in the reduced proof corresponding to $Q, P(c) \overset{\{2\}}{\vdash} \exists y P(y)$. (Still, the missing Q is added by $(iw\text{-}l)$ in $R(\sigma')$ to make the application of (com) possible.)

From the cut-relevant (and therefore underlined) $(\vee\text{-}l)$-inference one obtains an additional pair $\langle \rho_2, P(x) \overset{\{4\}}{\vdash} \rangle$ from its right premise, where ρ_2 is the derivation of $P(x) \overset{\{4\}}{\vdash} \exists y P(y)$ from the axiom.

For the succeeding cut-irrelevant application of $(\vee\text{-}l)$, the pair $\langle \rho_2, P(x) \overset{\{4\}}{\vdash} \rangle$ remains unchanged, as the left disjunct $P(x)$ does not occur at the left side in the end-hypersequent $\overset{\{4\}}{\vdash} \exists y P(y)$ of ρ_2. (This is case $(\vee\text{-}l)$/(b)/2 in the proof of Theorem 9.2.4.) The reduced proof ρ_3 of the final pair is formed by applying $(\vee\text{-}l)$ as indicated to the end-hypersequent of $R(\sigma')$ and to $Q \overset{\{5\}}{\vdash} Q$ as right and left premises, respectively. The corresponding d-hyperclause arises from merging $Q \overset{\{2\}}{\vdash} \mid \overset{\{3\}}{\vdash}$ and $\overset{\{5\}}{\vdash}$ into $Q \overset{\{2\}}{\vdash} \mid \overset{\{3,5\}}{\vdash}$.

For the final application of cut we have to take the union of the sets of pairs constructed for its two premises. Therefore the characteristic set of pairs for σ is

$$\{\langle \rho_1, \overset{\{1\}}{\vdash} Q \rangle,\ \langle \rho_2, P(x) \overset{\{4\}}{\vdash} \rangle,\ \langle \rho_3, Q \overset{\{2\}}{\vdash} \mid \overset{\{3,5\}}{\vdash} \rangle\}\,.$$

It is easy to check that conditions (1) and (2) of Theorem 9.2.4 are satisfied. $\diamond$

9.2.5 Hyperclause Resolution

By a *hyperclause* we mean a hypersequent in which only atomic formulas occur. Remember that, from the proof of Theorem 9.2.4, we obtain *d-hyperclauses*, which are like hyperclauses, except for allowing *disjunctions* of atomic formulas at the right hand sides of their components. However, using the derivable rule (*distr*) (see Section 9.2.1) it is easy to see that an **HG**-derivation of, e.g., the d-hyperclause

$$A \vdash B \vee C \mid\vdash D \vee E \vee F$$

can be replaced by an **HG**-derivation of the hyperclause

$$A \vdash B \mid A \vdash C \mid\vdash D \mid\vdash E \mid\vdash F.$$

Also the converse holds: using the rules ($\vee_i$-r), and (ec) we can derive the mentioned d-hyperclause from the latter hyperclause. Therefore we can refer to hyperclauses instead of d-hyperclauses in the following.
We also want to get rid of occurrences of $\bot$ in hyperclauses. Since $\bot \vdash$ is an axiom, any hyperclause which contains an occurrence of $\bot$ at the left hand side of some component is valid. But such hyperclauses are redundant, as our aim is to construct *refutations* for unsatisfiable sets of hyperclauses. On the other hand, any occurrence of $\bot$ at the right hand side of a component is also redundant and can be deleted. In other words: we can assume without loss of generality that $\bot$ does not occur in hyperclauses. (Note that this does not imply that occurrences of $\bot$ are removed from **HG**-proofs.)
In direct analogy to classical resolution, the combination of a cut-inference and most general unification is called a *resolution* step. The lower hyperclause in

$$\frac{\mathcal{H} \mid \Gamma_1 \vdash A \quad \mathcal{H}' \mid A', \Gamma_2 \vdash \Delta}{\theta(\mathcal{H} \mid \mathcal{H}' \mid \Gamma_1, \Gamma_2 \vdash \Delta)}\ (res)$$

where θ is the most general unifier of the atoms A and A', is called *resolvent* of the premises, that have to be variable disjoint. If no variables occur, and thus θ is empty, (res) turns into (cut) and we speak of *ground resolution*. The soundness of this inference step is obvious. We show that hyperclause resolution is also refutationally complete. It is convenient to view hyperclauses as sets of atomic sequents. This is equivalent to requiring that external contraction is applied whenever possible. Consequently, there is a unique unsatisfiable hyperclause, namely the *empty hyperclause*. A derivation of the empty hyperclause by resolution from initial hypersequents contained in a set Σ of hyperclauses is called a *resolution refutation* of Σ.

As usual for resolution, we focus on inferences on ground hyperclauses and later transfer completeness to the general level using a corresponding *lifting lemma.*

Theorem 9.2.5 *For every unsatisfiable set of ground hyperclauses Ψ there is a ground resolution refutation of Ψ.*

Proof: We proceed by induction on $e(\Psi) = \|\Psi\| - |\Psi|$, where $\|\Psi\|$ is the total number of occurrences of atoms in Ψ, and $|\Psi|$ is the cardinality of Ψ. If $e(\Psi) \leq 0$ then either Ψ already contains the empty hyperclause, or else Ψ contains exactly one atom per hyperclause. In the latter case, as Ψ is unsatisfiable, there must be hyperclauses $C_1 = (\vdash A)$ and $C_2 = (A \vdash)$ in Ψ. Obviously the empty clause is a ground resolvent of C_1 and C_2.
$e(\Psi) \geq 1$: Ψ must contain a hyperclause C that has more than one atom occurrence. Without loss of generality let $C = (\mathcal{H} \mid \Gamma \vdash A)$, where Γ may be empty. (The case where all atoms in C occur only on the left hand side of sequents is analogous.) Since Ψ is unsatisfiable also the sets $\Psi' = (\Psi - \{C\}) \cup \{\mathcal{H} \mid \Gamma \vdash\}$ and $\Psi'' = (\Psi - \{C\}) \cup \{\vdash A\}$ must be unsatisfiable. Since $e(\Psi') < e(\Psi)$ and $e(\Psi'') < e(\Psi)$ we obtain ground resolution refutations ρ' of Ψ' and ρ'' of Ψ'', respectively. By adding in ρ' an occurrence of A to the right side of the derived empty hyperclause and likewise to all other hyperclauses in ρ' that are on a branch ending in the initial hyperclause $\mathcal{H} \mid \Gamma \vdash$, we obtain a resolution derivation ρ'_A of $\vdash A$ from Ψ. By replacing each occurrence of $\vdash A$ as initial hyperclauses in ρ'' by a copy of ρ'_A we obtain the required ground resolution refutation of Ψ. □

Remark: Note that our completeness proof does not use any special properties of **G**. Only the polarity between left and right hand side of sequent and the disjunctive interpretation of "|" at the meta-level are used. For any logic $\mathcal{L}$: whenever we can reduce $\mathcal{L}$-validity (or $\mathcal{L}$-unsatisfiability) of a formula F to $\mathcal{L}$-unsatisfiability of a corresponding set of atomic hyperclauses, we may use hyperclause resolution to solve the problem.

To lift Theorem 9.2.5 to general hyperclauses, one needs to add (the hypersequent version of) *factorization*:

$$\frac{\mathcal{H} \mid \Gamma \vdash \Delta}{\theta(\mathcal{H} \mid \Gamma' \vdash \Delta)}\ (factor)$$

where θ is the most general unifier (see, e.g., [61]) of some atoms in Γ and where $\theta\Gamma'(\theta)$ is $\theta(\Gamma)$ after removal of copies of unified atoms. The lower hyperclause is called a *factor* of the upper one.

Lemma 9.2.1 *Let C_1' and C_2' be ground instances of the variable disjoint hyperclauses C_1 and C_2, respectively. For every ground resolvent C' of C_1' and C_2' there is a resolvent C of factors of C_1 and C_2.*

The proof of Lemma 9.2.1 is exactly as for classical resolution (see, e.g., [61]) and thus is omitted here. Combining Theorem 9.2.5 and Lemma 9.2.1 we obtain the refutational completeness of general resolution.

Corollary 9.2.1 *For every unsatisfiable set of hyperclauses Σ there is a resolution refutation of Σ.*

We will make use of the observation that any general resolution refutation of Σ can be *instantiated* into (essentially) a ground resolution refutation of a set Σ' of instances of hyperclauses in Σ, whereby resolution steps turn into cuts and factorization turns into additional contraction steps. (Note that additional contractions do not essentially change the structure of a ground resolution refutation.)

9.2.6 Projection of Hyperclauses into HG-Proofs

Remember that from Theorem 9.2.4 (in Section 9.2.4) we obtain a characteristic set of pairs $\{\langle R_1(\hat{\sigma}), D_1\rangle, \ldots \langle R_n(\hat{\sigma}), D_n\rangle\}$ for the proof $\hat{\sigma}$ of $\mathcal{H}^S$. As described in Section 9.2.5, we can construct a resolution refutation γ of the hyperclause set $\{C_1, \ldots, C_n\}$ corresponding to the d-hyperclauses $\{D_1, \ldots, D_n\}$ (this is step 3 of hyperCERES). Forming a ground instantiation γ' of γ yields a derivation of the empty hypersequent that consists only of atomic cuts and contractions. (Step 4 of hyperCERES.) Each leaf node of γ' is a ground instance $\theta(C_i)$ of a hyperclause in $\{C_1, \ldots, C_n\}$. From Theorem 9.2.4 we also obtain, for each $i \in \{1, \ldots, n\}$ a cut-free proof $R_i(\hat{\sigma})$ of $\mathcal{G}_i \odot D_i$, where $\mathcal{G}_i$ is a sub-hypersequent of the cut-irrelevant part of $\mathcal{H}_{\hat{\sigma}}$ and D_i is the d-hyperclause corresponding to C_i. We instantiate $R_i(\hat{\sigma})$ using θ and finally apply $(distr)$, as indicated in Section 9.2.5, to obtain a cut-free proof $\hat{\sigma}_i^\theta$ of $\theta(\mathcal{G}_i) \odot \theta(C_i)$.
To get a proof $\gamma'[\hat{\sigma}]$ of a linked Skolem instance of the original hypersequent $\mathcal{H}$ (cf. Section 9.2.3) we replace each leaf node $\theta(C_i)$ of γ' with the proof $\hat{\sigma}_i^\theta$ of $\theta(\mathcal{G}_i) \odot \theta(C_i)$, described above, and transfer the instances $\theta(\mathcal{G}_i)$ of cut-irrelevant formulas in $\mathcal{H}$ also to the inner nodes of γ' in the obvious way, That is, to regain correct applications of atomic cuts. As mentioned in Section 9.2.2, the remaining atomic cuts can easily be removed from $\gamma'[\hat{\sigma}]$. The resulting proof is subjected to de-Skolemization as described in

Theorem 9.2.3. This final step 7 of hyperCERES yields the desired cut-free proof of $\mathcal{H}$.

Example 9.2.2 We continue Example 9.2.1, where we have obtained the characteristic set of pairs $\{\langle \rho_1, \overset{\{1\}}{\vdash} Q\rangle,\ \langle \rho_2, P(x)\overset{\{4\}}{\vdash}\rangle,\ \langle \rho_3, Q\overset{\{2\}}{\vdash} \mid \overset{\{3,5\}}{\vdash}\rangle\}$ for the proof σ of the (trivially) Skolemized prenex hypersequent $Q \overset{\{1,2,4\}}{\vdash} \exists y P(y) \mid P(c) \vee Q \overset{\{3,5\}}{\vdash} Q$.
The obtained d-hyperclauses are in fact already hyperclauses. Moreover, one can immediately see that the hyperclauses $\overset{\{1\}}{\vdash} Q$ and $Q\overset{\{2\}}{\vdash} \mid \overset{\{3,5\}}{\vdash}$ can be refuted by a one-step resolution derivation γ:

$$\dfrac{\overset{\{1\}}{\vdash} Q \qquad Q\overset{\{2\}}{\vdash} \mid \overset{\{3,5\}}{\vdash}}{\overset{\{1,2\}}{\vdash} \mid \overset{\{1,3,5\}}{\vdash}}\ (res)$$

Note that $P(x)\overset{\{4\}}{\vdash}$ and the corresponding reduced proof ρ_2 are redundant. In our case, γ is already ground. Therefore no substitution has to be applied to the reduced proofs ρ_1 and ρ_3. By replacing the two upper (d-)hyperclauses in γ with ρ_1 and ρ_3, respectively we obtain the desired proof $\gamma[\sigma]$ that only contains an atomic cut:

$$\dfrac{Q \overset{\{1\}}{\vdash} Q \qquad \dfrac{\dfrac{\dfrac{Q, P(c) \overset{\{2\}}{\vdash} P(c) \mid \overset{\{3,5\}}{\vdash}}{Q, P(c) \overset{\{2\}}{\vdash} \exists y P(y) \mid \overset{\{3,5\}}{\vdash}}\,(\exists\text{-}r) \qquad \dfrac{Q\overset{\{2\}}{\vdash} \mid \overset{\{3,5\}}{\vdash} Q}{Q\overset{\{2\}}{\vdash} \mid P(c) \overset{\{3,5\}}{\vdash} Q}\,(iw)\text{-}l}{Q \overset{\{2\}}{\vdash} \exists y P(y) \mid P(c) \overset{\{3,5\}}{\vdash} Q}\,(com) \qquad Q \overset{\{5\}}{\vdash} Q}{Q \overset{\{2,4\}}{\vdash} \exists y P(y) \mid P(c) \vee Q \overset{\{3,5\}}{\vdash} Q}\,(\vee\text{-}l)}{Q \overset{\{1,2,4\}}{\vdash} \exists y P(y) \mid P(c) \vee Q \overset{\{3,5\}}{\vdash} Q}\,(cut)$$

◇

The results of this chapter are easily extendable to larger fragments **G**: (de-)Skolemization is sound already for intuitionistic logic **I** without positive occurrences of universal quantifiers, if an additional existence predicate is added [14]. Therefore hyperCERES applies after incorporation of the mentioned existence predicate. Other classes where Skolemization is sound for **I** are described by Mints [64].
The most interesting question however is whether hyperCERES can be extended to intuitionistic logic itself. Note that we obtain a calculus for **I** by dropping the communication rule from **HG**. It turns out that hyperCERES

is applicable to the class of (intuitionistic) hypersequents not containing negative occurrences of $\vee$ or positive occurrences of $\forall$, as the distribution rule (*distr*) is still sound for this fragment of **I**. This fragment actually is an extension of the Harrop class [43] with weak quantifiers.

The extendability of hyperCERES to full intuitionistic logic depends on the development of an adequate (de-)Skolemization technique, together with a concept of parallelized resolution refutations, that takes into account the disjunctions of atoms at the right hand side of clauses without using (*distr*).

From a more methodological viewpoint, it should be mentioned that hyper-CERES uses the fact that "negative information" can be treated classically in intermediate logics like **G**, and that cuts amount to entirely negative information in our approach. In this sense, global cut elimination, as presented in this paper, is more adequate for intermediate logics than stepwise reductions, which treat cuts as positive information.

Chapter 10

Related Research

10.1 Logical Analysis of Mathematical Proofs

Proof transformations for the analysis of proofs can roughly be classified according to the degree to which they eliminate the structure of the given proof.

Functional interpretations extract the desired information without changing the proof structure. At the moment they are the most widespread tool for analyzing proofs and truly deserve the description "proof mining" – if we think of mining in the sense of Agricola, where it is not intended to remove the mountain to obtain the ore. A concise overview can be found in [54].

Methods of proof analysis based on the first epsilon theorem or the epsilon substitution method dismiss the propositional structure in terms of the quantificational kernel of the proof [56–58].

The method of Herbrand analysis frees the proof from its quantificational aspects [63]. This holds to a lesser degree also for methods based on the no-counterexample-interpretation, the generalization of Herbrand's theorem.

Cut-elimination, the focus of this book, deletes both propositional and quantificational information when it is not related to the result of the proof.

Of course there is the proviso that the methods mentioned above can be combined: The no-counterexample-interpretation has been established originally in connection with epsilon calculus, and Herbrand disjunctions can be obtained from a proof using functional interpretations [39]. The final results of the application of the mentioned methods respect however more or less the classification above.

Another perimeter for a classification is the flexibility of the methods usually connected to the degree of formalization necessary. Very flexible methods

DOI 10.1007/978-94-007-0320-9_10,

such as the functional interpretation allow a greater range of proof improvements. On the other hand, less flexible methods such as cut elimination allows better comparisons of proofs, especially in the sense of negative statements, e.g. that a certain Herbrand disjunction cannot be obtained by any reasonable transformation from a given proof.

10.2 New Developments in CERES

During our work on this book the CERES-method was extended, modified and improved in several ways. The computation of the characteristic clause set from the characteristic clause term as defined in via the semantics of clause terms Section 6.3 is simple but inefficient. Another refined model, the profile, was defined by Stefan Hetzl [45] (see also [46]) and used for an improved analysis of cut-elimination via subsumption as defined in Section 6.8. Several different methods of computing clause sets from characteristic terms were developed by Bruno Woltzenlogel Paleo [80]. The clause sets obtained by these methods are smaller and much less redundant and pave the way for an improved complexity analysis of CERES. In this thesis the computation of canonic refutations (see Definition 6.7.2) and (therefore) the complexity results in Section 6.3 were substantially improved. The new form of clause computation made it also possible to modify CERES to a cut-introduction method. By application of this method exponential compressions of LK-proof sequences can be obtained by cut-introduction (see [80]). In the CERES method as described in Chapter 6 the resolution refutation can be any refutation obtained either by unrestricted or refined resolution (for the standard refinements of automated deduction see [61]). But more is possible: the characteristic clause sets encode structural properties of cuts and are always unsatisfiable. In [80] structural information from the cuts is encoded into resolution refinements, leading to a substantial restriction of the search space. These refinements, which are incomplete in general – but complete for characteristic clause sets, can contribute to a substantial improvement of the performance of CERES. Proof theoretically they yield a tool to encode reductive methods via resolution demonstrating the generality of the CERES approach.
The CERES-method presented in this book was defined for first-order logic (classical, finitely-valued and Gödel logic). Though first-order language is capable of expressing important mathematical concepts in a natural way, an extension of CERES to higher-order logic is essential to the analysis of more rewarding mathematical proofs which require complex types (like function-

als and operators) and induction. An extension to a weak second-order logic with quantifier-free comprehension was defined in [49]. Due to quantifier-free comprehension it was possible to Skolemize the proofs and extend the resolution calculus to an Andrews-type resolution calculus for higher-order logic (see [2]). An extension to full type theory is work in progress and is documented on the CERES-web page http://www.logic.at/ceres. The method CERES$^\omega$ substantially differs from CERES in first-order logic. In CERES$^\omega$ the proofs are no longer Skolemized (indeed Skolemization of formulas introduced by weak quantifier-rules is impossible in general). Proof projections and the construction of an atomic cut normal form are more involved than in the first-order case. The relative completeness of the clausal calculus of CERES$^\omega$ to Andrew's resolution in type theory is still an open problem. An example of a higher-order proof analysis by CERES$^\omega$, the transformation of an induction proof to a proof using the least number principle, is shown on http://www.logic.at/ceres. As a full automation of proof search in higher-order resolution is unrealistic, a semi-automated theorem proving enviroment is currently under development. Intgrating methods from Isabelle [52] and Coq [32] might be fruitful in future investigations.

For nonclassical (especially intermediate) logics the problem of Skolemization can be solved by choosing unusual concepts of Skolemization (see, e.g. [14]). The main problem here lies in the development of an adequate resolution calculus. For intermediary first-order logics we expect to overcome this problem because the cut-relevant information is negative and is therefore expected to behave, in some way, classically.

For the analysis of more advanced mathematical proofs the representation of induction is vital. As a first step we suggest to investigate schematic representations of proofs being closest to first-order formalisms. Using these intended improvements we will try to analyze Witt's proof of the theorem of Wedderburn [1] which is one of the most surprising examples of a composition of two elementary proofs; here the aim would be to eliminate the arguments concerning complex numbers and to check whether Euler's truncated argumentation can be obtained as suggested by André Weil.

References

[1] M. Aigner, G.M. Ziegler: Proofs from THE BOOK. Springer, Berlin, 1998.

[2] P.B. Andrews: Resolution in Type Theory. *Journal of Symbolic Logic*, 36, pp. 414–432, 1971.

[3] A. Avron: Hypersequents, Logical Consequence and Intermediate Logics for Concurrency. *Annals of Mathematics and Artificial Intelligence*, 4, pp. 225–248, 1991.

[4] A. Avron: The Method of Hypersequents in Proof Theory of Propositional Non-Classical Logics. In *Logic: From Foundations to Applications.* Clarendon Press, Oxford, pp. 1–32, 1996.

[5] M. Baaz, A. Ciabattoni: A Schütte-Tait Style Cut-Elimination Proof for First-Order Gödel Logic. Proceedings of *Tableaux 2002*. LNAI 2381, pp. 24–38, 2002.

[6] M. Baaz, A. Ciabattoni, C.G. Fermüller: Herbrand's Theorem for Prenex Gödel Logic and Its Consequences for Theorem Proving. Proceedings of *LPAR'2001*. LNAI 2250, pp. 201–216, 2001.

[7] M. Baaz, A. Ciabattoni, C.G. Fermüller: Cut Elimination for First Order Gödel Logic by Hyperclause Resolution. Proceedings of LPAR'2008. *Lecture Notes in Computer Science* 5330, pp. 451–466, 2008.

[8] M. Baaz, C. Fermüller: Resolution-Based Theorem Proving for Many-Valued Logics. *Journal of Symbolic Computation*, 19(4), pp. 353–391, 1995.

[9] M. Baaz, C. Fermüller, G. Salzer: Automated Deduction for Many-Valued Logics. In: J.A. Robinson, A. Voronkov (eds.), *Handbook of Automated Reasoning 2*, Elsevier and MIT Press, pp. 1356–1402, 2001.

M. Baaz, A. Leitsch, *Methods of Cut-Elimination*, Trends in Logic 34,
DOI 10.1007/978-94-007-0320-9,

[10] M. Baaz, C. Fermüller, R. Zach: Elimination of Cuts in First-Order Finite-Valued Logics. *Journal of Information Processing Cybernetics (EIK)*, 29(6), pp. 333–355, 1994.

[11] M. Baaz, S. Hetzl, A. Leitsch, C. Richter, H. Spohr: Cut-Elimination: Experiments with CERES. Proceedings of LPAR 2004, *Lecture Notes in Computer Science* 3452, pp. 481–495, 2005.

[12] M. Baaz, S. Hetzl, A. Leitsch, C. Richter, H. Spohr: Proof Transformation by CERES. In: J.M. Borwein, W.M. Farmer (eds.), *MKM 2006*, LNCS (LNAI), vol. 4108, pp. 82–93. Springer, Heidelberg, 2006.

[13] M. Baaz, S. Hetzl, A. Leitsch, C. Richter, H. Spohr: CERES: An Analysis of Fürstenberg's Proof of the Infinity of Primes. *Theoretical Computer Science*, 403 (2–3), pp. 160–175, 2008.

[14] M. Baaz, R. Iemhoff: The Skolemization of Existential Quantifiers in Intuitionistic Logic. *Annals of Pure and Applied Logic*, 142, pp. 269–295, 2006.

[15] M. Baaz, A. Leitsch: Complexity of Resolution Proofs and Function Introduction. *Annals of Pure and Applied Logic*, 57, pp. 181–215, 1992.

[16] M. Baaz, A. Leitsch: On Skolemization and Proof Complexity. *Fundamenta Informaticae*, 20(4), pp. 353–379, 1994.

[17] M. Baaz, A. Leitsch: Cut Normal Forms and Proof Complexity. *Annals of Pure and Applied Logic*, 97, pp. 127–177, 1999.

[18] M. Baaz, A. Leitsch: Cut-Elimination and Redundancy-Elimination by Resolution. *Journal of Symbolic Computation*, 29, pp. 149–176, 2000.

[19] M. Baaz, A. Leitsch: CERES in Many-Valued Logics. Proceedings of *LPAR'2005*. LNAI 3452, pp. 1–20, 2005.

[20] M. Baaz, A. Leitsch: Towards a Clausal Analysis of Cut-Elimination. *Journal of Symbolic Computation*, 41, pp. 381–410, 2006.

[21] M. Baaz, A. Leitsch, R. Zach: Incompleteness of an Infinite-Valued First-Order Gödel Logic and of Some Temporal Logic of Programs. Proceedings of *CSL'95*, LNCS 1092, pp. 1–15, 1996.

[22] M. Baaz, N. Preining, R. Zach: First-Order Gödel Logics. *Annals of Pure and Applied Logic*, 147, pp. 23–47, 2007.

[23] M. Baaz, P. Pudlak: Kreisel's Conjecture for LE1. In: P. Clote, J. Krajicek (eds.), *Arithmetic Proof Theory and Computational Complexity*, Oxford University Press, Oxford, pp. 30–49, 1993.

[24] M. Baaz, P. Wojtylak: Generalizing Proofs in Monadic Languages. *Annals of Pure and Applied Logic*, 154(2): 71–138, 2008.

[25] M. Baaz, R. Zach: Hypersequents and the Proof Theory of Intuitionistic Fuzzy Logic. Proceedings of *CSL'2000*, pp. 187–201, 2000.

[26] A. Beckmann, N. Preining: Linear Kripke Frames and Gödel Logics. *Journal of Symbolic Logic*, 71(1), pp. 26–44, 2007.

[27] E. Börger, E. Grädel, Y. Gurevich: The Classical Decision Problem. *Perspectives in Mathematical Logic*. Springer, Berlin, 1997.

[28] W.S. Brianerd, L.H. Landweber: Theory of Computation. Wiley, New York, NY, 1974.

[29] P. Brauner, C. Houtmann, C. Kirchner: Principles of Superdeduction. Proceedings of LICS, pp. 41–50, 2007.

[30] W.A. Carnielli: Systematization of Finite Many-Valued Logics through the Method of Tableaux. *Journal of Symbolic Logic*, 52(2), pp. 473–493, 1987.

[31] C.L. Chang, R.C.T. Lee: Symbolic Logic and Mechanical Theorem Proving. Academic Press, New York, NY, 1973

[32] Coq: Project Home Page. URL: http://pauillac.inria.fr/coq/.

[33] A. Degtyarev, A. Voronkov: Equality Reasoning in Sequent-Based Calculi. In: A. Robinson, A. Voronkov (eds.), *Handbook of Automated Reasoning*, vol. I, chapter 10, Elsevier Science, Amsterdam, pp. 611–706, 2001.

[34] G. Dowek, T. Hardin, C. Kirchner: Theorem Proving Modulo. *Journal of Automated Reasoning*, 31, pp. 32–72, 2003.

[35] G. Dowek, B. Werner: Proof Normalization Modulo. *Journal of Symbolic Logic*, 68(4), pp. 1289–1316, 2003.

[36] E. Eder: Relative Complexities of First-Order Calculi. Vieweg, Braunschweig, 1992.

[37] P. Erdös: On a New Method in Elementary Number Theory which Leads to an Elementary Proof of the Prime Number Theorem. *Proceedings of the National Academy of Sciences of the United States of America*, 35, pp. 374–384, 1949.

[38] G. Gentzen: Untersuchungen über das logische Schließen. *Mathematische Zeitschrift*, 39, pp. 405–431, 1934–1935.

[39] P. Gerhardy, U. Kohlenbach: Extracting Herbrand Disjunctions by Functional Interpretations. *Archive for Mathematical Logic*, 44, pp. 633–644, 2005.

[40] J.Y. Girard: Proof Theory and Logical Complexity. In *Studies in Proof Theory*, Bibliopolis, Napoli, 1987.

[41] K. Godden: Lazy Unification. Proceedings of the 28th Annual Meeting of the Association for Computational Linguistics, pp. 180–187, 1990.

[42] P. Hájek: Metamathematics of Fuzzy Logic. Kluwer, Dordrecht, 1998.

[43] R. Harrop: Concerning Formulas of the Types $A \Rightarrow B \vee C$, $A \Rightarrow (\exists x)B(x)$ in Intuitionistic Formal Systems. *Journal of Symbolic Logic*, 25, pp. 27–32, 1960.

[44] J. Herbrand: Sur le problème fondamental de la logique mathématique. *Sprawozdania z posiedzen Towarzysta Naukowego Warszawskiego, Wydzial III*, 24, pp. 12–56, 1931.

[45] S. Hetzl: Proof Profiles, Characteristic Clause Sets and Proof Transformations. PhD Thesis, Vienna University of Technology, 2007.

[46] S. Hetzl, A. Leitsch: Proof Transformations and Structural Invariance. In Algebraic and Proof Theoretic Aspects of Non-classical Logics. *Lecture Notes in Computer Science*, 4460, pp. 201–230, 2007.

[47] S. Hetzl, A. Leitsch, D. Weller, B. Woltzenlogel Paleo: Herbrand Sequent Extraction. AISC/Calculemus/MKM 2008, LNAI 5144, pp. 462–477, 2008.

[48] S. Hetzl, A. Leitsch, D. Weller, B. Woltzenlogel Paleo: Proof Analysis with HLK, CERES and ProofTool: Current Status and Future Directions. Proceedings of the CICM Workshop on Empirically Successful Automated Reasoning in Mathematics, CEUR Workshop Proceedings, vol 378, ISSN 1613-0073, 2008.

[49] S. Hetzl, A. Leitsch, D. Weller, B. Woltzenlogel Paleo: A Clausal Approach to Proof Analysis in Second-Order Logic. Proceedings of the LFCS 2009, *Lecture Notes in Computer Science*, 5407, pp. 214–229, 2009.

[50] D. Hilbert: Grundlagen der Geometrie. Mit Supplementen von Paul Bernays. 14. Auflage. Herausgegeben und mit Anhängen versehen von Michael Toepell. Mit Beiträgen von Michael Toepell, Hubert Kiechle, Alexander Kreuzer und Heinrich Wefelscheid. B. G. Teubner Stuttgart - Leipzig 1999.

[51] D. Hilbert, P. Bernays: Grundlagen der Mathematik Band 2. Springer, Berlin, 1939.

[52] Isabelle: System Home Page. URL: http://www.cl.cam.ac.uk/Research/HVG/Isabelle.

[53] W.H. Joyner: Resolution Strategies as Decision Procedures. *Journal of the ACM*, 23(1), pp. 398–417, 1976.

[54] U. Kohlenbach: Applied Proof Theory: Proof Interpretations and their Use in Mathematics. In *Springer Monographs in Mathematics*, Springer, Berlin, 2009.

[55] C.J. Krajicek, P. Pudlak: The Number of Proof Lines and the Size of Proofs in First Order Logic. *Archive for Mathematical Logic*, 27, pp. 69–84, 1988.

[56] G. Kreisel: On the Interpretation of Nonfinitist Proofs, Part I. *Journal of Symbolic Logic* 16, pp. 241–267, 1951.

[57] G. Kreisel: On the Interpretation of Nonfinitist Proofs, Part II: Interpretation of Number Theory, Applications. *Journal of Symbolic Logic*, 17, pp. 43–58, 1952.

[58] G. Kreisel: Mathematical Significance of Consistency Proofs. *Journal of Symbolic Logic*, 23, pp. 155–182, 1958.

[59] G.W. Leibniz: Philosophical Writings. M. Morris, G.H.R. Parkinson (ed. and trans.), J.M. Dent and Sons Ltd, London, 1973.

[60] G.W. Leibniz: New Essays on Human Understanding, 1703-5, first publ. 1765, P. Remnant, J. Bennett (ed. and trans.), Cambridge University Press, Cambridge, 1981.

[61] A. Leitsch: The Resolution Calculus. *EATCS Texts in Theoretical Computer Science*. Springer, Berlin, 1997.

[62] D. Loveland: Automated Theorem Proving – A Logical Basis. North-Holland, Amsterdam, 1978.

[63] H. Luckhardt: Herbrand-Analysen zweier Beweise des Satzes von Roth: polynomiale Anzahlschranken. *Journal of Symbolic Logic*, 54, pp. 234–263, 1989.

[64] G. Mints: The Skolem Method in Intuitionistic Calculi. *Proceedings of the Steklov Institute of Mathematics*, 121, pp. 73–109, 1974.

[65] G. Mints: Quick Cut-Elimination for Monotone Cuts. In *Games, Logic, and Constructive Sets*, The University of Chicago Press, Chicago, IL, 2003.

[66] R. Nieuwenhuis, A. Rubio: Paramodulation-Based Theorem Proving. In: J.A. Robinson, A. Voronkov (eds.), *Handbook of Automated Reasoning*, Elsevier, Amsterdam, pp. 371–443, 2001.

[67] V.P. Orevkov: Lower Bounds for Increasing Complexity of Derivations after Cut Elimination. *Journal of Soviet Mathematics*, 20, pp. 2337–2350, 1982.

[68] P. Pudlak: The Lengths of Proofs. In: S.R. Buss (ed.), *Handbook of Proof Theory*, Elsevier, Amsterdam, 1998.

[69] J.A. Robinson: A Machine Oriented Logic Based on the Resolution Principle. *Journal of the ACM* 12(1), pp. 23–41, 1965.

[70] K. Schütte: Beweistheorie. Springer, Berlin, 1960.

[71] H. Schwichtenberg: Proof Theory: Some Applications of Cutelimination. In: J. Barwise (ed.), *Handbook of Mathematical Logic*, North Holland, Amsterdam, 1977.

[72] R. Statman: Lower Bounds on Herbrand's Theorem. *Proceedings of the American Mathematical Society* 75, pp. 104–107, 1979.

[73] W.W. Tait: Normal Derivability in Classical Logic. In: J. Barwise (ed.), *The Syntax and Semantics of Infinitary Languages*, Springer, Berlin, pp. 204–236, 1968.

[74] G. Takeuti: Proof Theory. North-Holland, Amsterdam, 2nd edition, 1987.

[75] G. Takeuti, T. Titani: Intuitionistic Fuzzy Logic and Intuitionistic Fuzzy Set Theory. *Journal of Symbolic Logic*, 49, pp. 851–866, 1984.

[76] T. Tao: Soft Analysis, Hard Analysis, and the Finite Convergence Principle. http://terrytao.wordpress.com/2007/05/23/soft-analysis-hardanalysis-and-the-finite-convergence-principle/

[77] A.S. Troelstra, H. Schwichtenberg: Basic Proof Theory. Cambridge University Press, Cambridge, 2nd edition, 2000.

[78] C. Urban: Classical Logic and Computation. PhD Thesis, University of Cambridge Computer Laboratory, 2000.

[79] B. Woltzenlogel Paleo: Herbrand Sequent Extraction. VDM Verlag Dr.Müller e.K., Saarbrücken, , ISBN-10: 3836461528, February 7, 2008.

[80] B. Woltzenlogel Paleo: A General Analysis of Cut-Elimination by CERes. PhD Thesis, Vienna University of Technology, 2009.

Index

$|\ \ |$, 111
$\|X\|$, 26
$\wedge_3$, 167
$\wedge_n$, 164
$\mathcal{A}_{\top\bot}$, 190
$\circ$, 13
CL( ), 115
comp, 11
$\leq_{cp}$, 133
cutcomp, 19, 86
$def(P){:}l$, 172
$def(P){:}r$, 172
depth(ν), 21
DIFF(t_1, t_2), 29
$dm{:}r$, 167
$\rhd$, 73, 113
$\mathcal{A}_e$, 41
e, 51
$\gamma \cdot D$, 139
H_i, 55
HC$_\mathcal{A}$(), 50
$H^\star$(), 46
$\leq_s$, 11, 112
LKDe, 173
LKe, 170
LK', 167
$\Phi^\mathcal{A}$, 19
Φ^s, 110
Φ^s_i, 110
$\Phi^s_\emptyset$, 110
$l(\varphi)$, 23
$<_d$, 183
N_c, 178
nmc, 87
$\|\ \|_l$, 107
$|\,t\,|$, 165
$=:l$, 170
$=:r$, 170
PC$_\mathcal{A}$, 50
PES(φ), 125, 207
$\varphi[C]$, 123
$\Phi[\mathbf{K}]$, 234
$\pi()$, 39
$\oplus^n$, 165
$\oplus$, 111
$q(,)$, 46
rc, 175
$>_\mathcal{R}$, 71
RES(ψ), 144
$>_G$, 72
RH, 179
$\rho_{<_d}$, 183
ρ_H, 179
RH^*, 179
RH^*_+, 179
$R_{<_d}$, 183
$R^*_{<_d}$, 183
$>_T$, 86
$\mathcal{R}$, 65
$S^\star$(), 46
s, 51
$\bar{S}(\nu, \Omega)$, 120
$\sim_p$, 190
sk, 106

$S(\nu, \Omega)$, 22
$\leq_{ss}$, 133
$\sqsubseteq$, 13, 112
T, 52
τ, 183
$\tau_{\max}(x, A)$, 183
Ax_T, 55
$\Theta(\ )$, 114
$\otimes^n$, 165
$\otimes$, 111
$\times$, 111
V_b, 9
V_f, 9
$\varphi(\gamma)$, 126
Abstraction, 200
ACNF, 64
Ancestor, 21
$\{\wedge, \vee\}$-combination, 206
    immediate, 22
    path, 22
$\mathcal{A}$-valid, 45
AXDC, 182
Axiom set, 14
    atomic, 14
    standard, 15

C
Calculus
    hypersequent, 252
    LM-type, 232
CERES
    fast on $\mathcal{K}$, 176
CERES normal form, 126
Characteristic clause set, 115, 166
Characteristic hyperclauses, 259
Characteristic term, 114, 165
Clause, 35
    hyper, 265
    W-, 236
Clause term, 111, 165
Complexity
    Herbrand, 50
    logical, 11
    proof, 50
Composition
    of sequents, 13
Condensation, 178
    normal form, 178
Condensed, 178
Context product, 117
Contraction normalization, 35
$\mathrm{CORR}(t_1, t_2)$, 28
Corresponding pairs, 28
Cut
    essential, 23
Cut complexity, 19
Cut derivation, 23
Cut rule, 17
Cut-complexity, 86
Cut-derivation
    simple, 72
    strict, 86
    uppermost, 72
Cut-elimination
    relation, 64
    sequence, 64

D
Definition rule, 172
Depth, 21
Depth ordering, 183
De-Skolemization, 256
DIFF(,) difference set, 29
Difference set, 29
Distribution, 231

E
Elementary in, 60

F
Factor, 36, 266

Formula, 11
  $\{\top, \bot\}$, 190
  monotone, 178
Function
  elementary, 51

G
Gödel logic, 251
Grade, 23
Ground projection, 37, 237
  minimal, 130

H
Hauptsatz, 73
Herbrand complexity, 50
Herbrand sequent, 45
H**G**, 252
HyperCERES, 255
Hyperclause, 265
  resolution, 265
Hyperresolution, 179
Hyperresolvent, 179
Hypersequent, 252

I
Instantiation sequent, 45
Interpolant, 190
  weak, 190
Interpolation, 190
  derivation, 190
Interpolation theorem, 203

K
K_{mon}, 185

L
Lattice, 222
Lattice proof, 224
Lifting theorem, 136
LK, 15
  derivation, 18
  proof, 19
LM-type calculus, 232

M
m.g.u.
  total, 128
MC, 184
m.g.u., 24
Mid-hypersequent, 257
Minimal ground projection, 130
Mix rule, 17

N
NE-improvement, 93, 160
Nonelementary in, 60

P
Pair
  corresponding, 28
  irreducible, 28
  unifiable, 29
Partition
  of a sequent, 190
Permutation
  of sequents, 13
Polarity, 14
Position, 10
Preproof, 209
Pre-resolution
  derivation, 128
Pre-resolvent, 128
p-resolvent, 36
PRF-resolvent, 36
Projection, 123
Projection derivation, 206
Proof
  generalized, 209
  length, 23
  LK, 19
  path, 20
  pre-, 209

regular, 21
Skolemized, 107
Proof complexity, 50
Pseudo-cut, 164

Q

QMON, 189
QM-sequent, 189
Quantifier
strong, 14, 231
weak, 14, 231
Quasi-monotone, 188

R

Rank, 23
Reduction
relation, 64
sequence, 64
Regularity, 21
Resolution
complexity, 175
hyperclause-, 265
ordered, 183
W-, 236
Resolution deduction, 36
Resolution refutation, 36
canonic, 142
Resolved atom, 36
Resolvent, 36
ordered, 183
pre-, 128
PRF-, 36
Rule
logical, 15
structural, 16

S

Semi-formula, 11
Semi-lattice, 221
Semi-term, 9
Seq, 18
Sequent, 12
atomic, 12
Herbrand, 45
prenex, 46
weakly quantified, 14
Skolemization, 238
of formulas, 106
of hypersequent, 256
of proofs, 107
of sequents, 107
Statman's sequence, 54
Subderivation, 20
Subsequent, 13
Substitution, 10
more general, 11
Subsumption, 133
of clause terms, 133
of proofs, 138

T

Tape proof, 216
Term, 9
critical, 200
Term basis, 212
Total m.g.u., 128

U

UAL, 31
UIE, 176
UILM, 178
UIRM, 181
Unification
algorithm, 31
theorem, 31
Unifier, 24
most general, 24
simultaneous, 24

V

Variant
of a clause, 36

W

W-clause, 236
Weight, 87
W-resolution, 236